# RECENT DEVELOPMENTS
# IN CHEMICAL PROCESS
# AND PLANT DESIGN

# RECENT DEVELOPMENTS IN CHEMICAL PROCESS AND PLANT DESIGN

Editors

## Y. A. LIU
Vilbrandt Professor of Chemical Engineering
Virginia Polytechnic Institute and State University

## HENRY A. McGEE, JR.
Professor of Chemical Engineering
Virginia Polytechnic Institute and State University

## W. ROBERT EPPERLY
Executive Vice President
Fuel Tech, Inc.

A Wiley-Interscience Publication

## JOHN WILEY & SONS
New York • Chichester • Brisbane • Toronto • Singapore

*To the Memory of Professor Frank C. Vilbrandt*
*on the Golden Anniversary of His Founding*
*of the Department of Chemical Engineering,*
*Virginia Polytechnic Institute and State University*

*Library of Congress Cataloging in Publication Data:*
Recent developments in chemical process and plant design/editors,
   Y. A. Liu, Henry A. McGee, Jr., W. Robert Epperly.
      p.   cm.
     "A Wiley-Interscience publication."
     Bibliography: p.
     Includes index.
     ISBN 0-471-84780-1
     1. Chemical processes.  2. Chemical engineering.  I. Liu, Y. A.
  (Yih-An)  II. McGee, Henry A. Jr.  III. Epperly, W. R.
  TP155.7.R43 1987
  660.2'81—dc19

Printed in the United States of America

10  9  8  7  6  5  4  3  2  1

# CONTRIBUTORS

**Kenneth J. Bell,** Regents Professor of Chemical Engineering, Oklahoma State University, Stillwater, Oklahoma and Senior Chemical Engineer, Argonne National Laboratory, Argonne, Illinois

**W. Robert Epperly,** General Manager of Corporate Research, Exxon Research and Engineering Company, Annandale, New Jersey; presently Executive Vice President and General Manager of Research and Development, Fuel Tech, Inc., Stamford, Connecticut

**Lawrence B. Evans,** Professor of Chemical Engineering, Massachusetts Institute of Technology, and Chairman of the Board, ASPEN Technology, Inc., Cambridge, Massachusetts

**James R. Fair,** John J. McKetta Centennial Energy Chair in Engineering, and Director, Separations Research Program, Center for Energy Studies, University of Texas, Austin, Texas

**Fun-Gau Ho,** Senior Engineer, Research and Development Department, Union Carbide Corporation, South Charleston, West Virginia

**George E. Keller II,** Corporate Research Fellow, Research and Development Department, Union Carbide Corporation, South Charleston, West Virginia

**Y. A. Liu,** Frank C. Vilbrandt Professor of Chemical Engineering, Virginia Polytechnic Institute and State University, Blacksburg, Virginia

**Thomas J. McAvoy,** Professor of Chemical Engineering, University of Maryland, College Park, Maryland

**T. W. Fraser Russell,** Allan P. Colburn Professor and Chairman of Chemical Engineering, and Director, Institute of Energy Conversion, University of Delaware, Newark, Delaware

**Henry G. Schwartzberg,** Professor of Food Engineering, University of Massachusetts, Amherst, Massachusetts

**John H. Sinfelt,** Senior Scientific Advisor, Exxon Research and Engineering Company, Annandale, New Jersey

**Geoffrey L. Wells,**  Senior Lecturer, Department of Chemical Engineering and Fuel Technology, University of Sheffield, Sheffield, United Kingdom

**Arthur W. Westerberg,**  John E. Swearingen Professor of Chemical Engineering, Carnegie-Mellon University, Pittsburgh, Pennsylvania

# FOREWORD

It is difficult to do justice through words to the memory of Dr. Frank C. Vilbrandt, for he was a man who chose his words carefully. His energy was directed toward deeds rather than rhetoric. Yet I can think of no finer tribute to "Doc" than to dedicate this book to his memory and establish the Frank Vilbrandt Memorial Fund to benefit chemical engineering students at Virginia Tech.

I am sure Doc would appreciate a volume dedicated to his honor and a fund to assist engineering students, for they perpetuate one of his basic beliefs: that engineers, particularly chemical engineers, are vital to our culture. By traditional design, engineers solve problems. They create; they design; they discover; they improve. And this same basic belief that Doc made so much a part of his teachings is even truer today.

New directions in the chemical process industries such as medical prostheses and appliances will depend on chemical engineering technology. At the same time, the manufacture of materials will rely greatly on chemical engineering as we develop organo-ceramic and organo-metallic polymers, composite materials, and synthetic ceramics. As we create these new applications, the future of chemical engineering education will expand the application of chemical engineering principles to the design of systems that support and enhance physiology (artificial organs); space exploration (life support systems); energy conversion via solar, fuel cell, and marine conversion systems; and biochemical and electromechanical production techniques in industry. A review of the subject matter of this memorial edition indicates that many of these topics are explored in detail.

The young people Doc taught over a long and dedicated career were fortunate indeed! He opened a door into a promising future while he threatened, cajoled, and enticed us to enter the field he loved so much. I think Doc would be pleased that chemical engineering continues to be the dynamic, innovative, and exciting profession that he promised us.

ALEXANDER F. GIACCO
CHAIRMAN, PRESIDENT AND
CHIEF EXECUTIVE OFFICER
HERCULES INCORPORATED

# PREFACE

This book describes the latest developments in and applications of chemical process- and plant-design methodology. Written by internationally recognized, award-winning engineers and scholars, this self-contained sourcebook provides beginners and experts with state-of-the-art reviews of a variety of important topics in chemical process and plant design. The topics covered include computer-aided design, safety, catalyst design, process synthesis, heat integration, process exchangers, energy-efficient separations, integration of design and control, synthetic-fuel plants, food and biochemial production, and electronic materials processing.

This volume is prepared in celebration of the 50th anniversary of the founding of the Department of Chemical Engineering at Virginia Polytechnic Institute and State University (''Virginia Tech'') by the late Professor Frank Carl Vilbrandt in 1935. Professor Vilbrandt's McGraw-Hill textbook *Chemical Engineering Plant Design*, first published in 1939, was a pioneering volume on the subject, and it was used in chemical engineering departments nationally and abroad for over three decades. In recognition of Professor Vilbrandt's significant professional contributions, the editors and contributors have agreed to donate all royalties from book sales to establish the *Frank Vilbrandt Memorial Fund*. The fund will be used to promote and recognize excellence in performing process- and plant-design projects by chemical engineering students at Virginia Tech.

Because of its broad coverage and in-depth review of recent developments in chemical process and plant design, this book provides both an introduction to the subject as well as a fairly comprehensive status report. It should be of value and interest to all chemical engineers, process engineering managers, and chemical engineering faculty. It should also be beneficial to chemical engineering seniors and graduate students as a textbook or reference volume for design courses.

Y. A. LIU
HENRY A. MCGEE, JR.
W. ROBERT EPPERLY

*Blacksburg, Virginia*
*Annandale, New Jersey*
*September 1987*

ix

# ACKNOWLEDGMENTS

It is a pleasure to thank the individuals and organizations who contributed directly and indirectly to the preparation of this volume.

We express our deepest appreciation to the authors (and their institutions and companies), who contributed their precious time in preparing the manuscripts and who relinquished the tangible rewards of their efforts so that the Frank Vilbrandt Memorial Fund could be established.

We are indebted to Professor William L. Conger, Department Head of Chemical Engineering at Virginia Tech, who encouraged the preparation of this book and provided the necessary staff and financial support. Our thanks go to James L. Smith and the rest of the staff of Wiley-Interscience for their continued assistance.

We recognize also the following individuals, who reviewed the manuscripts for this volume: Seymour B. Alpert, Electric Power Research Institute; James D. Batchelor, U.S. Department of Energy; Lorenz J. Biegler, Carnegie-Mellon University; J. Peter Clark, Epstein Process Engineering, Inc.; Richard D. Colberg, California Institute of Technology; James Doss, Tennessee Eastman Company; James M. Douglas, University of Massachusetts; Heinz Heinmann, Lawrence Berkeley Laboratory; Ernest J. Henley, University of Houston; Davis E. James, Asheville, North Carolina; C. Judson King, University of California at Berkeley; Trevor A. Kletz, Loughborough University of Technology, United Kingdom; Walter C. Kohfeldt, Exxon Chemical Company; Pierre R. Latour, Set Point, Inc.; Kenneth N. McKelvey, E. I. du Pont de Nemours Company; Manfred Morari, California Institute of Technology; Alex G. Oblad, University of Utah; Richard Pollard, University of Houston; Mark A. Stadtherr, University of Illinois; Larry T. Thompson, AT&T Bell Laboratories; M. Albert Vannice, Pennsylvania State University; and Irving Wender, University of Pittsburgh.

We are grateful to Diane Cannaday and Diane Haden at Virginia Tech and Kathy Greeley at Exxon Research and Engineering Company, who have patiently and skillfully handled the extensive correspondence and typing generated by the preparation of this volume. Finally, the first editor (YAL) acknowledges a special debt to his wife, Hing-Har, who so graciously undertook what must have been the seemingly endless chores associated with the preparation of this volume.

Y.A.L.
H.A.M.
W.R.E.

# CONTENTS

## 11. PROCESS AND PLANT DESIGN FOR FOOD AND BIOCHEMICAL PRODUCTION 383

**Henry G. Schwartzberg**

## 12. CHEMICAL ENGINEERING IN ELECTRONIC MATERIALS PROCESSING: SEMICONDUCTOR REACTION AND REACTOR ENGINEERING 457

**T. W. Fraser Russell**

# 1

# CATALYST DESIGN: SELECTED TOPICS AND EXAMPLES

JOHN H. SINFELT

1.1. Introduction

1.2. Transport Considerations
  A.  Pore Diffusion—Influence on Catalytic Activity
  B.  Influence of Pore Diffusion on Selectivity

1.3. Extent of Catalyst Surface
  A.  Determination of Total Surface Area—Physical Adsorption
  B.  Dispersion of Supported Metals—Selective Chemisorption

1.4. Catalyst Structure
  A.  Structure of Pt-Ir Catalysts
  B.  Surface Composition of Metal-Alloy Catalysts
  C.  Bimetallic Systems of Immiscible Components

1.5. Intrinsic Activity and Selectivity
  A.  Hydrogenolysis of C-C Bonds in Alkanes
  B.  Selective Inhibition of Hydrogenolysis

1.6. Bifunctional Catalysis
  A.  Nature of Platinum-on-Alumina Catalysts
  B.  Hydrocarbon Reactions on $Pt/Al_2O_3$ Catalysts

1.7. Catalyst Poisoning and Deactivation
  A.  Poisoning by Impurities in a Feed Stream

## 1.1.  INTRODUCTION

Catalytic processes have been known and practiced by man since ancient times. An early example is the production of wine and vinegar by fermentation. While the nature of the fermentation process was not appreciated by the ancients, it is now known that enzymes are involved. By the early 1800s, many diverse observations had been made of processes in which the transformation of one kind of chemical species into another apparently depended on the presence of an additional agent, although the agent itself was not consumed in the process. Examples included (1) the production of ether from alcohol by the action of sulfuric acid, (2) the conversion of starch to sugar in the presence of dilute acids, and (3) the reaction between oxygen and hydrogen that occurs when a platinum wire is introduced into a mixture of the two gases.

In 1836, the Swedish chemist Berzelius critically examined these apparently unrelated observations and recognized the existence of a general kind of phenomenon, which he called ''catalysis.'' The word catalysis is of Greek origin and means ''loosening.'' Berzelius believed that catalysis was a force that various agents called catalysts could exert on substances to bring about chemical change, the catalysts themselves remaining unaltered in the process. In choosing the name catalysis for the phenomenon, Berzelius perhaps thought that the catalytic force loosened the binding between the atoms of a chemical species, thus making it possible for the atoms to enter into a new configuration characteristic of a different chemical species.

Although the concept of a catalytic force was eventually discarded, the recognition of catalysis as a general phenomenon was itself an important milestone in chemistry. It led to a search for new examples of the phenomenon and stimulated scientists to develop a better understanding of catalytic action. Around 1900, a major fundamental advance in the understanding of catalytic phenomena was made by the German chemist Wilhelm Ostwald, who proposed that the influence of a catalyst is limited to an effect on the rate of a chemical reaction. Inherent in this proposal was the idea that a chemical change occurring in the presence of a catalyst must also be capable of occurring in the absence of the catalyst, although at a lower rate. The rate of chemical change in the absence of a catalyst may be so low that it escapes detection. There are numerous examples of reactions that should proceed to a very large extent on the basis of thermodynamics, but which are virtually undetectable in the absence of suitable catalysts. Thus a mixture of hydrogen and oxygen in a molecular ratio of $2:1$ can be maintained in a vessel at room temperature for years with no evidence of formation of water, although thermodynamic calculations indicate that the reaction should proceed to virtually 100%. If a finely divided platinum catalyst is introduced into the mixture of hydrogen and oxygen, the formation of water occurs readily. The role of the catalyst is to increase the reaction rate, an aspect of the reaction not addressed by thermodynamics. The extent of reaction possible at equilibrium is not affected by the catalyst. Only the rate of approach to equilibrium is influenced.

Thus Ostwald identified catalysis as a phenomenon in chemical kinetics. His proposal provided a basis for scientific inquiry into catalysis and paved the way for the widespread investigation and application of catalytic phenomena (Sinfelt, 1984a,b; 1985). Since the time of Ostwald, the rates of many different kinds of catalytic reactions have been measured as a function of reaction conditions and of catalyst properties. Such rate data have been extremely important for designing reactors used in industrial catalytic processes and for obtaining insight into catalytic phenomena. The impact of catalysis during the twentieth century has been enormous—it now provides the basic technology for the manufacture of a host of vitally important materials, ranging from fertilizers, to synthetic fibers, to petroleum products such as gasoline and heating oil.

Catalytic processes are commonly divided into two categories, homogeneous and heterogeneous (see, for example, Sinfelt, 1979). The former are processes in which the catalyst and reactants are present in a single phase, as in a solution. In the latter, the reactants and catalyst are present in separate phases, a common example comprising reactants in a vapor phase in contact with a solid catalyst. There are many examples of both types of catalysis; I restrict my attention to heterogeneous catalysis, which is especially important in large-scale industrial applications.

In heterogeneous catalysis, the reaction occurs at the boundary between the phases. For example, molecules in a vapor may undergo reaction on the surface of a solid employed as a catalyst. In a typical situation, a vapor is passed through a bed of solid catalyst particles. In one form of metal catalyst which is widely used commercially, the catalyst particles consist of a porous refractory material and small metal crystallites or clusters dispersed throughout the particles. The term "carrier," or "support," is commonly used in referring to the refractory material, and the catalyst is known as a supported metal catalyst. In a small laboratory reactor, the particles could be granules approximately 0.5 mm in size. In a commercial reactor, the particle size would be somewhat larger, perhaps 2 or 3 mm. The refractory material constituting the bulk of the particles is frequently an oxide such as alumina ($Al_2O_3$) or silica ($SiO_2$), with a structure consisting of a network of pores. A schematic drawing of a catalyst particle illustrating this network is given in Figure 1.1, which also includes an enlarged view of a single pore (Sinfelt, 1985). The metal clusters or crystallites reside on the walls of the pores. In a catalytic reaction, molecules from the vapor stream diffuse into the pores of the catalyst and are adsorbed on the active metal clusters. The adsorbed molecules

TYPICAL CATALYST PARTICLE
NETWORK OF PORES

SINGLE PORE
ENLARGED VIEW

METAL
CLUSTERS

**FIGURE 1.1** Schematic drawing of a catalyst particle with a structure comprising a network of pores with an average diameter of perhaps 100 Å. The particle consists of a refractory material such as alumina or silica, with small metal clusters residing on the walls of the pores (Sinfelt, 1985). In some commonly used catalysts, the metal clusters have sizes on the order of 10 Å.

then undergo chemical transformations on the clusters to yield molecules of a different chemical species, which are subsequently desorbed to yield molecules of product in the pores. The product molecules must then diffuse out through the network of pores into the gas stream flowing through the void space between the particles. The composition of the gas stream changes as it is depleted of molecules of reactant and enriched in molecules of product in its passage through the catalyst bed.

In the foregoing paragraphs, I presented a very general introduction to the subject of catalysis before addressing the topic of catalyst design. In moving ahead to this topic, I begin with a few general comments intended to put the problem of catalyst design in some perspective. Some scientists have expressed the view that the design of catalysts is just now becoming possible, presumably because of recent breakthroughs in areas such as surface science, organometallic chemistry, and quantum chemistry. I do not agree with this view. For a long time, scientists and engineers with the responsibility for developing catalytic processes have applied sound physicochemical principles in the formulation and manufacture of suitable catalysts. The design of catalysts, in common with the design of virtually anything, is an ever-evolving activity. When new information becomes available, it is utilized in the design process, to the extent that it is useful to do so. With the accumulation of such knowledge, it is possible to produce more sophisticated designs. As the science of catalysis has steadily developed during the twentieth century, we have indeed witnessed its gradual incorporation in the design of catalysts and catalytic processes. Some elements of design that have played a role in the development of catalysts are considered in this review. No attempt is made to present a comprehensive treatment. The intent is to provide perspective and to impart some of the flavor of the design process by a choice of selected topics and examples. While much emphasis is given to metal catalysts, which are of primary interest to the author, the issues considered are relevant to the design of catalysts in general.

## 1.2. TRANSPORT CONSIDERATIONS

In a reactor in which a fluid is passed through a bed of catalyst granules or pellets, the transport of molecules between the bulk-fluid phase and the active surface of the catalyst is one of the crucial steps in the conversion of reactants to products. It is customary to divide the process into two categories: (a) transport between the bulk fluid and the external surface of the catalyst pellets and (b) diffusion in the network of pores within the catalyst pellets. The first step is often called external mass transfer, while the term pore diffusion is used for the second. In the design of catalysts and catalytic reactors, transport effects are commonly handled by considering such factors as the fluid velocity in the reactor and the size and porosity of the catalyst pellets. Because of the heat of reaction generally associated with chemical change, it is also necessary to consider problems involving the transport of heat and the possibility of nonisothermal conditions in a reactor.

The topic of transport effects in heterogeneous catalysis has been considered extensively by other authors (Thiele, 1939; Weisz and Prater, 1954; Wheeler, 1955) and is not treated in any detail in this chapter. This discussion is very general, and is included to emphasize that consideration of transport effects is an important aspect of the design of catalysts and catalytic reactors. Moreover, it is an aspect which is generally well understood and under control (Wheeler, 1955). This fact is well appreciated by chemical engineers concerned with the development of industrial catalytic processes. However, the substantial progress in this area may not be apparent to others.

### A. Pore Diffusion—Influence on Catalytic Activity

Pore diffusion effects are commonly discussed with reference to the Thiele modulus $h$ defined by the expression

$$h = d\left(\frac{k}{rD}\right)^{1/2} \tag{1.1}$$

in which $d$ is the size of the catalyst pellet, $k$ is the intrinsic reaction rate constant, $r$ is the pore radius, and $D$ is the diffusion coefficient (Thiele, 1939). The possibility that a reaction will be limited by pore-diffusion effects increases as the value of $h$ increases. If $h$ is low enough, pore diffusion is not a factor in determining the rate of reaction, and the concentration of a reactant has the same value throughout a catalyst pellet. When $h$ becomes sufficiently large, the reaction is limited strongly by pore diffusion and a marked gradient in concentration of reactant from the external surface to the center of a pellet is apparent. It is customary to define a quantity known as the effectiveness factor, which is simply the ratio of the reaction rate actually obtained with a catalyst to that which would be obtained in the absence of any diffusional limitation. The effectiveness factor increases as the value of $h$ decreases, approaching unity when $h$ is very small.

From Equation 1.1 we see that diffusional effects can be minimized by increasing the size of the pores in the catalyst and by decreasing the size of the catalyst pellets. A greater extent of reaction can then be obtained with a given mass or volume of catalyst, so that the catalyst is being used more effectively. In the specification of a catalyst for a particular industrial reaction, pore size and particle size are variables that should be considered. In general, the choice of particle size for a catalyst is not made solely on the basis of diffusional effects. Other factors, such as reactor pressure drop, must also be considered.

## B.  Influence of Pore Diffusion on Selectivity

The diffusional properties of a catalyst can have an important influence on the selectivity of conversion of reactants to products, as well as the overall extent of reaction (Weisz and Prater, 1954; Wheeler, 1955). As an example, we consider a sequential reaction, $A \rightarrow B \rightarrow C$, where $B$ is the desired product and $C$ is a minor by-product resulting from the further reaction of $B$. In this situation, the intrinsic rate constant for the conversion of $A$ to $B$ is high compared with that for the conversion of $B$ to $C$. If there is no limitation due to pore diffusion (i.e., the pores are large and the catalyst particles are small), good yields of $B$ relative to $C$ can be obtained over a wide range of conversion of the reactant $A$. However, if there is a strong diffusional limitation (i.e., the pores are small and the catalyst particles are large), the product $B$ is held up in the pores and the probability of its conversion to $C$ increases. The yield of $B$ relative to $C$ in the external fluid stream is therefore lower. An example of a reaction system where the foregoing considerations would apply is the hydrogenation of acetylene to ethylene, since ethylene can undergo further reaction with hydrogen to yield ethane. A second example is the partial oxidation of hydrocarbons to various oxygen-containing products which can be degraded to carbon dioxide and water through further reaction with oxygen.

Another situation of interest with regard to selectivity is the simultaneous reaction of two different compounds $A$ and $B$ in a mixture to give products $C$ and $D$; that is, $A \rightarrow C$ and $B \rightarrow D$. If the intrinsic rate of conversion of $A$ to $C$ is much higher than that of $B$ to $D$, a strong limitation due to pore diffusion will retard the formation of $C$ to a larger extent than the formation of $D$. A reasonable example would be the simultaneous hydrogenation of a mixture of 1-hexene and benzene to $n$-hexane and to cyclohexane, respectively. The olefin hydrogenation reaction is retarded more extensively than the hydrogenation of the aromatic. If the objective is to remove the olefin from the mixture as selectively as possible by hydrogenation, it is desirable to eliminate the influence of pore diffusion by selecting a catalyst with large pores and by using particles of catalyst that are as small as possible.

In recent years, the phenomenon of shape-selective catalysis with zeolites has attracted much attention (Venuto and Landis, 1968; Weisz, 1973). Zeolites are aluminosilicates that are crystalline. The primary structural units of aluminosilicates are $SiO_4$ and $AlO_4$ tetrahedra, whether or not they are crystalline. In zeolites, however, there are larger secondary units which are clearly discernible. The secondary units are composed of the primary tetrahedra, and exist in the form of regular polyhedra, rings, or chains. Various types of zeolite structures are obtained by linking these secondary units together in different ways (Barrer, 1964; Breck, 1964). There are internal cavities of uniform size within the structures. The cavities are interconnected through well-defined openings or windows, giving rise to an intracrystalline pore structure within which catalysis can occur. The catalytic sites within this pore structure may arise from the zeolite itself or from very small metal clusters introduced into the structure. The sizes of the openings to the cavities are in the range of molecular dimensions. Consequently, there is the possibility of selectivity in the admission of different molecules into the cavities on the basis of differences in molecular size or shape.

There have been a variety of industrial applications of shape-selective catalysis (Heinemann, 1981). In an example employing a zeolite known as erionite (with a pore size of about 5 Å), normal alkanes are selectively hydrocracked out of a gasoline fraction rich in aromatic hydrocarbons and also containing branched alkanes. The normal alkanes are admitted through the pore openings, but the aromatics and branched alkanes are not. In the cavities, the normal alkanes are hydrocracked to lower-molecular-weight alkanes, which are not retained in the gasoline fraction. Since the normal alkanes are the lowest octane-number components, their removal increases the octane rating of the remaining mixture of aromatics and branched alkanes.

An improvement in the shape-selective hydrocracking process was made by replacing the erionite catalyst with a zeolite known as ZSM-5, which has a larger pore opening (6–7 Å). As a consequence of the larger pore opening, ZSM-5 can admit singly branched alkanes, and also benzene and toluene. The singly branched alkanes, as well as the normal alkanes, are therefore hydrocracked. Since the singly branched alkanes are the next lowest octane-number components of the gasoline fraction, their removal leads to additional octane-number improvement. Furthermore, olefins formed in the cracking process can alkylate the benzene and toluene. The alkylaromatics formed have higher octane numbers than the benzene and toluene, and hence contribute to additional improvement in the octane rating of the gasoline fraction. Since cracked fragments are utilized in the alkylation reaction, the yield of valuable liquid products relative to gas is also increased.

Shape-selective catalysis has also been utilized in hydrocracking processes for the selective removal of waxy components (long-chain n-alkanes) from distillate fuels and lubricating oils. This type of processing, using ZSM-5 or mordenite type zeolites, leads to a dramatic reduction in the pour point of the hydrocarbon fraction. As a result, more high-boiling hydrocarbons can be included in distillate oil products, thus increasing the yield of valuable products from a given amount of crude oil. The shape-selective features of the ZSM-5 zeolite have also been applied in the isomerization of xylenes and for the conversion of methanol to gasoline. The various applications of zeolites as shape-selective catalysts provide intriguing examples of the influence of diffusional phenomena in catalysis. Moreover, they illustrate the use of such phenomena in the design of catalysts for meeting very specific objectives.


## 1.3. EXTENT OF CATALYST SURFACE

The quantity of catalyst required to achieve a given amount of reaction depends on the extent of active surface per unit mass or volume of catalyst. Information on this property of a catalyst is commonly obtained from gas-adsorption measurements.


### A. Determination of Total Surface Area—Physical Adsorption

The determination of the surface area of a finely divided or porous solid by the low-temperature physical adsorption of a gas is a classical method in surface chemistry and catalysis (Emmett, 1954). Since physical adsorption is nonspecific with respect to the nature of the surface, the method gives information on the total surface area of a solid. Consequently, it is most useful for a catalyst consisting of a single component, such as a porous alumina or a nickel powder. The method involves the well-known BET (Brunauer, Emmett, and Teller) equation

$$\frac{x}{v(1-x)} = \frac{1}{v_m c} + \frac{(c-1)x}{v_m c} \tag{1.2}$$

In this equation $v$ is the volume of gas adsorbed at relative pressure $x$, where $x$ is equal to $p/p_o$, the ratio of the equilibrium gas pressure to the saturation vapor pressure at the temperature of the measurement. The quantity $v_m$ is the volume of gas adsorbed on completion of the first monolayer. The constant $c$ is a function of the heat of adsorption in the first layer and of the heat of liquefaction of the gas. A plot of the left-hand side of the equation versus $x$ is a straight line, the slope and intercept of which yield the quantities $v_m$ and $c$. From $v_m$ one can determine the number of molecules adsorbed in a monolayer, which, when multiplied by a suitable value for the cross-sectional area of the adsorbed molecule, yields the surface area of the solid. In the application of the BET method, the adsorbate is frequently nitrogen or one of the rare gases, and the isotherm is often measured at the boiling point of nitrogen, 77K. For nitrogen, the cross-sectional area of the adsorbed molecule is generally taken as 16.2 $\text{Å}^2$ in computing surface areas. The BET method has been widely used by catalytic chemists for almost half a century for the characterization of catalytic materials. It is among the foremost developments in surface science during the twentieth century.


### B. Dispersion of Supported Metals—Selective Chemisorption

For a catalyst consisting of several components, one is often interested in determining the extent of surface associated with a particular component. A good example is a supported metal catalyst, in which metal clusters or crystallites are dispersed on the surface of a carrier such as alumina or silica. Frequently, the metal surface area is a small fraction of the total surface area of the material. If a reaction occurs exclusively on the metal component, one is more concerned with the extent of metal surface than with the total surface associated with the catalyst mass. Since nonspecific physical adsorption can only give information on the total surface area of the composite material, it is inadequate for the purpose.

In the case of a supported metal catalyst, the property of interest is the ratio of surface atoms S to total atoms M in the metal clusters or crystallites present. This ratio, $S/M$, is generally a function of M. If we consider a face-centered cubic metal such as platinum, and make the simplifying assumption that the metal clusters are cubes, we can calculate the relation between $S/M$ and M. The results of the calculation are given in Figure 1.2. The value of $S/M$, which is commonly called the metal dispersion (Boudart, 1969), increases as M decreases, approaching unity when M becomes sufficiently small. Metal dispersions close to unity are commonly realized in precious-metal catalysts used for the production of high antiknock quality components for gasoline. Such high dispersions imply that the metal clusters are extremely small, of the order of 10 Å in size. Alternatively, metal dispersions of unity would be obtained if the metal clusters were "raft-like" or "plate-like" structures consisting of single layers of metal atoms on the carrier. This alternative would require significant interaction between the metal and carrier to impart stability to clusters with such shapes.

An experimental estimate of $S/M$ can be made from a measurement of the amount of a gas chemisorbed on the metal clusters. In contrast with physical adsorption, the chemisorption must be very selective, readily saturating the surfaces of the metal clusters with a monolayer but not occurring on the metal-free surface of the carrier. The chemisorption of a gas such as hydrogen, carbon monoxide, or oxygen at room temperature has been used very effectively for this purpose (Sinfelt, 1975). Use of the method requires knowledge of the stoichiometry of the chemisorption process, that is, the number of molecules chemisorbed per surface

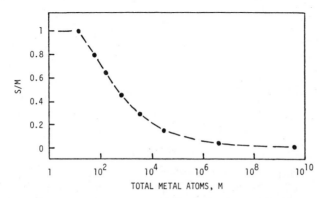

**FIGURE 1.2** The ratio of surface atoms S to total atoms M in a metal crystal, as a function of M. For simplicity, the metal crystals are assumed to be cubes with the face-centered cubic structure.

metal atom. A measurement of the ratio of molecules chemisorbed to total number of metal atoms present in the catalyst, coupled with a knowledge of the chemisorption stoichiometry, makes it possible to determine the metal dispersion $S/M$.

In the chemisorption of hydrogen on the Group VIII metals, the hydrogen molecule dissociates into atoms. Consequently, it is common to express chemisorption data in terms of the quantity $H/M$, which represents the ratio of chemisorbed hydrogen atoms H to total metal atoms M present in the catalyst. The determination of the value of $H/M$ corresponding to monolayer coverage of the metal clusters is made from a chemisorption isotherm such as that shown in Figure 1.3 for a rhodium-on-alumina catalyst (Via et al., 1983). In the right-hand ordinate of the figure, the quantity $H/Rh$ corresponds to $H/M$. The amount adsorbed is only slightly pressure-dependent over the range of pressures employed in obtaining the data. The small pressure dependence is associated with a weakly bound fraction of the total chemisorption. An estimate of the strongly bound fraction is commonly made by extrapolating the data back to zero pressure, since saturation with regard to the strongly chemisorbed component is attained at equilibrium pressures very much lower than those corresponding to the measured isotherm. If the adsorption stoichiometry for the strongly bound fraction is one hydrogen atom per surface rhodium atom, the extrapolated value of $H/Rh$ is equal to the dispersion $S/M$.

In Figure 1.4, data on $H/M$ are shown for a series of rhodium catalysts of widely varying rhodium concentration (Yates and Sinfelt, 1967; Sinfelt, 1972b), including a rhodium powder (100% Rh) and a number of silica-supported rhodium catalysts with rhodium concentrations varying from 0.1 to 10%. The value of $H/M$

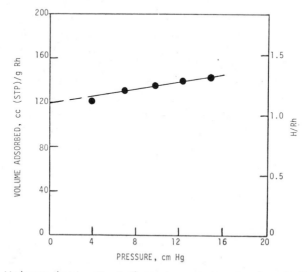

**FIGURE 1.3** Hydrogen chemisorption isotherm at room temperature for a rhodium-on-alumina catalyst containing 0.5 wt % rhodium. The quantity $H/Rh$ in the right-hand ordinate represents the ratio of the number of hydrogen atoms adsorbed to the number of rhodium atoms in the catalyst (Via et al., 1983). Reprinted with permission from the American Institute of Physics.

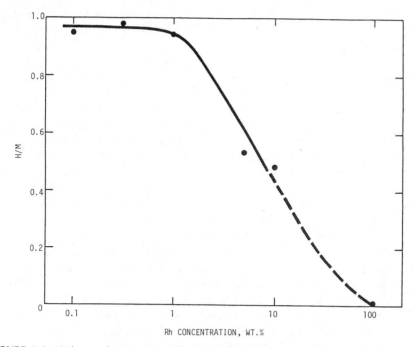

**FIGURE 1.4** Hydrogen chemisorption data at room temperature on a series of rhodium catalysts of varying rhodium concentration. The catalysts containing 0.1–10% rhodium consist of rhodium clusters dispersed on silica, while the 100% rhodium catalyst is a metal powder. The quantity $H/M$ refers to the number of hydrogen atoms adsorbed per atom of rhodium in the catalyst. All of the catalysts were contacted with hydrogen at 723K in their preparation (Yates and Sinfelt, 1967; Sinfelt, 1972b). Figure prepared from data tabulated in the paper by Yates and Sinfelt (1967).

increases continuously with decreasing rhodium concentration, approaching a limiting value close to 1 at the lowest rhodium concentrations. The size of the rhodium clusters in the catalysts decreases as the rhodium concentration decreases. At a rhodium concentration of 1 wt %, the clusters are so small that virtually all of the rhodium atoms are surface atoms. Decreasing the rhodium concentration below 1 wt % therefore has little effect on the dispersion $S/M$, thus accounting for the approach of $H/M$ to a limiting value. The fact that the limiting value is close to 1 is consistent with a stoichiometry of one adsorbed hydrogen atom per surface metal atom. Hydrogen chemisorption data on dispersed platinum catalysts also appear to be consistent with such a stoichiometry. Moreover, this stoichiometry may be closely approached for hydrogen chemisorption on other Group VIII metals dispersed on carriers such as silica and alumina. A notable exception is encountered with dispersed iridium catalysts, where a stoichiometry of two adsorbed hydrogen atoms per surface metal atom is suggested by data obtained by various workers (McVicker et al., 1980; Carter et al., 1982).

The development of selective chemisorption methods that provide a satisfactory estimate of metal dispersion has been one of the important advances in metal catalysis during the past couple of decades. It has been important for fundamental studies of catalysis by dispersed metals. It has also been very valuable in the characterization of metal catalysts used in industrial processes and has been employed widely for monitoring the quality of metal catalysts manufactured on a large scale for industrial use. It has also been a valuable tool for assessing changes in catalyst properties resulting from exposure to conditions encountered in commercial practice (e.g., in the regeneration and reactivation of metal catalysts).

In the design of a supported metal catalyst for an industrial application, metal dispersion is a very important consideration. There is the matter of determining how the catalyst should be prepared to obtain high dispersion. Frequently, the metal is incorporated in the catalyst by contacting the carrier with an aqueous solution of a metal precursor. After the material is dried, it is commonly heated to a higher temperature to decompose the original metal precursor. This heat treatment is often called calcination. The choice of temperature for this step is an important, and frequently critical, consideration.

In the preparation of a platinum-on-alumina catalyst, the first step generally involves contact of alumina with a chloroplatinic acid solution (Sinfelt, 1972b). After a drying step, the material is exposed to air at a temperature of 775–875K for several hours. Subsequent exposure to hydrogen at a temperature in the vicinity of 725–775K yields a material consisting of very small clusters of platinum dispersed on alumina. If exactly the same procedure is used to prepare an iridium-on-alumina catalyst, with the substitution of chloroiridic acid for chloroplatinic acid in the first step, the dispersion of the iridium in the final material is very low (Garten and Sinfelt, 1980). In this case, exposure to air at a temperature of 775–875K is deleterious, leading to formation of large crystallites of $IrO_2$ which are readily identified in an X-ray diffraction pattern. Upon reduction of the material, large iridium crystallites are obtained, so that the metal dispersion is an order of magnitude lower than that obtained for a platinum catalyst. To obtain a high iridium dispersion, it is vital to avoid exposure of the material to air at temperatures as high as those used in the preparation of a platinum catalyst. Decreasing the calcination temperature from 775–875K to 525–575K avoids the formation of large crystallites of $IrO_2$ and leads to a well-dispersed iridium catalyst.

## 1.4. CATALYST STRUCTURE

The structure of a catalyst, or one of its primary components, can have a major influence on its performance. As already noted, the pore structure of a catalyst is important from the standpoint of transport of molecules to and from the active sites in a catalyst. Both the activity and selectivity of a catalyst can be affected. In the case of zeolite catalysts, the dimensions of the openings into the cavities within the crystals can be critical with regard to the types of molecules which can be admitted to undergo reaction. Considerations of structure other than pore structure are also frequently important, as illustrated in this section for bimetallic catalyst systems.

## A. Structure of Pt-Ir Catalysts

For metal catalysts, we have seen that the conditions of preparation can have a crucial effect on the nature of the catalyst, as was illustrated for a supported iridium catalyst. Depending on the calcination temperature used, a catalyst with large iridium crystallites or very small iridium clusters can be obtained. An important extension of this phenomenon is observed in the preparation of supported platinum–iridium bimetallic catalysts (Sinfelt, 1976; Sinfelt and Via, 1979). Catalysts containing platinum–iridium clusters are of interest for reforming petroleum naphtha fractions for the production of high-octane-number components of gasoline. The reforming process, and the use of platinum–iridium catalysts, are discussed in another section. Here we are concerned strictly with the question of the influence of preparative conditions on the structure of a platinum–iridium catalyst.

In a typical preparation of highly dispersed platinum–iridium bimetallic clusters on a silica or alumina carrier, the carrier is first coimpregnated with chloroplatinic and chloroiridic acids and then dried at a temperature of 383K prior to calcination in air at a temperature in the approximate range of 525–650K. The material is exposed to hydrogen at a temperature of 725–775K in situ in a reactor. The platinum and iridium in the catalyst are then present in the form of platinum–iridium clusters. If the calcination temperature used in the preparation of the catalyst is too high (e.g., 773K), the iridium undergoes oxidation and agglomeration to form large crystallites of $IrO_2$. On subsequent treatment in hydrogen at 773K, the $IrO_2$ is reduced to yield large crystallites of iridium, as discussed earlier for a catalyst containing iridium alone. At this point, the material consists of a mixture of highly dispersed platinum or platinum-rich clusters and large crystallites of iridium. A portion of an X-ray diffraction pattern in a region for a (220) reflection is shown in the upper field of Figure 1.5 for a silica-supported platinum–iridium sample heated in air at 500°C (773K) and then contacted with hydrogen at the same temperature (Sinfelt and Via, 1979). The diffraction profile is unsymmetrical and contains a narrow line on the high-angle side corresponding to large iridium crystallites. On the low-angle side of the profile, a broader line corresponding to highly dispersed platinum or platinum-rich clusters is evident. The contrast between this profile and the profile for platinum–iridium bimetallic clusters in the lower field of Figure 1.5 is clear. The latter profile is symmetrical and is centered at a diffraction angle about midway between the angles at which lines for the pure metals are observed. These results demonstrate the importance of preparative conditions in the formation of bimetallic clusters in the platinum–iridium system. In general, exposure to air at temperatures below about 650K does not appear to be harmful, but contact with air at temperatures above about 720K should be avoided.

While calcination of a platinum–iridium catalyst at 773K leads to formation of large $IrO_2$ crystallites, and subsequently to large iridium or iridium-rich crystallites on reduction, the deleterious effect is not so pronounced as it is with a catalyst containing only iridium (Carter et al., 1982). The presence of the platinum inhibits the oxidative agglomeration of the iridium at 773K, as a consequence of the interaction between the two elements in the bimetallic clusters. This stabilizing effect

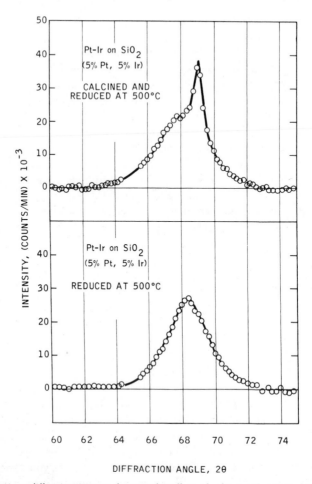

**FIGURE 1.5**  X-ray diffraction pattern showing the effect of calcining (heating) a silica-supported platinum–iridium sample in air at 500°C prior to exposure to hydrogen at 500°C, illustrating the importance of the preparative conditions in the formation of highly dispersed platinum–iridium bimetallic clusters (Sinfelt and Via, 1979). Reprinted with permission from Academic Press, Inc.

of the platinum is particularly evident in preparations containing low contents of the metals (of the order of 0.5 wt %), and has the important practical feature that the platinum–iridium catalyst is more stable than a pure iridium catalyst in the oxygen-containing gases employed for the regeneration of the catalysts. Such regeneration is periodically necessary for the removal of carbonaceous residues deposited on the catalysts during reforming. In general, however, the regeneration procedures employed with platinum–iridium catalysts differ from those used with platinum catalysts because of the greater susceptibility of iridium to oxidative agglomeration.

## B. Surface Composition of Metal-Alloy Catalysts

Considerations of surface composition are frequently very important for bimetallic catalysts. In general, the surface composition is different from that in the bulk. This phenomenon is a consequence of Gibb's adsorption principle: If accumulation of one of the components in the surface serves to lower the surface energy of the system, the surface will then be enriched in this particular component (Sinfelt, 1975). Frequently, this component is the one with the lower heat of sublimation. In the case of combinations of a Group VIII and a Group IB metal, the Group IB metal has the lower heat of sublimation and would thus be expected to concentrate in the surface. Evidence for this in the case of catalysts consisting of nickel–copper alloys can be obtained from hydrogen chemisorption data, and is based on the observation that strong chemisorption of hydrogen does not occur on copper (van der Planck and Sachtler, 1967; Cadenhead and Wagner, 1968; Sinfelt et al., 1972). The addition of only a few percent of copper to nickel decreases the amount of strongly chemisorbed hydrogen severalfold, suggesting that the concentration of copper in the surface is much greater than in the bulk. Hydrogen chemisorption data on nickel–copper alloys are given in Figure 1.6, in which the amount of strongly chemisorbed hydrogen is shown as a function of the copper content of the alloy (Sinfelt et al., 1972; Sinfelt, 1974a). Here, strongly chemisorbed hydrogen refers to the amount that is not removed by evacuation at room temperature. The data of Figure 1.6 suggest that copper is the predominant component in the surface of nickel–copper alloys containing as little as 5 at. % copper overall.

A factor that should not be ignored in discussions of the surface composition of alloys is the effect of the gaseous atmosphere in contact with the alloy surface. The conclusions drawn for Group VIII–Group IB systems ignore the effect of chemisorbed gases. If the gas interacts very strongly and selectively with the Group VIII metal, for example, the Group IB metal will not be the predominant component in the surface. Thus, in an oxygen atmosphere, the surface of the nickel–gold system becomes rich in nickel rather than gold (Williams and Boudart, 1973). Similarly, in an atmosphere of carbon monoxide, the surface of palladium–silver alloys becomes enriched with palladium, whereas normally silver would be the predominant surface component (Bouwman et al., 1972). In attempting to generalize these comments on the effect of a particular chemisorbed gas on the surface composition of an alloy, we state simply that the component of the alloy with the greater affinity for the gas is drawn to the surface.

## C. Bimetallic Systems of Immiscible Components

In the nickel–copper alloys just considered, the two components were completely miscible. However, bimetallic systems of interest in catalysis are not limited to combinations of metallic elements that form solid solutions. A good example is provided by the ruthenium–copper system (Sinfelt, 1973b; 1977; Sinfelt et al., 1976). It is known that bimetallic aggregates can be prepared in which copper tends to cover the surface of ruthenium. Thus it appears that a significant inter-

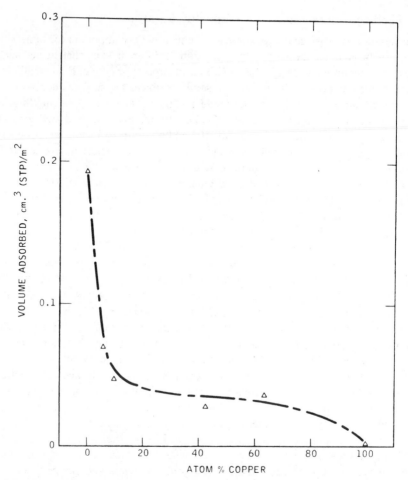

**FIGURE 1.6** Hydrogen chemisorption on nickel–copper alloys at room temperature (Sinfelt et al., 1972; Sinfelt, 1974a). The data are for strongly chemisorbed hydrogen, that is, the amount which cannot be desorbed at room temperature by evacuation for 10 min to a pressure of approximately $10^{-6}$ torr. Reprinted with permission from Academic Press, Inc.

action exists between the components at the surface. The copper can be considered as a species that forms chemical bonds with ruthenium at the interface, analogous to the bonds existing in chemisorption. The osmium–copper system behaves in a similar manner (Sinfelt, 1973b, 1977).

Two types of probes, chemical and physical, have been used in establishing the structures of these bimetallic catalysts. A catalytic reaction, the hydrogenolysis of ethane to methane ($C_2H_6 + H_2 \rightarrow 2CH_4$), is an example of one of the chemical probes. The catalytic activity of copper for the reaction is many orders of magnitude lower than that of ruthenium or osmium (Sinfelt, 1973a, 1974a). When copper is incorporated with either of these metals in a catalyst, the activity for ethane hydrogenolysis is decreased by orders of magnitude (Sinfelt, 1973b). This decrease would be impossible if there were no interaction of copper with the ruthenium or osmium in the catalysts. If we hypothesize that the copper is present as an adsorption layer on ruthenium or osmium, we can use ethane hydrogenolysis to test the hypothesis. Data on ruthenium–copper catalysts provide a good illustration (Sinfelt et al., 1976; Sinfelt, 1977). Suppose we have two different ruthenium–copper catalysts with widely different metal dispersions. One of them consists of unsupported bimetallic aggregates with a dispersion of 0.01, while the other consists of supported bimetallic clusters with a dispersion of 0.50. Thus the metal dispersion differs by a factor of 50 for the two catalysts. If the copper is present only on the surface of the ruthenium in both catalysts, the catalyst with the higher dispersion will require a 50-fold higher atomic ratio of copper to ruthenium for a given degree of coverage of ruthenium by copper, and hence for a given degree of suppression of the ethane hydrogenolysis activity of the ruthenium. This is clearly illustrated by the data in Figure 1.7, thus providing strong evidence for the hypothesis that copper is present as an adsorption layer on the ruthenium.

Physical probes that have provided additional support for this view of the structure of ruthenium–copper catalysts include electron spectroscopy and X-ray absorption spectroscopy (Sinfelt, 1983). These probes were applied somewhat later than the chemical probes, since they were not available during the early stages of the research on bimetallic catalysts. For bimetallic clusters, the extended fine

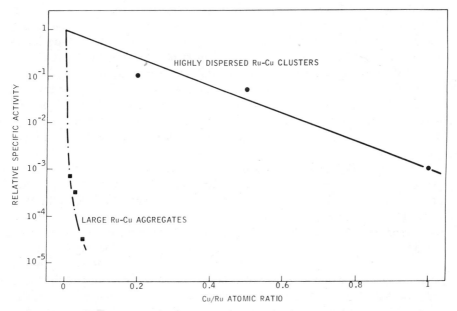

**FIGURE 1.7** Influence of the state of dispersion of ruthenium–copper catalysts on the relationship between ethane hydrogenolysis activity and catalyst composition (Sinfelt et al., 1976; Sinfelt, 1977). Reprinted with permission from Academic Press, Inc.

structure associated with X-ray absorption edges has been found to be especially useful for structural investigations (Sinfelt et al., 1984.)

## 1.5. INTRINSIC ACTIVITY AND SELECTIVITY

It is customary to use the term activity in place of reaction rate in referring to the performance of a catalyst. In general, people working in the field are interested in comparisons of activities of different catalysts or of modifications of a particular catalyst. Reaction rates depend on such variables as temperature and the concentrations or partial pressures of the various reactants, products, and other chemical species present in a reactor. Consequently, it is important to specify the reaction conditions when comparing the activities of catalysts.

At present, we do not have a sufficiently detailed understanding of catalysis to permit a calculation of intrinsic catalytic activity from a knowledge of the electronic structures of the chemical species involved. Indeed, the ability to make such calculations is still very limited even for the simplest chemical reactions not involving the complexities of heterogeneous catalysis. Our information on rates of chemical reactions in general is obtained experimentally. Significant progress in understanding chemical reactivity has evolved from experiments designed to yield information on the nature of the intermediate steps and chemical species involved in a reaction (i.e., the reaction mechanism), and by relating reaction rates to thermodynamic properties and structural features of the species involved.

Although catalyst design is not based on calculations of activities from first principles, useful deductions can be made from a knowledge of the general chemical properties of materials. It is customary for chemists to relate such properties to the positions of the elements in the periodic table. Elements in a given group tend to show similar chemical behavior, while those in a given period exhibit a systematic variation of properties across the period. For metals, as an example, it is instructive to consider how catalytic activity varies from one metal to another within such a period. Since Group VIII of the periodic table has three subgroups, designated as $VIII_1$, $VIII_2$, and $VIII_3$, a period contains three metals within this group.

For a number of reactions catalyzed by metals, maximum activity within a given period is found for a metal in one of the subgroups within Group VIII. The particular subgroup in which the maximum activity is observed depends on the reaction; and for a given reaction, it depends on which period is being considered. The patterns of variation of catalytic activity can be rationalized in a general way with the aid of a principle relating catalytic activity to the ease and strength of chemisorption of the reactants. According to this principle, maximum activity results when chemisorption of a reactant molecule is fast but not very strong. If the chemisorption bond is too strong, the surface complex is too stable to undergo reaction, or the product of the surface reaction does not desorb readily from the surface (Boudart, 1968; Sinfelt, 1975). For metal catalysts, this extreme is approached by metals immediately preceding the Group VIII metals in the periodic table, that is, by the metals in Groups VIA and VIIA. At the other extreme, which is approached by the Group IB metals immediately following the Group VIII metals, the chemisorption is significantly weaker and may be too weak to activate the reactant molecule in the manner required for the catalytic reactions to occur. Alternatively, the chemisorption may be prohibitively slow. For many reactions, the optimal situation in which chemisorption is fast but not too strong is observed for the metals in Group VIII of the periodic table, and consequently these metals are especially important in catalysis (Sinfelt, 1975).

The activities of metals for the hydrogenolysis of the carbon–carbon bond in alkanes provides a good illustration of the general ideas introduced in the previous paragraph (Sinfelt, 1973a; 1974a; 1985). This reaction is not a desirable one from the point of view of economics. In general, one is interested in suppressing hydrogenolysis in favor of other types of reactions leading to more valuable products. One might justify a study of the reaction on the basis that the knowledge gained could lead to ideas for moderating the hydrogenolysis activity of a metal, thereby rendering it more selective for reactions of greater value. In fact, the discovery of certain types of bimetallic catalysts for accomplishing this objective was influenced strongly by the results of studies on the hydrogenolysis activities of metals (Sinfelt, 1983; 1985).

## A.  Hydrogenolysis of C-C Bonds in Alkanes

Hydrogenolysis reactions of hydrocarbons involve the rupture of carbon–carbon bonds and the formation of carbon–hydrogen bonds. The simplest hydrogenolysis reaction of a hydrocarbon is the conversion of ethane to methane

$$C_2H_6 + H_2 \rightarrow 2CH_4$$

In this reaction, the available evidence indicates that ethane is chemisorbed with dissociation of carbon–hydrogen bonds (Cimino et al., 1954; Sinfelt, 1973a; 1974a). This yields a hydrogen-deficient surface species $C_2H_x$ which undergoes carbon–carbon bond scission. The following sequence of reaction steps leading to formation of $C_1$ surface fragments ($C$, $CH$, or $CH_2$) may be visualized, the symbol (ads) signifying an adsorbed species

$$C_2H_6 \rightleftarrows C_2H_x(ads) + aH_2$$

$$C_2H_x(ads) \rightarrow \text{adsorbed } C_1 \text{ fragments}$$

The adsorbed $C_1$ fragments are hydrogenated to methane to complete the reaction. The quantity $a$ is equal to $(6 - x)/2$.

A comparison of metals as hydrogenolysis catalysts reveals a striking variation in activity. Specific catalytic activities of all the metals in Group VIII and of rhenium in Group VIIA for the hydrogenolysis of ethane to methane are given in Figure 1.8 (Sinfelt, 1973a; 1974a). The metals were supported on silica. The figure has three separate fields representing the metals of the first, second, and third transition

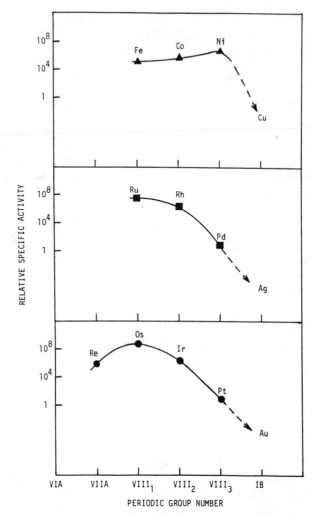

**FIGURE 1.8** Catalytic activities of metals for the hydrogenolysis of ethane to methane (Sinfelt, 1973a, 1974a, 1985). The activities were determined at a temperature of 478K and at ethane and hydrogen partial pressures of 0.030 and 0.20 atm, respectively. Reprinted with permission from Academic Press, Inc.

series. Activities of the Group IB metals (copper, silver, gold) are not shown, but they are much lower than the activities of the least active Group VIII metals (Sinfelt, 1973a). The most complete data for metals within a given period are those for the metals of the third transition series. Beginning with rhenium in Group VIIA, and proceeding in the direction of increasing atomic number to osmium, iridium, and platinum in Group VIII and on to gold in Group IB, the hydrogenolysis activity attains a maximum value at osmium. From osmium to platinum alone, the activity decreases by 8 orders of magnitude. A similar variation is observed from ruthenium to palladium in the second transition series.

In the first transition series, the Group VIII metals (iron, cobalt, and nickel) are much more active for hydrogenolysis than copper in Group IB. In this respect, the first transition series is very similar to the second and third transition series just discussed. However, maximum catalytic activity in the first transition series is observed for the metal in the third subgroup within Group VIII (i.e., nickel), whereas in the second and third transition series, the maximum activity is observed for the metal in the first subgroup within Group VIII, namely, ruthenium or osmium. Thus the pattern of variation of hydrogenolysis activity among the Group VIII metals of the first transition series is somewhat different from that observed for those of the second and third transition series. This tends to parallel known chemical differences between elements of the first transition series on the one hand, and the corresponding elements of the second and third transition series on the other.

We interpret the activity plots in Figure 1.8 in the following manner. The Group IB metals (Cu, Ag, Au) are inactive because of their inability to chemisorb ethane. This may be a consequence of very weak metal–carbon bonds, a high activation barrier for dissociative chemisorption of ethane, or both (activation barriers commonly increase with decreasing heats of adsorption). The Group VIII metals, in general, are not limited by their ability to chemisorb ethane. On platinum, for example, chemisorption occurs at temperatures far lower than those required for scission of the carbon–carbon bond. This is illustrated by data showing that platinum catalyzes the exchange reaction of ethane with deuterium at temperatures as low as 400K (Bond, 1962), while hydrogenolysis requires temperatures in the vicinity of 600K (Sinfelt, 1973a). For platinum, the rate-determining step in ethane hydrogenolysis is the scission of the carbon–carbon bond in the intermediate $C_2H_x$. On moving from platinum to iridium to osmium in Group VIII, the strength of the metal–carbon bond would be expected to increase, leading to higher rates of scission of the carbon–carbon bond. At some point the rate of scission becomes high relative to the rate of the subsequent process in which methane is formed from the chemisorbed monocarbon fragments. The latter process may consist of several steps, which are commonly lumped together to represent the product desorption step in the reaction. When this process constitutes the slow part of the reaction, one says that desorption of the product is rate limiting. An increase in metal–carbon bond strength then lowers the rate of desorption of methane, thus decreasing hydrogenolysis activity. A maximum activity for hydrogenolysis is therefore observed at the point where the rate-limiting step changes from carbon–carbon scission to methane desorption. For the metals of the third transition series, this change is observed on moving from osmium to rhenium. The situation is very similar for the metals in the second transition series, where maximum activity is observed for the metal in the same subgroup within Group VIII. For the metals of the first transition series, however, the shift in rate-determining step appears to occur at a point further to the right in the series, between nickel and cobalt.

The purpose of the foregoing discussion on rate-determining steps is to provide some rough guidance on how they might be expected to change when one considers ethane hydrogenolysis on a group of metals within a given series. It must be realized, however, that the rate-determining step for a given metal can change if the reaction is conducted over a wide enough range of conditions. Consequently, the details of such a discussion may be affected to some extent by the choice of conditions, particularly if one considers conditions substantially different from those ordinarily employed in studies of the reaction.

In the hydrogenolysis of alkanes with a number of nonidentical carbon–carbon bonds, there is the possibility of different rates of scission at different locations in the molecule. For platinum, such differences are not very pronounced, and the scission tends to be nonselective (Matsumoto et al., 1970; Carter et al., 1971). However, the situation is very different for the Group VIII metals of the first transition series (iron, cobalt, nickel). On these metals, the hydrogenolysis products are consistent with a reaction scheme involving successive demethylation of the hydrocarbon chain (Matsumoto et al., 1971). According to this scheme, scission occurs only at terminal carbon–carbon bonds. Thus one of the fragments is always a $C_1$ species which is hydrogenated to form methane, while the other fragment has one of two fates. It can undergo further scission at the terminal carbon–carbon bond to produce additional methane, or it can be hydrogenated and desorbed into the gas phase. On proceeding from nickel to cobalt to iron, one finds that the initial distribution of products shifts markedly in the direction of lower-carbon-number alkanes. On nickel, at temperatures of 450–525K, the products of $n$-hexane hydrogenolysis at low conversion levels are equimolar amounts of $n$-pentane and methane. On cobalt, by contrast, the molar ratio of methane to $n$-pentane in the products is approximately 10. On iron, $n$-pentane is not observed at all. For cobalt and iron, the results suggest that product desorption is the rate-limiting step in the reaction. For nickel, the reaction is probably limited by scission of carbon–carbon bonds. This is consistent with the suggestion made earlier in the discussion of ethane hydrogenolysis data on nickel, cobalt, and iron.

## B. Selective Inhibition of Hydrogenolysis

In the early 1960s, I became interested in bimetallic catalysts and initiated a research program in this area at the EXXON laboratories (Sinfelt, 1983). One major aspect of the research was an investigation of the possibility of altering the selectivity of a metal catalyst by incorporating a second metal component. The possibility of selectivity effects in hydrocarbon transformations on bimetallic catalysts may be considered from two different points of view. It is possible that the introduction of a second metal component will influence the strength of bonding of a reactant molecule to the surface, and that different types of transformations will be affected differently. In this kind of effect, one visualizes an electronic interaction between the metal components. It is also conceivable that selectivity effects can arise as a consequence of differences in the sensitivity of different reactions to surface structure. We consider the latter possibility first.

It is useful for discussion purposes to represent a catalytic site by the symbol

$S_n$, where the subscript $n$ refers to the number of active metal atoms in the site. If a reaction proceeds via a chemisorbed species requiring a site with a high value of $n$, it will be more sensitive to dilution of the surface with atoms of an inactive metal than will a reaction in which the chemisorbed species can be accommodated by a site with a low value of $n$. From the results of a variety of investigations of the catalytic activities of metal surfaces for hydrogenolysis and dehydrogenation reactions, we hypothesize that the value of $n$ is higher for the former. On this basis, we expect the hydrogenolysis activity of an active metal to be suppressed more strongly than the dehydrogenation activity when atoms of an inactive metal are randomly distributed in the surface.

For a specific example, it is useful to consider the catalytic activities of nickel–copper alloys for the hydrogenolysis of ethane to methane and the dehydrogenation of cyclohexane to benzene. In comparison with nickel, copper is inactive for the catalysis of either ethane hydrogenolysis or cyclohexane dehydrogenation. Before considering the activities of nickel–copper alloys for these two reactions, we recall that the surface composition of a nickel–copper alloy is in general different from the overall composition. Copper concentrates markedly in the surface of nickel–copper alloys. Thus, although a nickel–copper alloy may contain only a few percent of copper, it is the predominant component in the surface.

The activities of nickel–copper alloys for ethane hydrogenolysis and cyclohexane dehydrogenation have been investigated as a function of alloy composition (Sinfelt et al., 1972). The alloys used in the studies were metal powders (granules). Approximately one atom out of a thousand in the alloys was a surface atom. Activity is defined as the number of molecules of ethane or cyclohexane converted per second per square centimeter of alloy surface at a temperature of 589K. The activities for ethane hydrogenolysis were obtained with a reaction mixture of ethane and hydrogen diluted with an inert gas, helium. The partial pressures of ethane and hydrogen were 0.030 and 0.20 atm, respectively. In cyclohexane dehydrogenation, hydrogen is a product rather than a reactant. Nevertheless, the activities for this reaction were determined by admitting a mixture of cyclohexane and hydrogen to the reactor. This procedure is necessary to suppress side reactions in which the cyclohexane decomposes to form carbonaceous residues on the surface. Without addition of hydrogen, the rapid accumulation of such residues deactivates the catalyst. The introduction of hydrogen limits the concentration of the residues and is necessary to maintain a steady level of activity. Suppression of the dehydrogenation reaction itself by the hydrogen is insignificant at the temperature and pressure employed. The partial pressures of cyclohexane and hydrogen were 0.17 and 0.83 atm, respectively.

When a nickel–copper alloy is used as a catalyst for the aforementioned reactions, it may be regarded as a system comprising an active component (nickel) in combination with an inactive one (copper). From the hypothesis we advanced, one would expect the rate of ethane hydrogenolysis to decline more rapidly than the rate of cyclohexane dehydrogenation, when copper is introduced into a nickel surface. This expectation is clearly supported by the data in Figure 1.9 (Sinfelt et al., 1972).

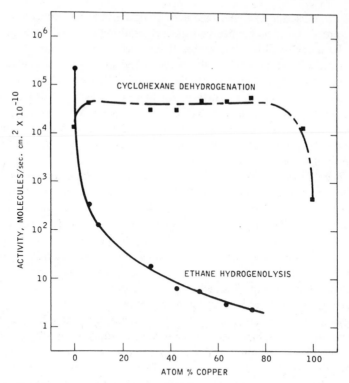

**FIGURE 1.9** Activities of nickel–copper alloys for the hydrogenolysis of ethane to methane and the dehydrogenation of cyclohexane to benzene (Sinfelt et al., 1972). The activities are reaction rates at 589K. Reprinted with permission from Academic Press, Inc.

In accordance with our earlier discussion on structure sensitivity, the effect of copper on the activity of nickel for ethane hydrogenolysis is indeed very different from the effect for cyclohexane dehydrogenation. For ethane hydrogenolysis, the presence of only 5 at.% copper decreases catalytic activity a thousandfold. That a small amount of copper has such a large inhibiting effect is associated with the marked tendency of copper to concentrate in the surface. As the copper content increases, the hydrogenolysis activity decreases continuously. In contrast, the activity of nickel for dehydrogenation of cyclohexane is affected very little over a wide range of composition, and actually increases on addition of the first increments of copper to nickel. Only as the catalyst composition approaches pure copper is a marked decline in dehydrogenation activity observed.

While a marked structure sensitivity can account for a large inhibiting effect of copper on the hydrogenolysis activity of nickel, the data on cyclohexane dehydrogenation indicate that some other factor must also be involved. Even if the reaction can occur on a single active nickel atom, an effect based purely on the dilution of the active nickel atoms would require the nickel–copper alloys to be less active than pure nickel. It does not explain why copper-rich alloys have dehydrogenation activities as high as, or higher than, that of pure nickel. The copper exerts an influence beyond that of a diluent; that is, it perturbs the properties of the surface in such a manner as to increase its intrinsic activity for the dehydrogenation reaction. The perturbation of the properties of the surface by the copper probably plays a role in the hydrogenolysis reaction too, although the effect on catalytic activity is directionally different. In view of the low ability of copper relative to nickel to chemisorb hydrocarbons, the presence of copper in nickel–copper alloys could decrease the strength of adsorption of hydrocarbon species. If the rate-limiting step in the reaction is accelerated by a decrease in the strength of adsorption (e.g., the desorption of the benzene product), the catalytic activity will then be higher when copper is present. This possibility appears reasonable for cyclohexane dehydrogenation on nickel–copper alloys with copper contents as high as 76%. When the catalytic activity finally exhibits a sharp decline at copper contents approaching 100%, it is likely that the rate-limiting step changes. For pure copper, the chemisorption of the cyclohexane itself may be limiting. However, if the rate-limiting step is inhibited by a decreased strength of adsorption, the catalytic activity will be lower when copper is present. This possibility appears reasonable in ethane hydrogenolysis, where the strength of bonding between the two carbon atoms in the chemisorbed intermediate might be expected to vary in an inverse manner with the strength of bonding of the carbon atoms to the metal.

Another example of a selectivity effect is observed with ruthenium–copper and osmium–copper bimetallic clusters. On pure ruthenium or osmium clusters, cyclohexane dehydrogenates to form benzene, but it also undergoes hydrogenolysis to yield small alkane molecules, primarily methane. When copper is incorporated with ruthenium or osmium, the hydrogenolysis reaction is strongly inhibited while the dehydrogenation reaction is affected to a much smaller degree. The selectivity of conversion of cyclohexane to benzene is therefore improved markedly, as shown by the data in Figure 1.10, where selectivity is defined as the ratio of the dehydrogenation activity D to the hydrogenolysis activity H (Sinfelt, 1973b, 1979). As noted earlier, the ruthenium–copper and osmium–copper systems are particularly interesting, because copper is essentially completely immiscible with either ruthenium or osmium in the bulk. In the bimetallic clusters, the copper is present on the surface of the ruthenium or osmium, where it forms bonds with atoms of these metals.

Thus we see that the presence of copper can influence the selectivity of a Group VIII metal whether or not it forms solid solutions with the latter in the bulk. Active arrays of Group VIII metal atoms responsible for hydrogenolysis activity are presumably broken up when copper atoms are introduced into, or on top of, the surface layer. Selective inhibition of the hydrogenolysis activity of a Group VIII metal has also been observed when gold or silver is substituted for copper (Sinfelt et al., 1969; 1971). In general, we have observed that a Group IB metal suppresses the hydrogenolysis activity of a Group VIII metal and improves its selectivity for catalyzing such reactions as the dehydrogenation and isomerization of hydrocarbons. Similar findings have been reported by other workers in the Netherlands (Roberti

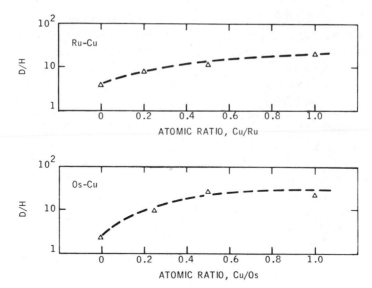

**FIGURE 1.10** Selectivity of conversion of cyclohexane over silica-supported bimetallic clusters of ruthenium–copper and osmium–copper at 589K, as represented by the ratio D/H (Sinfelt, 1973b; 1979). (D is the rate of dehydrogenation of cyclohexane to benzene, and H is the rate of hydrogenolysis to alkanes.) Reprinted with permission from Academic Press, Inc.

et al., 1973; Beelen et al., 1973). While there is still much to be learned about the detailed mechanisms of these reactions, we have seen how some simple concepts derived from systematic studies of the catalytic properties of pure metals can be useful in the design of bimetallic catalysts with improved selectivity.

## 1.6. BIFUNCTIONAL CATALYSIS

In general, a catalytic reaction involves a sequence of steps: (1) the initial chemisorption of reactants; (2) one or more surface transformations; and (3) the final desorption of products. It is conceivable that different steps in the sequence could occur on different kinds of sites. Thus, for certain types of reactions, the most effective type of catalyst may possess two quite different kinds of active sites.

### A. Nature of Platinum-on-Alumina Catalysts

For example, we consider a catalyst surface comprising a mixture of metal and acidic sites (Ciapetta and Hunter, 1953a, b; Heinemann et al., 1953; Mills et al., 1953). Such a surface can be obtained in the case of a supported metal catalyst by choosing a support that has acidic sites present at its surface. A catalyst containing small amounts of platinum dispersed on an acidic form of alumina (Haensel, 1949a, b) satisfies the requirements very well. The platinum clusters are very effective for catalyzing hydrogenation and dehydrogenation reactions, while the acidic sites on the alumina readily catalyze reactions involving a rearrangement of the carbon skeleton of hydrocarbon molecules containing carbon–carbon double bonds. The presence of the acidic sites is readily demonstrated by the affinity of the alumina surface for such basic molecules as ammonia, trimethylamine, *n*-butylamine, pyridine, and quinoline (Oblad et al., 1951).

The nature of the acidic sites on alumina has been the subject of much discussion. The surface is characterized by the presence of hydroxyl groups which could conceivably be a source of Bronsted (protonic) acidity. Chloride ions present at the surface of the alumina interact with the hydroxyl groups to enhance their acidity. The chloride ions are readily introduced in the original preparation of the catalyst by using chloroplatinic acid as the source of the platinum. They can, of course, also be incorporated by exposure of the catalyst to other chlorine-containing reagents such as HCl. Since the aluminas employed in the preparation of these catalysts are normally heated to high temperatures (775–875 K), there are Lewis acid sites present as a result of dehydroxylation reactions. A Lewis acid is defined as a species that can accept a pair of electrons from a base. At the surface of dehydroxylated alumina, there are incompletely coordinated aluminum atoms (i.e., aluminum atoms coordinated to three oxygen atoms instead of four) which could serve as electron acceptors. The interaction of these aluminum sites with a Lewis base such as ammonia, via bonding involving the lone electron pair on the nitrogen atom, is readily visualized.

## B.  Hydrocarbon Reactions on Pt/Al₂O₃ Catalysts

As an example of a reaction involving the participation of both metal and acidic sites, we consider the conversion of methylcyclopentane to benzene. The methylcyclopentane first dehydrogenates on platinum sites to yield methylcyclopentenes as intermediates (Mills et al., 1953). The latter then isomerize to cyclohexene on acidic sites. Cyclohexene subsequently returns to platinum sites, where it can either be hydrogenated to cyclohexane or dehydrogenated to benzene, the relative amounts of these products depending on reaction conditions. Isomerization of normal alkanes to branched alkanes involves the same kind of reaction sequence in which olefinic intermediates are transported between platinum and acidic sites on the catalyst. The mode of transport of the intermediates must be considered in this type of reaction scheme (Weisz, 1962). A sequence of steps involving transport between platinum and acidic sites via the gas phase gives a good account of much experimental data (Weisz and Swegler, 1957; Hindin et al., 1958; Sinfelt et al., 1960).

Kinetic investigations have played an important role in advancing the concept of bifunctional catalysis (Sinfelt, 1964; 1981). The isomerization of *n*-pentane on a platinum–alumina catalyst provides a good example (Sinfelt et al., 1960). The rate measurements were made at a temperature of 645 K. The *n*-pentane was passed over the catalyst in the presence of hydrogen at total pressures ranging from 7.7 to 27.7 atm at hydrogen to *n*-pentane ratios varying from 1.4 to 18. Over this range of conditions, the rate was found to be independent of total pressure and to increase with increasing $n$-pentane to hydrogen ratio.

The kinetic data were interpreted in terms of a mechanism involving $n$-pentene intermediates:

$$nC_5 \overset{Pt}{\rightleftarrows} nC_5^= + H_2$$

$$nC_5^= \overset{Acid}{\underset{Site}{\rightarrow}} iC_5^=$$

$$H_2 + iC_5^= \overset{Pt}{\rightleftarrows} iC_5$$

According to this mechanism the $n$-pentane dehydrogenates to $n$-pentenes on platinum sites. The $n$-pentenes are adsorbed on acid sites, where they are isomerized to isopentenes. The latter are then hydrogenated to isopentane on platinum sites, thus completing the reaction. The isomerization step on the acid sites has commonly been assumed to involve carbonium-ion type intermediates.

At the conditions used, equilibrium is readily established in the dehydrogenation and hydrogenation steps. The equilibrium concentration (or partial pressure) of $n$-pentenes in the gas phase is proportional to the molar ratio of $n$-pentane to hydrogen. The rate-limiting step is the isomerization of the $n$-pentenes to isopentenes. Consequently, the overall rate of isomerization of $n$-pentane is determined by the equilibrium partial pressure of $n$-pentenes in the gas phase, as demonstrated in Figure 1.11 by the circles. Also shown in the figure is a data point (the square) for the rate of isomerization of 1-pentene over a sample of catalyst containing no platinum. For this point, the pentene partial pressure is simply the partial pressure of the 1-pentene reactant. The point falls close to the line correlating $n$-pentane isomerization rate with the equilibrium partial pressure of $n$-pentenes, as one would expect from the foregoing discussion. Whether one starts with 1- or 2-pentene in this type of experiment should make little difference, since double-bond migration is fast compared with skeletal isomerization of olefins.

In this example of $n$-pentane isomerization, the rate of reaction is determined by the catalytic activity of the acid sites in the catalyst, since the rate-limiting step in the reaction occurs on these sites. This situation is the usual one in the isomerization of alkanes on platinum–alumina catalysts at the conditions employed in catalytic reforming. Increasing the dehydrogenation activity of the catalyst by increasing the platinum content has no significant effect on the rate of isomerization (Sinfelt et al., 1962). Data on the isomerization of $n$-heptane to 2-methylhexane and 3-methylhexane show no change in the reaction rate when the platinum content is increased from 0.1 to 0.6 wt. %. In this range of composition, the dehydrogenation activity is proportional to platinum content. When the platinum content is as high as 0.1 wt. %, the dehydrogenation activity is already high enough to maintain the equilibrium concentration of alkene (olefin) intermediates in the gas phase. Hence there is no effect of increasing the dehydrogenation activity further. The same type of result is found for the isomerization-dehydroisomerization of alkyl-

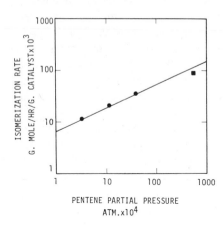

ISOMERIZATION RATE
G. MOLE/HR/G. CATALYST$\times10^3$

PENTENE PARTIAL PRESSURE
ATM.$\times10^4$

**FIGURE 1.11** Isomerization rate as a function of pentene partial pressure at a temperature of 645K. The circles are data for the rate of isomerization of *n*-pentane on a platinum-on-alumina catalyst. For these points the pentene partial pressure represents the equilibrium value, which is determined by the molar ratio of *n*-pentane to hydrogen in the reactor. The square is a data point for the isomerization of 1-pentene on the acidic component of the catalyst alone (i.e., on a sample of catalyst that contains no platinum). For this point, the pentene partial pressure is simply the pressure of the 1-pentene reactant (Sinfelt et al., 1960). Reprinted with permission from J. H. Sinfelt, H. Hurwitz, and J. C. Rohrer, *J. Phys. Chem.*, **64**, 892 (1960). Copyright 1960 American Chemical Society.

cyclopentanes (Sinfelt et al., 1962), as typified by the conversion of methylcyclopentane to cyclohexane and benzene (mostly the latter at typical reforming conditions). In this reaction, methylcyclopentenes are intermediates.

In the range of hydrogen pressures commonly used in reforming, the rates of isomerization of alkanes and isomerization–dehydroisomerization of alkylcyclopentanes decrease with increasing hydrogen pressure. However, when a much wider range of hydrogen pressures is considered, the directional effect of this variable changes. In the complete absence of hydrogen, the reactions are not observed (Sinfelt, 1981). In the presence of hydrogen, the rates of reaction increase with increasing hydrogen pressure up to a given point, beyond which they decrease with further increase in hydrogen pressure. The change in the nature of the hydrogen pressure effect on the rates has been interpreted as follows: In the absence of hydrogen, or at low hydrogen pressures, the platinum sites become heavily covered with hydrogen-deficient hydrocarbon residues. As a result, the reaction rates are limited by the dehydrogenation activity of the catalyst, that is, the rate of formation of intermediate olefins. The role of hydrogen is then one of keeping the platinum sites free of surface residues. The extent of coverage of the surface by such residues decreases with increasing hydrogen pressure, thus accounting for the beneficial effect of hydrogen pressure on the overall rate of isomerization. At sufficiently high hydrogen pressures, however, coverage of platinum sites by surface residues ceases to be a limiting factor, and the rate does not continue to increase with increasing hydrogen pressure. This means that the formation of intermediate olefins is no longer the rate-controlling step, and the overall reaction becomes limited by the isomerization of olefins on the acidic centers of the catalyst. The decrease in the rate of isomerization with further increase in hydrogen pressure is attributed to the effect of hydrogen pressure in limiting the concentration of olefin intermediates attainable at equilibrium, as can be seen from our discussion of *n*-pentane isomerization kinetics.

The development of bifunctional systems for the effective catalysis of such hydrocarbon transformations as the isomerization of alkanes and the dehydroisomerization of alkylcyclopentanes provides an excellent example of catalyst design in practice. The design was accomplished through a knowledge of the chemistry of metal and acid catalysis. Research in this area led to a major industrial process known as catalytic reforming. The process has been widely applied in the petroleum industry for the production of high-octane-number components for automotive fuels (Haensel and Donaldson, 1951; Sinfelt, 1964, 1981). In addition to the reactions discussed here, there are two other major kinds of hydrocarbon transformations in reforming, namely, the dehydrocyclization of alkanes to aromatics and fragmentation reactions (hydrogenolysis and hydrocracking). The latter produce low-carbon-number alkanes and are far less desirable than the reactions yielding products in the gasoline boiling range. The major reactions in catalytic reforming, with specific examples, are illustrated in Figure 1.12.

## 1.7. CATALYST POISONING AND DEACTIVATION

In practice, the activity and/or selectivity of a catalyst may change with time. There are a variety of reasons for such a change. The presence of a small amount of an impurity in a feed stream to a reactor is one possibility. Amounts of impurities as low as 1 ppm in a feed stream can have marked effects in a catalytic process. The impurity may gradually accumulate on the surface of the catalyst and thereby inactivate the sites responsible for the catalytic reaction.

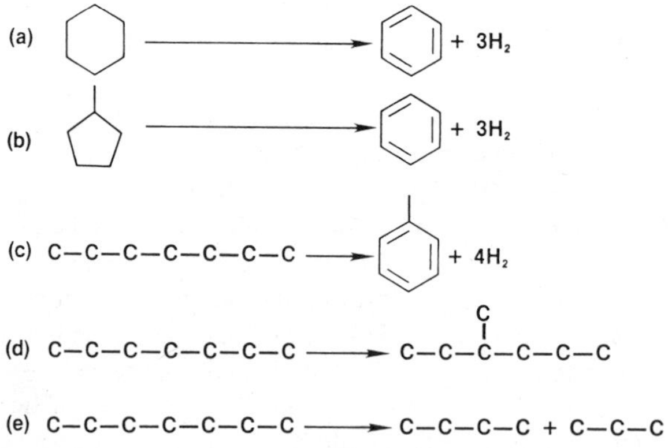

**FIGURE 1.12** Major reactions in catalytic reforming illustrated with specific examples: (a) dehydrogenation of cyclohexanes to aromatic hydrocarbons; (b) dehydroisomerization of alkylcyclopentanes to aromatic hydrocarbons; (c) dehydrocyclization of alkanes to aromatic hydrocarbons; (d) isomerization of n-alkanes to branched alkanes; (e) fragmentation reactions (hydrocracking and hydrogenolysis) yielding low-carbon-number alkanes (Sinfelt, 1983).

## A.  Poisoning by Impurities in a Feed Stream

In the hydrogenation of unsaturated hydrocarbons (e.g., olefins or aromatics) over a nickel catalyst, the presence of small amounts of sulfur compounds in the feed stream leads to a decline in catalytic activity over a period of time. A sulfur compound may be strongly chemisorbed itself on the active catalytic sites, or it may decompose in such a manner that sulfur alone is retained by the surface. In essence, such a decomposition process leads to irreversible chemisorption of sulfur on the nickel surface, which may be viewed as the formation of a surface sulfide. Nickel sites useful for the hydrogenation reaction are thereby eliminated, and the catalytic activity declines.

The contemplated use of a nickel catalyst for such a hydrogenation reaction could depend upon several considerations. First, one might investigate the possibility of removing the sulfur compounds from the feed stream by some sort of chemical pretreatment. If a cheap reagent were available for the quantitative removal of the sulfur compounds, so that the material could simply be discarded after use, or if the reagent could be regenerated easily and cheaply, this approach might be feasible. Another consideration would be the use of a nickel catalyst with as high a nickel surface area as possible. As the surface area per unit mass of nickel in the catalyst increases, the capacity for sulfur increases. This lengthens the period of time over which a given fraction of the nickel surface is inactivated. The life of the catalyst is thereby extended. As indicated earlier in this chapter, the formulation and development of methods for extending the active surface area of a catalyst is an important aspect of catalyst design.

## B.  Poisoning Due to Reactions—Carbonaceous Residues

Apart from the deactivating effect of impurities in a feed stream, there is the possibility that reactant molecules themselves will to some extent be converted to species that poison the catalyst surface. In hydrocarbon reactions over a variety of catalysts, the formation of carbonaceous residues on the surface is an undesirable side reaction accompanying the reactions of primary interest. Such residues are formed via extensive dehydrogenation of chemisorbed hydrocarbons to highly unsaturated species, which in turn readily undergo condensation or polymerization reactions. In the reforming of petroleum naphtha fractions over a bifunctional platinum-on-alumina catalyst, for example, both the metal sites and the acidic sites on the alumina may be deactivated by the residues. The formation of the carbonaceous residues is suppressed greatly by operating a reforming unit at elevated pressure (20–30 atm) with recycle of hydrogen-rich gas from the reformer effluent (Sinfelt, 1964, 1981, 1983). The important feature here is the maintenance of a suitably high partial pressure of hydrogen throughout the reaction zone. Reforming units employing this feature with platinum-on-alumina catalysts were introduced widely in petroleum refining during the 1950s and 1960s. The hydrogen inhibits the formation of highly unsaturated species on the surface and also removes hydrocarbon residues via hydrogenolysis reactions. Since increasing hydrogen pressure also has the effect of decreasing the yield of aromatic hydrocarbons, the choice of hydrogen pressure for a reformer is a balance of product yields against catalyst deactivation rate. This is clearly a key element in the design of a catalytic reforming unit.

While the formation of carbonaceous residues on a reforming catalyst is retarded by operating at high hydrogen partial pressure, this approach to the problem has the disadvantage that the process is not being operated at optimal conditions for maximizing the yields of aromatic hydrocarbons, which are the desired high-octane-number components for gasoline. To maximize yields, one would prefer to operate the process at hydrogen partial pressures as low as possible. The modification of a reforming catalyst to permit such operation therefore presents a good challenge in catalyst design. The challenge was met through the development of bimetallic catalyst systems for reforming (Jacobsen et al., 1969; Sinfelt 1972a, 1976).

## C.  Development of Bimetallic Reforming Catalysts

A research program on bimetallic clusters, which I initiated in the EXXON laboratories during the early 1960s (Sinfelt, 1983), led to the development of a catalyst containing platinum–iridium clusters dispersed on alumina. This catalyst exhibited much better activity and activity maintenance than the platinum-on-alumina catalysts used previously. Consequently, a liquid product with a higher octane number could be obtained, and the decline in octane number with time was much less pronounced. The octane number is largely determined by the aromatic hydrocarbon content of the product. A comparison of alumina-supported platinum–iridium and platinum catalysts showing the octane number of the liquid product as a function of time-on-stream is given in Figure 1.13 for the reforming of a 323–473K boiling-range Venezuelan naphtha at 760K and 14.6 atm. The naphtha-weight hourly space velocity (grams of naphtha per hour per gram of catalyst) was 2.1, and the mole ratio of hydrogen to naphtha at the reactor inlet was approximately 6.

In commerical practice a reformer is operated to produce a product of constant octane number. As the catalyst deactivates, the temperature of the system may be increased to compensate for the lower activity. In this way, the octane number can be maintained at the desired level. Illustrative data for platinum and platinum-iridium catalysts in this type of operation are given in Figure 1.14. The temperature required to produce a 100-octane-number product is shown as a function of time on stream in the reforming of a 372–444K boiling-range naphtha. The reactor pressure was 14.6 atm, and the naphtha-weight hourly space velocity was 3. The molar ratio of hydrogen to hydrocarbon at the reactor inlet was approximately 6. As shown in the figure, the platinum–iridium catalyst required a lower temperature than the platinum catalyst to produce the 100-octane-number product, which is indicative of its higher activity. Furthermore, the rate of increase of temperature required to maintain the octane number over an extended period of time was much lower for the platinum–iridium catalyst. This means that it can function for a much longer period of time before regeneration is required.

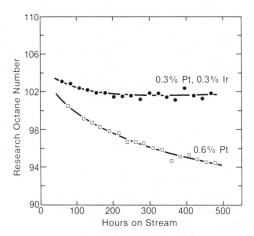

**FIGURE 1.13** Comparison of alumina-supported platinum–iridium and platinum catalysts showing the research octane number of the liquid reformate as a function of time-on-stream in the reforming of a 323–473K boiling-range Venezuelan naphtha at 760K and 14.6 atm (Carter et al., 1982). Reprinted with permission from Elsevier Scientific Publishing Company.

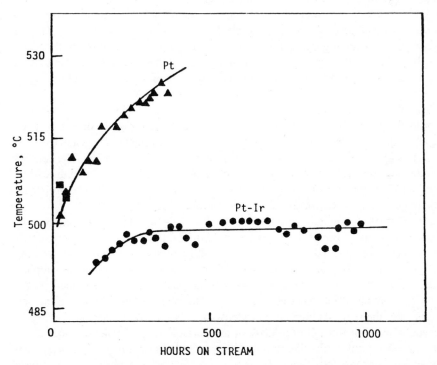

**FIGURE 1.14** Data on the reforming of a 372–444K boiling-range naphtha showing the temperature required to produce a 100-octane-number product as a function of time-on-stream for alumina-supported platinum and platinum–iridium catalysts at 14.6 atm pressure (Sinfelt, 1976; 1981).

At a given temperature, pressure, and hydrogen-to-hydrocarbon mole ratio, the extent of formation of carbonaceous residues is much lower for a platinum–iridium catalyst than for a platinum catalyst. The improved activity maintenance of platinum–iridium catalysts can be attributed, at least in part, to the lower accumulation of carbonaceous residues on the surface. This, in turn, may be related to increased hydrogenolysis activity resulting from the presence of iridium. Since iridium is much more active than platinum for hydrogenolysis, the platinum–iridium clusters have higher hydrogenolysis activity than clusters containing platinum alone. As indicated earlier, hydrogenolysis provides a mechanism for limiting the accumulation of carbonaceous residues on the surface.

In exploiting the hydrogenolysis properties of iridium for suppressing formation of carbonaceous residues, one must be aware of a possible adverse effect on the product distribution in reforming, since high hydrogenolysis activity leads to increased rates of conversion of hydrocarbons to light gaseous products. A catalyst containing iridium alone on alumina is limited in this manner. In the case of platinum–iridium catalysts, however, the interaction between the platinum and iridium moderates the hydrogenolysis activity of the latter. At the conditions employed in reforming (temperatures near 775K and pressures of 15–30 atm), the inclusion of iridium with the platinum increases the rate of conversion of naphtha hydrocarbons to valuable gasoline components to about the same extent that it increases the rate of conversion to light gaseous products. Consequently, one obtains a satisfactory distribution of reforming products and simultaneously realizes large advantages in activity and activity maintenance over catalysts containing platinum alone on alumina.

The platinum–iridium reforming catalyst was developed in the EXXON laboratories within the framework of a fundamental research program on catalysis by metals. The novel design feature is the concept of a bimetallic cluster (Sinfelt, 1972a; 1973b), in this case a platinum–iridium cluster (Sinfelt, 1976). The concept was developed with the aid of systematic studies of the kinetics of hydrocarbon reactions on metals, coupled with investigations of catalyst structure and properties utilizing a variety of chemical and physical probes (Sinfelt, 1983).

Since its first commercial application in 1971, the platinum–iridium catalyst has been used in many Exxon reforming units worldwide. It has also been used by other companies, under license from Exxon. Simultaneously, another reforming catalyst consisting of platinum and rhenium supported on alumina has been applied widely in the petroleum industry. This catalyst was originally developed by the Chevron Company (Jacobsen et al., 1969). The platinum–rhenium combination differs from the types of bimetallic systems already considered in that rhenium is a metallic element from Group VIIA of the periodic table. In some respects, reforming operations with platinum–iridium and platinum–rhenium catalysts are similar; in other respects, there are significant differences. Both types of catalysts are exposed to a sulfur-containing gas stream in situ prior to use in commercial reformers. This treatment provides an additional moderation of the hydrogenolysis activity of the catalyst, over and above any effect achieved as a result of interaction of the iridium or rhenium with the platinum. In reforming operations, platinum–

iridium catalysts are generally about twice as active as platinum–rhenium catalysts, but the yields of liquid product of a given octane number are usually about 1 vol % lower (Carter et al., 1982; Sinfelt, 1983). However, in some operations, liquid yields as well as reforming activities are higher for platinum–iridium catalysts (Carter et al., 1982; Sinfelt, 1983). The lower rates of activity decline of platinum–iridium catalysts relative to platinum catalysts are also characteristic of platinum–rhenium catalysts. However, the rates of accumulation of carbonaceous residues on platinum–rhenium catalysts in reforming operations are about the same as on platinum catalysts and are typically higher than those on platinum–iridium catalysts by a factor of two or more. Thus the activity maintenance of a platinum–rhenium catalyst is better than one might have expected on the basis of the observed rate of accumulation of carbonaceous residues on the surface. The differences between platinum-iridium and platinum-rhenium catalysts in reforming are in part attributable to differences in sulfur retention (Carter et al., 1982).

The attractive features of platinum–rhenium and platinum–iridium catalysts can be combined in a reforming operation (Sinfelt, 1974b, 1981, 1983, 1985). Data on the reactions of selected hydrocarbons on platinum–rhenium and platinum–iridium catalysts indicate that the former is more selective for the conversion of cycloalkanes to aromatic hydrocarbons, while the latter is more selective for the aromatization of noncyclic alkanes (Carter et al., 1982). Since the first of these aromatization reactions occurs much more readily than the second, a reforming system can be divided into two separate reaction zones. The aromatization of cycloalkanes occurs primarily in the first zone, while the conversion of noncyclic alkanes to aromatics occurs predominantly in the second zone. Consequently, it is reasonable to use a platinum–rhenium catalyst in the first zone and a platinum–iridium catalyst in the second zone. In practice, a reformer consists of a number of reactors in series. Thus the platinum–rhenium catalyst can be charged to the initial reactors and the platinum–iridium catalyst to the tail reactor or reactors (Sinfelt, 1974b). When the combined catalyst system is employed, one obtains the high activity characteristic of a platinum–iridium catalyst. Simultaneously, the yield of liquid reformate is more nearly equivalent to that obtained with a platinum–rhenium catalyst in cases where the latter has a yield advantage over a platinum–iridium catalyst (Carter et al., 1982). Combined catalyst systems of the type described here are in use in a number of commercial reforming units. The choice of a catalyst system for a reforming unit is affected by a number of factors, including the activity of the catalyst system and the yields of the various products obtained with it. The relative importance of the different factors, and hence the economic analysis, depends on the particular application.

## 1.8.  CONCLUDING REMARKS

In this chapter, some basic aspects of catalyst design have been discussed and illustrated with selected examples. As indicated in the Introduction, no attempt has been made to present a detailed account of the many diverse issues one encounters in industrial practice. Instead, the author has concentrated on providing some perspective about the design process.

Although our understanding of catalysis has not progressed to an extent that enables us to make quantitative predictions of catalytic activities from first principles, catalysts can be designed and developed with a rational, scientific basis. In the time that has elapsed since Ostwald clearly recognized catalysis as a phenomenon in chemical kinetics, much has been achieved in this extremely important field.

## REFERENCES

Barrer, R. M., "Molecular Sieves," *Endeavour*, **23,** 122 (1964).

Beelen, J. M., V. Ponec, and W. M. H. Sachtler, "Reaction of Cyclopropane on Nickel and Nickel-Copper Alloys," *J. Catal.*, **28,** 376 (1973).

Bond, G. C., *Catalysis by Metals*, Academic, London, 1962, p. 195.

Boudart, M., *Kinetics of Chemical Processes*, Prentice-Hall, Englewood Cliffs, NJ, 1968, pp. 198–203.

Boudart, M., "Catalysis by Supported Metals," *Adv. Catal.*, **20,** 153 (1969).

Bouwman, R., G. J. M. Lippits, and W. M. H. Sachtler, "Photoelectric Investigation of the Surface Composition of Equilibrated Ag-Pd Alloys in Ultrahigh Vacuum and in the Presence of CO," *J. Catal.*, **25,** 350 (1972).

Breck, D. W., "Crystalline Molecular Sieves," *J. Chem. Educ.*, **41** (12), 678 (1964).

Cadenhead, D. A. and N. J. Wagner, "Low Temperature Hydrogen Adsorption on Copper-Nickel Alloys," *J. Phys. Chem.*, **72,** 2775 (1968).

Carter, J. L., J. A. Cusumano, and J. H. Sinfelt, "Hydrogenolysis of *n*-Heptane Over Unsupported Metals," *J. Catal.*, **20,** 223 (1971).

Carter, J. L., G. B. McVicker, W. Weissman, W. S. Kmak, and J. H. Sinfelt, "Bimetallic Catalysts: Application in Catalytic Reforming," *Appl. Catal.*, **3,** 327 (1982).

Ciapetta, F. G. and J. B. Hunter, "Isomerization of Saturated Hydrocarbons in the Presence of Hydrogenation-Cracking Catalysts," *Ind. Eng. Chem.*, **45,** 147 (1953a).

Ciapetta, F. G. and J. B. Hunter, "Isomerization of Saturated Hydrocarbons. Normal Pentane, Isohexanes, Heptanes, and Octanes," *Ind. Eng. Chem.*, **45,** 155 (1953b).

Cimino, A., M. Boudart, and H. S. Taylor, "Ethane Hydrogenation-Cracking on Iron Catalysts with and without Alkali," *J. Phys. Chem.*, **58,** 796 (1954).

Emmett, P. H., "Measurement of the Surface Area of Solid Catalysts," in *Catalysis*, Vol. 1, Paul H. Emmett (ed.), Reinhold, New York, 1954, pp. 31–74.

Garten, R. L. and J. H. Sinfelt, "Structure of Pt-Ir Catalysts: Mössbauer Spectroscopy Studies Employing $^{57}$Fe as a Probe," *J. Catal.*, **62,** 127 (1980).

Haensel, V., "Alumina-Platinum-Halogen Catalyst and Preparation Thereof," U.S. Patent 2,479,109 (1949a).

Haensel, V., "Process of Reforming a Gasoline with an Alumina-Platinum-Halogen Catalyst," U.S. Patent 2,479,110 (1949b).

Haensel, V. and G. R. Donaldson, "Platforming of Pure Hydrocarbons," *Ind. Eng. Chem.*, **43,** 2102 (1951).

Heinemann, H., "Technological Applications of Zeolites in Catalysis," *Catal. Rev. Sci. Eng.*, **23** (1&2), 315 (1981).

Heinemann, H., G. A. Mills, J. B. Hattman, and F. W. Kirsch, "Houdriforming Reactions. Studies with Pure Hydrocarbons," *Ind. Eng. Chem.*, **45,** 130, (1953).

Hindin, S. G., S. W. Weller, and G. A. Mills, "Mechanically Mixed Dual Function Catalysts," *J. Phys. Chem.*, **62,** 244 (1958).

Jacobson, R. L., H. E. Kluksdahl, C. S. McCoy, and R. W. Davis, "Platinum–Rhenium Catalysts: A Major New Catalytic Reforming Development," *Proceedings of the American Petroleum Institute, Division of Refining*, **49,** 504 (1969).

Matsumoto, H., Y. Saito, and Y. Yoneda, "Contrast Between Nickel and Platinum Catalysts in Hydrogenolysis of Saturated Hydrocarbons," *J. Catal.*, **19,** 101 (1970).

Matsumoto, H., Y. Saito, and Y. Yoneda, "The Classification of Metal Catalysts in Hydrogenolysis of Hexane Isomers," *J. Catal.*, **22,** 182 (1971).

McVicker, G. B., R. T. K. Baker, R. L. Garten, and E. L. Kugler, "Chemisorption Properties of Iridium on Alumina Catalysts," *J. Catal.*, **65,** 207 (1980).

Mills, G. A., H. Heinemann, T. H. Milliken, and A. G. Oblad, "Houdriforming Reactions. Catalytic Mechanism," *Ind. Eng. Chem.*, **45,** 134 (1953).

Oblad, A. G., T. H. Milliken, and G. A. Mills, "Chemical Characteristics and Structure of Cracking Catalysts," *Adv. Catal.*, **3,** 199 (1951).

Roberti, A., V. Ponec, and W. M. H. Sachtler, "Reactions of Methylcyclopentane on Nickel and Nickel-Copper Alloys," *J. Catal.*, **28,** 381 (1973).

Sinfelt, J. H., "Bifunctional Catalysis," *Adv. Chem. Eng.*, **5,** 37 (1964).

Sinfelt, J. H., "Esso Catalyst Based on Multimetallic Clusters," *Chem. Eng. News*, **50,** 18 (3 July 1972a).

Sinfelt, J. H., "Highly Dispersed Catalytic Materials," *Ann. Rev. Mater. Sci.*, **2,** 641 (1972b).

Sinfelt, J. H., "Specificity in Catalytic Hydrogenolysis by Metals," *Adv. Catal.*, **23,** 91 (1973a).

Sinfelt, J. H., "Supported Bimetallic Cluster Catalysts," *J. Catal.*, **29,** 308 (1973b).

Sinfelt, J. H., "Catalysis by Metals: The P. H. Emmett Award Address," *Catal. Rev.*, **9,** (1), 147 (1974a).

Sinfelt, J. H., "Combination Reforming Process," U. S. Patent 3,791,961 (1974b).

Sinfelt, J. H., "Heterogeneous Catalysis by Metals," *Prog. Solid-State Chem.*, **10** (2), 55 (1975).

Sinfelt, J. H., "Polymetallic Cluster Compositions Useful as Hydrocarbon Conversion Catalysts," U. S. Patent 3,953,368 (1976).

Sinfelt, J. H., "Catalysis by Alloys and Bimetallic Clusters," *Acc. Chem. Res.*, **10,** 15 (1977).

Sinfelt, J. H., "Structure of Metal Catalysts," *Rev. Mod. Phys.*, **51** (3), 569 (1979).

Sinfelt, J. H., "Catalytic Reforming of Hydrocarbons," in *Catalysis-Science and Tech-*

*nology*, Vol. 1, John R. Anderson and Michel Boudart (eds.), Springer-Verlag, Berlin, Heidelberg, 1981, pp. 257–300.

Sinfelt, J. H., *Bimetallic Catalysts: Discoveries, Concepts and Applications*, Wiley, New York, 1983.

Sinfelt, J. H., "Some Reflections on Catalysis" (Perkin Medal Address), *Chem. Ind.*, 403 (4 June 1984a).

Sinfelt, J. H., "The Evolution of Catalytic Science and Technology" (Gold Medal Address, American Institute of Chemists), *The Chemist*, **61** (10), 6 (1984b).

Sinfelt, J. H., "Bimetallic Catalysts," *Sci. Am.*, **253** (3), 90 (1985).

Sinfelt, J. H., A. E. Barnett, and J. L. Carter, "Inhibition of Hydrogenolysis," U. S. Patent 3,617,518 (1971).

Sinfelt, J. H., A. E. Barnett, and G. W. Dembinski, "Isomerization Process Utilizing a Gold-Palladium Alloy in the Catalyst," U. S. Patent 3,442,973 (1969).

Sinfelt, J. H., J. L. Carter, and D. J. C. Yates, "Catalytic Hydrogenolysis and Dehydrogenation over Copper-Nickel Alloys," *J. Catal.*, **24,** 283 (1972).

Sinfelt, J. H., H. Hurwitz, and J. C. Rohrer, "Kinetics of *n*-Pentane Isomerization over Pt-Al$_2$O$_3$ Catalyst," *J. Phys. Chem.*, **64,** 892 (1960).

Sinfelt, J. H., H. Hurwitz, and J. C. Rohrer, "Role of Dehydrogenation Activity in the Catalytic Isomerization and Dehydrocyclization of Hydrocarbons," *J. Catal.*, **1,** 481 (1962).

Sinfelt, J. H., Y. L. Lam, J. A. Cusumano, and A. E. Barnett, "Nature of Ruthenium-Copper Catalysts," *J. Catal.*, **42,** 227 (1976).

Sinfelt, J. H. and G. H. Via, "Dispersion and Structure of Platinum-Iridium Catalysts," *J. Catal.*, **56,** 1 (1979).

Sinfelt, J. H., G. H. Via, and F. W. Lytle, "Application of EXAFS in Catalysis. Structure of Bimetallic Cluster Catalysts," *Catal. Rev. Sci. Eng.*, **26** (1), 81 (1984).

Thiele, E. W., "Relation Between Catalytic Activity and Size of Particle," *Ind. Eng. Chem.*, **31,** 916 (1939).

van der Planck, P. and W. M. H. Sachtler, "Surface Composition of Equilibrated Copper-Nickel Alloy Films," *J. Catal.*, **7,** 300 (1967).

Venuto, P. B. and P. S. Landis, "Organic Catalysis over Crystalline Aluminosilicates," *Adv. Catal.*, **18,** 259 (1968).

Via, G. H., G. Meitzner, F. W. Lytle, and J. H. Sinfelt, "Extended X-Ray Absorption Fine Structure (EXAFS) of Highly Dispersed Rhodium Catalysts," *J. Chem. Phys.*, **79,** 1527 (1983).

Weisz, P. B., "Polyfunctional Heterogeneous Catalysis," *Adv. Catal.*, **13,** 137 (1962).

Weisz, P. B., "Zeolites-New Horizons in Catalysis," *Chem. Technol.*, **3** (8), 498 (1973).

Weisz, P. B. and C. D. Prater, "Interpretation of Measurements in Experimental Catalysis," *Adv. Catal.*, **6,** 143 (1954).

Weisz, P. B. and E. W. Swegler, "Stepwise Reaction on Separate Catalytic Centers: Isomerization of Saturated Hydrocarbons," *Science*, **126,** 31 (1957).

Wheeler, A., "Reaction Rates and Selectivity in Catalyst Pores," in *Catalysis*, Vol. 2, P. H. Emmett (ed.) Reinhold, New York, 1955, pp. 105–165.

Williams F. L. and M. Boudart, "Surface Composition of Nickel-Gold Alloys," *J. Catal.*, **30,** 438 (1973).

Yates, D. J. C. and J. H. Sinfelt, "The Catalytic Activity of Rhodium in Relation to Its State of Dispersion," *J. Catal.*, **8,** 348 (1967).

# 2

# PROCESS HEAT-EXCHANGER DESIGN: QUALITATIVE FACTORS IN SELECTION AND APPLICATION

KENNETH J. BELL

2.1. Introduction

2.2. Quantitative Aspects of Heat-Exchanger Design
    A. Basic Correlations of Heat-Transfer and Fluid-Flow Phenomena
    B. Thermal–Hydraulic Modeling of Heat Exchangers
    C. Basic Equations of Heat-Exchanger Design
    D. Criteria for Heat-Exchanger Selection
    E. Logical Structure of the Heat-Exchanger Design Process
    F. Further Aspects of Computer Design of Heat Exchangers
    G. Mechanical Design of Heat Exchangers

2.3. Problem Identification
    A. Process Design and Heat-Exchanger Design
    B. The "Context" of the Design Problem
    C. Proper Goals for Optimization of Heat Recovery
    D. Pulling the Design Together

2.4. Uncertainty in Heat-Exchanger Design
    A. Sources of Uncertainty
    B. Analyzing Uncertainty
    C. Designing Around Uncertainty

41

## 2.1.  INTRODUCTION

The central problem of process heat-exchanger design may perhaps be stated as "the creation of equipment to accomplish desired thermal changes within allowable pressure drops and with proper attention to the mechanical integrity and operational flexibility of the heat exchanger in service." Clearly, such a charge calls forth a variety of problems to be addressed, ranging from very fundamental areas in fluid flow and heat transfer to very applied considerations of, for example, gasket selection. No one person, nor any one reference, is able to deal competently with all of these problems, but there are some brave attempts.

Kern (1950) wrote the classic work *Process Heat Transfer*, which largely defined the field. This book, though technically obsolete now, established the basic structure of the thermal–hydraulic calculations and was until recently the only book seriously addressing the practical questions of heat-exchanger design for the process industries. The *Heat Exchanger Design Handbook* (Schlünder, 1983) is undoubtedly the most comprehensive effort both physically (ca. 2500 pages in five sections and seven loose-leaf binders) and in subject-matter coverage (from the most fundamental fluid-mechanics and heat-transfer equations and correlations to design methods, international codes for mechanical design, and physical properties of common fluids). Singh and Soler (1984) have produced in their *Mechanical Design of Heat Exchangers and Pressure Vessel Components* the best available reference on mechanical design.

Several other sources are less comprehensive, but contain modern material directly relating to thermal–hydraulic design of heat exchangers. Among them are Kakac et al., (1981), *Heat Exchangers: Thermal–Hydraulic Fundamentals and Design* (1981); Marto and Nunn (1981), *Power Condenser Heat Transfer Technology*; and Taborek et al. (1983), *Heat Exchangers: Theory and Practice*.

These books, and the hundreds of references given in them, provide a strong quantitative base for heat-exchanger design and application, though chasms of ignorance still exist in many areas together with needs and opportunities for new work on almost any topic. But what seems to be missing from the literature, and particularly from the teaching material, is much discussion on what I have chosen to call the "qualitative factors" in the selection and design of heat exchangers. This chapter attempts to introduce some of these ideas into the teaching of design.

These qualitative factors are of diverse natures and may in fact have only two features in common. First, they are essential to the proper application of heat exchangers in the process industries (and probably elsewhere); second, they will never be brought to the engineer's attention by the equations he/she is using, though these factors may have a great deal to do with the proper selection and use of those equations in design. Description of these factors tends to be by way of example, or even anecdotes, though there are attempts to bring some of them to a more analytical form. Practicing engineers in the field will regard many of these matters as self-evident. My own experience suggests that most are lessons learned the hard way—by being involved in a foul-up and learning from the experience, hopefully in generalizable terms.

This chapter opens with a very brief statement of the state of the quantitative base of heat-exchanger design and the structure of the design process. This section contains the only equations in the chapter, and these are included only for the purpose of establishing the proper starting point. The next section deals with the need for "Problem Identification"—probably the point at which more heat-exchanger designs go sour than any other. Then we discuss "Uncertainty in Heat-Exchanger Design"—omnipresent but overlooked, ignored, and generally an embarrassment. The last sections on "Flow Maldistribution" and "Fouling" could be similarly described, but do represent specific causes of uncertainty, and offer fertile fields for research as well as the need for good judgment in design.

In looking ahead to what has been promised, the author feels rather daunted at the dimensions of the task. But this sort of material is indisputably part of the problem of design, and it seems high time some effort was made to put it into the teaching literature.

## 2.2. QUANTITATIVE ASPECTS OF HEAT-EXCHANGER DESIGN

### A. Basic Correlations of Heat-Transfer and Fluid-Flow Phenomena

The starting point for any thermal–hydraulic design method for heat exchange is the collection of basic correlations for fluid flow (especially pressure drop or fluid friction) and heat transfer for the flow geometries existing in the heat exchanger. For example, for a shell-and-tube exchanger, one would expect that friction factor and Colburn $j$-factor (or Nusselt number) relationships for flow inside tubes and across banks of tubes would be required, since the shell-side flow geometry can be approximated as an ideal tube bank. These correlations can sometimes be obtained by theoretical analysis alone [e.g., laminar-flow heat transfer inside a tube

(Shah and London, 1978)], but for most geometries experimental data are essential [e.g., transition and turbulent flow in tube banks (Bergelin et al., 1952)].

The literature on these basic correlations is enormous and one should ordinarily start with one of the handbooks, such as Rohsenow and Hartnett (1985) or Volume 2 of Schlünder (1983). These handbooks generally give some recommendations as to the preferred correlations, as well as key references to the original literature. But one may also discover that there are substantial gaps in the literature, especially for newer geometries such as enhanced tubes or plate heat exchangers. In these cases, one has the choice of making a best guess (which is done a good deal more often than the engineering student may realize) or going into the laboratory to obtain the required data (a lengthy and expensive process, which accounts for the popularity of guessing). Actually, the "best-guess" alternative is really more a matter of extrapolating existing information into new ranges of application, and making use of conservative engineering judgment and an understanding of the fundamental transport processes that are expected to control.

A new technique that is being increasingly used is to generate by computer-specific numerical solutions (over a range of flow parameters) for fluid-flow and heat-transfer quantities, starting from the most basic available equations (e.g., the Navier–Stokes equations, with some sort of turbulence model and possibly temperature-dependent physical properties) and to present the results as fitted functions of conventional dimensionless numbers for actual use in design calculations. But the process cannot be any better—and may be much worse—than its starting assumptions, particularly for the turbulence model. Furthermore, the method will not signal internal errors except in obvious cases. Insofar as possible, the results should be checked against any available experimental data. In any case, one should use only as prescribed, and with a strong dose of skepticism.

As a general statement, I would say that the basic correlations are adequate for most heat-exchanger design at this time. Specific critical problems may require extended analysis, extreme conservatism, and even proof; but this is not the area of greatest concern or difficulty in the field.

## B. Thermal–Hydraulic Modeling of Heat Exchangers

Having the thermal–hydraulic correlations for the "building block" geometries of a heat exchanger is a necessary, but not sufficient, condition for a design method. A "real" heat exchanger is more than the sum of the idealized geometries that go into it. Using the shell-and-tube exchanger as an example again, we can say that the tube side is not too far removed from the textbook geometry of plain round tubes, which are long enough that entrance effects do not significantly alter the mean values of pressure gradient and heat-transfer coefficient for the entire tube. (In fact, this is not true for laminar or transition flow, for unusual header designs that may result in maldistributed flow, or for fluids that undergo significant property changes—especially viscosity—with temperature.)

But the shell side is a very different story. The first effect is the combination of longitudinal and cross-flow in the baffle windows (or turnarounds), and this is compounded by flow bypassing between the outermost tubes and the shell, and by fluid leaking through the clearances between the baffles and the shell and between the tubes and the baffles. So we very quickly get away from the idealized geometry of flow perpendicular to a rectangular bank of tubes.

Tinker (1951) gave the first reasonably complete description of the shell-side flow patterns, and used that as a basis for a rating procedure to calculate pressure drop and heat transfer for a shell-and-tube exchanger. Unfortunately, he lacked both the basic data correlations for such effects as shell-to-baffle leakage streams, and the computational power to carry out the calculations required by his model. These came along later, resulting in fairly good hand-design methods [the Delaware method, in a number of revisions of which Section 3.3 in Schlünder (1983) is the most comprehensive] and computer methods, a typical case of which is named ''Stream Analysis'' (Palen and Taborek, 1969).

The main point is that a great deal of insight is required to model fluid-flow and heat-transfer processes inside real exchangers, and then to pull together the basic correlations into these models. This task can only be done by someone reasonably knowledgable about the construction details and the real conditions of application of a heat exchanger. It cannot be done successfully by those whose interest, knowledge, and experience do not extend beyond the basic differential equations of non-isothermal fluid flow. The final test of the methods comes when their predictions are compared to high-quality test data obtained on typical heat exchangers.

Probably more work needs to be done on modeling heat exchangers and verifying these models experimentally than on additional fundamental studies. The conceptual aspects are highly demanding and require imagination as well as solid technical capabilities, but almost always the result is a computer program with rigorous requirements of programming precision. Very few people are actually engaged in this activity on a full-time basis. Nonetheless, these efforts proceed apace, and the level of sophistication in the models is increasing about as rapidly as the computer capabilities can handle.

## C. Basic Equations of Heat-Exchanger Design

The basic design equation for a two-fluid steady-state heat exchanger is:

$$A_T^* = \int_0^{Q_T} \frac{dQ}{U^*(T - t)} \tag{2.1}$$

where $A_T^*$ is the total heat-transfer area required in the exchanger, $Q_T$ is the total heat transferred in the exchanger, $U^*$ is the overall heat-transfer coefficient based on the reference area $A_T^*$, and $T$ and $t$ are the local temperatures of the hot and cold streams, respectively. The overall heat-transfer coefficient is related to the individual film coefficients $h_o$ and $h_i$, the fouling resistances $R_{f_o}$ and $R_{f_i}$, and the wall resistance $\Delta X_w / k_w$ by

$$U^* = \frac{1}{A_T^*} \left[ \frac{1}{h_o A_o} + \frac{R_{fo}}{A_o} + \frac{\Delta X_w}{k_w A_w} + \frac{R_{fi}}{A_i} + \frac{1}{h_i A_i} \right]^{-1} \qquad (2.2)$$

where $A_o$, $A_i$, and $A_w$ are the heat transfer areas of the two convective processes and the conduction through the wall, respectively, $\Delta X_w$ is the wall thickness, and $k_w$ is the wall thermal conductivity.

An explanation of the area terms is in order. As long as virtually all heat exchangers used plain tube (or pipe, in double-pipe heat exchangers), one could take $A_o$ and $A_i$ as the outside and inside surface areas without ambiguity and $A_T^*$ was usually taken equal to $A_o$. However, more and more, enhanced surfaces are being used—first fins (radial and longitudinal), then grooves, flutes, corrugations, and bumps, and whatever else some manufacturers thought might improve the heat transfer process without causing unacceptable pressure losses. The geometrical area of some of these surfaces is very difficult to determine, and so "effective" areas of one sort or another are often used—"total outside surface including fins," "rubber band area," "projected area," "area based on mean inside diameter," and so on. There is no difficulty in using these areas as long as they are defined *clearly* and used *consistently*, including in the data reduction process from which the heat-transfer and pressure-loss correlations are deduced from experimental data. Equations (2.1) and (2.2) are written to accommodate this flexibility.

An explanation of the area terms is in order. As long as virtually all heat exchangers used plain tube (or pipe, in double-pipe heat exchangers), one could take $A_o$ and $A_i$ as the outside and inside surface areas without ambiguity and $A_T^*$ was usually taken equal to $A_o$. However, more and more, enhanced surfaces are being used—first fins (radial and longitudinal), then grooves, flutes, corrugations, and bumps, and whatever else some manufacturers thought might improve the heat transfer process without causing unacceptable pressure losses. The geometrical area of some of these surfaces is very difficult to determine, and so "effective" areas of one sort or another are often used—"total outside surface including fins," "rubber band area," "projected area," "area based on mean inside diameter," and so on. There is no difficulty in using these areas as long as they are defined *clearly* and used *consistently*, including in the data reduction process from which the heat-transfer and pressure-loss correlations are deduced from experimental data. Equations (2.1) and (2.2) are written to accommodate this flexibility.

As indicated above, certain basic correlations can be used to predict $h_o$ and $h_i$ with various corrections applied for the nonidealities of the flow arrangement. The wall term is usually a small resistance compared to the others, and generally poses no difficulties to the designer. Fouling is often a major resistance to heat transfer and, simultaneously, is highly unpredictable; it has been termed "the major unresolved problem in heat transfer" (Taborek, et al., 1972) and more recently "the major unsolved problem in heat transfer." Section 2.6 of this chapter is devoted to describing its importance and a few ways of getting around this problem.

The integrand of Equation (2.1) represents the area required to transfer an increment of heat $dQ$ at a point in the heat exchanger where the hot and cold stream temperatures are $T$ and $t$, respectively. Since ordinarily at least one of the stream temperatures varies from point to point in the exchanger (an exception is a pure vapor condensing isobarically on one side to vaporize a pure liquid isobarically on the other), and since the individual film coefficients may vary with temperature or local flow conditions, it is necessary to integrate over the entire required heat-transfer duty. Most heat exchangers are in fact designed using computer programs that are capable of carrying out this integration numerically if the flow arrangements and requisite physical properties are supplied.

However, Equation (2.1) can be reduced to an analytically integrable form for many heat-exchanger configurations if a small set of assumptions is satisfied. These assumptions include such items as constant specific heats or isothermal phase transition for each stream, constant overall heat transfer coefficient, and steady state. The result is the "Mean Temperature Difference Concept":

$$A_T^* = \frac{Q_T}{U^*(\text{MTD})} \tag{2.3}$$

where MTD stands for "Mean-Temperature Difference," a quantity computable from the four terminal temperatures of the fluids. The MTD can be written as

$$\text{MTD} = F(\text{LMTD}) \tag{2.4}$$

where LMTD is the "Logarithmic-Mean Temperature Difference" for countercurrent flow*:

$$\text{LMTD} = \frac{(T_1 - t_2) - (T_2 - t_1)}{\ln\left(\dfrac{T_1 - t_2}{T_2 - t_1}\right)} \tag{2.5}$$

where $T_1$ and $t_1$ are the inlet temperatures, and $T_2$ and $t_2$ are the outlet temperatures of the hot and cold streams, respectively.

The parameter $F$ is a configuration correction factor which corrects the LMTD to flow configurations that are not purely countercurrent. It depends only upon the four terminal temperatures and the specific arrangements of flows in the heat ex-

---

*The LMTD for the special case of pure cocurrent flow is

$$(\text{LMTD})_{\text{cocurrent}} = \frac{(T_1 - t_1) - (T_2 - t_2)}{\ln\left[(T_1 - t_1)/(T_2 - t_2)\right]} \tag{2.6}$$

Pure cocurrent flow is rarely used (though it does occasionally solve some very tricky problems) and Equation (2.6) has no application to any other situation.

changer, and charts of $F$ in these terms are widely available in the chemical engineering literature. The best detailed treatment of the subject is given in Part 1 of Schlünder (1983). Here $F$ varies from 1.00 (for purely countercurrent flow) downward to 0, but any situation in which the value of $F$ is less than 0.8, or possibly 0.75 at the lowest, should be very carefully considered by the designer—there is almost always a better way to handle a problem so that $F$ falls between 0.8 and 1.0 (see also Section 6.3.H.1.c.).

Having set the stage with a few basic equations, let us now take a look at the proper way to use them.

## D.  Criteria for Heat-Exchanger Selection

The criteria that a process heat-exchanger must satisfy can be easily stated if we confine ourselves for the moment to rather broad statements.

First, of course, the heat exchanger must meet the process requirements. That is, it must achieve the desired change in the thermal condition of the process stream (or streams) within the allowable pressure drops, and it must continue to do this until the next scheduled shutdown of the plant for maintenance.

Second, the exchanger must withstand the service conditions of the plant environment. This includes the mechanical stresses of transportation, installation, and operation, and the thermal stresses induced by the temperature differences. It must also resist corrosion by the process and service streams (as well as by the environment); this is usually mainly a matter of choice of construction materials, but mechanical design does have some effect. Desirably, the exchanger should also resist fouling, but there is not much the designer may do with confidence in this regard except for keeping the velocities as high as pressure drop, erosion, and vibration limits permit.

Third, the exchanger must be maintainable, which usually implies choosing a configuration that permits cleaning—tube and/or shell side, as may be indicated—and replacement of tubes, gaskets, and any other components that may be especially vulnerable to corrosion, erosion, vibration, or aging. This requirement may also place limitations on positioning the exchanger and providing clear space around it.

Fourth, the exchanger should cost as little as is consistent with the requirements stated above. In the present listing, this refers to first cost or installed cost, since the operating cost and the cost of lost production due to exchanger unavailability have already been considered by implication in the earlier and more important criteria. The true cost of an exchanger can often be best evaluated by considering how much money will be lost if the exchanger must be shut down prematurely for repair or cleaning, or if it "bottlenecks" the plant production rate.

Finally, there may be limitations on exchanger diameter, length, weight, and/or tube specifications due to site requirements, lifting and servicing capabilities, or inventory considerations.

It is sometimes considered desirable that the exchanger design be selected with an eye to possible alternative uses in other applications. This has disturbing implications. Most heat exchangers are intended for projects having an expected life of 5–20 yr—equal to or greater than the probable life of the exchanger. To suggest that a heat exchanger might become available sooner implies either that the exchanger or the process will prove unsatisfactory in its role. It is far better to labor under the positive compulsion that the only hope for success is by designing each item uniquely for the best performance in the task at hand.

## E.  Logical Structure of the Heat-Exchanger Design Process

The basic logical structure of the design procedure for a process heat-exchanger is shown in Figure 2.1. The basic structure is the same whether we use hand- or computer-design methods; the only difference is the replacement of the very subtle and implicit human thought processes by a logical structure suited to a fast but inflexible computer program. In fact, all of the most important decisions are made outside the dashed line bounding the computer's contribution to the solution.

First, the problem must be identified as completely and unambiguously as pos-

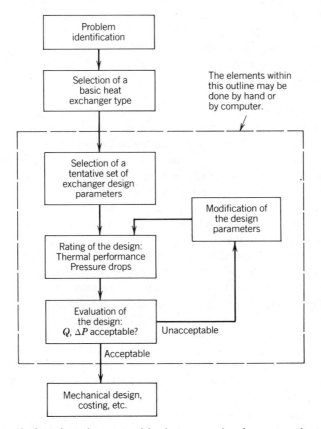

**FIGURE 2.1**  The basic logical structure of the design procedure for a process heat exchanger.

sible. This includes not only required flow rates and temperatures and compositions, but also the important questions: ''What is the process engineer really asking for?'' and ''Is that what he really needs?'' and ''Has he told me everything I need to know?'' It is necessary to distinguish here between *what* must be done, which falls in the domain of the process engineer, and *how* it is to be done, which is primarily the responsibility of the heat-exchanger designer. Not infrequently, the process engineer seeks to impose unwise design choices upon the exchanger designer. If there is a chance that the choice will seriously adversely affect plant cost or performance, the exchanger designer has no choice but to object strenuously and with solid technical arguments. We develop this idea of ''problem identification'' in greater length in Section 2.3.

The designer determines at the start of each problem the critical points of the problem—those that exercise a disproportionate effect upon the operational success, reliability, and/or cost of the exchanger. Each case is a little different in this respect. In one, the critical point is the low coefficient associated with one stream; in another, a close internal temperature approach between the streams. These points are the ones upon which the designer lavishes his attention and skill.

By this time the exchanger designer is making some very basic decisions. From the moment the problem hits his desk, the designer has some feel for whether or not this is a big exchanger (or exchanger system), which directly translates into the amount of money involved. The designer must draw some early conclusions about the levels of engineering effort required and justified. The designer employed by an exchanger construction company may make a very different decision than one employed by a process company, to take the two extreme examples. The vendor cannot afford to invest a great deal of engineering effort in an inexpensive heat exchanger (except in the unusual situation that his engineering work is separately contracted for), so the designer judges his level of effort by the selling price of the exchanger. The process company must make a product, so the criterion for an engineering effort is the cost of the exchanger failing to work; a small but critical exchanger may warrant more design effort than a much larger but less vital one. Hence most processors now make detailed checks of the design of critical components in their new plants, even though they have engaged an engineering construction company to do the complete design.

Next, the single most important design decision is made: What basic configuration of heat exchanger will be used: double-pipe, shell-and-tube, plate, etc.? Usually the decision is even more detailed; for example, given that a shell-and-tube exchanger is chosen for a vaporizing service, the design may be further specified among the kettle, vertical or horizontal thermosiphon, forced circulation, or falling-film types. Actually, the decision may have been made much earlier in the process engineer's mind, who recalls that a given configuration was used in the last instance of a similar application. The process flowsheet from which the exchanger designer is working may already indicate by word or symbol a specific configuration. In many applications (such as using atmospheric air to cool a process stream), there is really very little choice available as far as basic configurations are concerned. But, if the exchanger designer is to play his proper role, he should have control over the choice of configuration within very broad guidelines and such physical constraints as exist in fact.

The next decision is about what design method to use. Basically, these fall into two categories: hand- and computer-design methods. Hand-design methods as they exist in the most recent literature and as applied by a competent designer are still valid for at least half and perhaps as many as 90% of all heat-exchanger problems. However, anyone who tries to design a vertical thermosiphon reboiler by hand is asking for trouble. If one chooses to use a computer-design method, one still has the task of selecting the level of the method. There are short-cut and detailed computer-design methods available for most exchanger types.

Whether the chosen design method is by hand or by computer, the next step is to select a tentative set of geometrical parameters of the exchanger. Any experienced designer has a set of rules and procedures by which he can select a plausible set of candidate parameters for his first attempt; the better he is at estimating the starting design, the sooner he will come to the final design, and this is very important for hand-calculation methods. A method for shell-and-tube applications is given in Section 3.1 of Schlünder (1983). However, it is usually faster to let a computer select a starting point—usually a very conservative case—and use its enormous computational speed to move toward the desired design.

In either case, the initial design will be "rated"; that is, the thermal performance and the pressure drops for both streams will be calculated for this design. The rating program is described in more detail later in this section.

If the calculation shows that the required amount of heat cannot be transferred or if one or both allowable pressure drops are exceeded, it is necessary to select a different, usually larger, heat exchanger, and to rerate. Alternatively, if one or both pressure drops are much smaller than allowable, it is probably possible to find a smaller and less costly heat exchanger, while utilizing more of the available pressure drop. The selection of the modified design requires a very elegant piece of logic; a simple example is given later in this section.

Once an acceptable thermal–hydraulic design is found, the design process continues with detailed mechanical design, shop drawings, material requirements, cost estimates, and so on.

The rating program shown schematically in Figure 2.2 is the core of the entire heat-exchanger design program. In the rating program, the problem specification and a preliminary estimate of the exchanger configuration are used as input data; the exchanger configuration given is tested for its ability to effect the required temperature changes on the process streams within the pressure drop limitations specified.

The rating process carries out basically three kinds of calculations. First, it computes a number of internal geometry parameters–surface and flow areas, leakage areas, by-pass area and so on (quantities that are required as further input into the heat-transfer and pressure-drop calculations)—are calculated for each stream in the configuration specified. The individual heat-transfer coefficients are combined via Equation (2.2) to give an overall coefficient, which is then used in Equation (2.1) or (2.3).

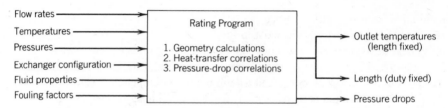

**FIGURE 2.2**  The rating program—the core of the heat-exchanger design process.

The results from the rating program are either the outlet temperatures of the streams if the length (and therefore area) of the heat exchanger has been fixed, or the length of the heat exchanger required to effect the necessary thermal change if the duty (inlet and outlet temperatures) of the heat exchanger has been fixed. In either case, the rating program also calculates the pressure drops for both streams in the exchanger.

The greatest amount of technical effort is required in writing the rating program because all of the correlations for heat transfer and pressure drop must be put in a quantitative form by the heat-exchanger designer. As mentioned above, it is usually necessary to do a great deal of model building to correct the basic correlations to the real geometry factors and interaction effects that actually exist in such a complicated geometry as a shell-and-tube heat exchanger.

The configuration modification program* takes the output from the rating program and modifies the configuration in such a way that the new configuration will do a ''better'' job of solving the heat-transfer problem.

The configuration modification program is typically an extremely complex one logically because it must determine what limits the performance of the heat exchanger and what can be done to remove that limitation without adversely affecting either the cost of the exchanger or the operational characteristics of the exchanger that are satisfactory. If, for example, it finds that the specified configuration cannot transfer enough heat, the program will try either to increase the heat-transfer coefficient or to increase the area of the heat exchanger. To increase the tube-side coefficient, one can increase the number of tube passes, thereby increasing the tube-side velocity. If the shell-side heat transfer is limiting, one can try decreasing baffle spacing or the baffle cut. To increase the area, one can increase the length of the exchanger or the shell diameter, or go to multiple shells in series or in parallel.

If the exchanger is limited by either a very low mean-temperature difference or value of the correction factor on the mean-temperature difference, the modification program can go to a countercurrent configuration (rarely, because of thermal stress

---

*In general, wherever the words ''modification program'' are used or implied, the reader may substitute the words ''thought processes of the exchanger designer.'' They are trying to accomplish the same ends, though the largely intuitive thought processes of the human designer are much harder to ''flow-chart.''

problems implicit in the configurations that permit countercurrent flow) or to multiple shells in series (commonly).

If the exchanger is limited by pressure drop on the tube side, the program can decrease the number of tube passes or it can increase the tube diameter; alternatively, it can decrease the tube length and increase the shell diameter and the number of tubes. If the exchanger is limited by pressure drop on the shell side, the program may increase the baffle cut, the baffle spacing, or the tube pitch, or go to double or triple segmental baffles. Other mechanical modifications could be made, but these are most commonly tested in the configuration modification programs. While changes are being made in the design for thermal–hydraulic reasons, the program must also be operating within the limits imposed by various codes, standards, and input restrictions specified by the customer. For example, baffle spacings should meet TEMA Standards (1978) as well as avoiding tube vibration.

A very simple but illustrative configuration modification program is diagrammed in Figure 2.3. In this particular program, it is assumed that the duty of the exchanger has been specified and the length required for the exchanger has been calculated by the rating program. The rating program has as its initial output the length of the shell of the largest acceptable diameter, the fewest tube passes consistent with the shell configuration, and the greatest allowable baffle spacing permitted by mechanical construction standards. The quantities with asterisks are the maximum allowable values of that parameter as specified in the input data. This choice gives the lowest tube-side and shell-side pressure drops consistent with the shell-size limitations and the exchanger type.

The first question that arises is: Is the length calculated by the rating program less than the maximum allowable length of the exchanger that can be accommodated? If the answer to this question is "No," it is necessary either to add another shell in parallel to split the flow or to adjust internal parameters in the heat exchanger to increase that heat-transfer coefficient (and inevitably also the pressure drop). The decision is made on the basis of the calculated pressure drops. The parameter adjusted depends upon the circumstances. For example, if the tube-side coefficient and pressure drop are low compared to the limit allowed, one would increase the number of tube passes. In any case, after the adjustments have been made, it is necessary to go back to the rating program and rerate the new design. The output of the rating program then comes back into the block on the left side of Figure 2.3 again.

If, however, the length of the exchanger is calculated to be less than the maximum allowable length (i.e., $L < L^*$), it is necessary to determine first whether the pressure-drop limitations have been exceeded. If either pressure-drop limitation is exceeded, it is necessary to add a shell in parallel and re-rate the new configuration. If both of the pressure drops are very much less than those allowed, it will probably be possible to decrease the shell diameter and come up with a smaller and less expensive exchanger than the one previously rated. If the pressure drops calculated are only slightly less than the pressure drop allowed, the exchanger designed is very close to the smallest acceptable exchanger (i.e., one that will meet all the operational requirements). One then adjusts the design parameters to

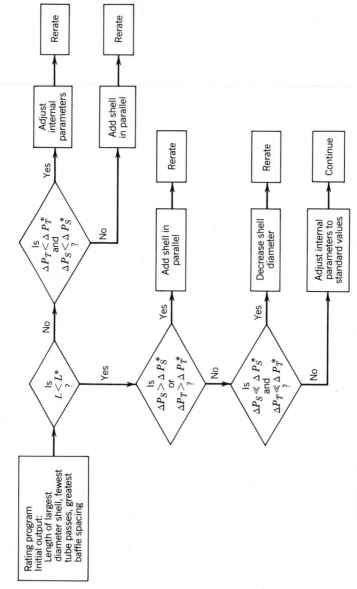

**FIGURE 2.3** A simple example of a configuration modification program. The asterisk denotes the largest value of that parameter permitted in the final design.

standard values for exchanger design and continues with the rest of the exchanger-design program.

## F. Further Aspects of Computer Design of Heat Exchangers

Every key output item from a computer program should be personally inspected to verify its basic rationality. For heat-exchanger designs, this inspection should be done by the heat-exchanger designer who can very quickly determine whether the computer-generated design is a reasonable one. Short-cut methods are available (Schlünder, 1983, Section 3.1) which can give an approximate design by hand within a few minutes. If this design agrees with the computer reasonably well, the computer design is likely a very good one. But if there are disagreements, the designer should reconsider the computer result.

It is important to do this every time. While any heat-exchanger design program prepared by a competent designer or programmer has been verified by running it against a number of test cases (typically 10 or 12), the logical structure of a large program is so complex that there is no possibility of ever testing all of the possible logical circuits that might be encountered. For example, even a modest-sized design program might have 40 logical decision points in its structure, resulting in $2^{40}$ or $1.1 \times 10^{12}$ distinct ways through the logic. If we were to run a different logical path every second around the clock, it would take just under 35,000 yr to check out all the possibilities. Therefore, in any given design problem, it is possible that a particular logical loop will be entered for the first time. If the logic of this loop is faulty, the results from it may be completely unrealistic or lead to further illogical steps.

Any questionable item in the output should be verified by analysis at whatever level of detail is required to establish its basic validity. Individual heat-transfer coefficients that seem to be too high or too low can be verified by going back to the basic correlation and doing a calculation. Geometry factors can be checked very quickly against plausible values by simple geometrical relationships. Particular attention should be paid to those items that can substantially affect the performance and operating life of the plant. For example, if an overall heat transfer coefficient turns out to be especially sensitive to the choice of a fouling resistance, it is worth taking additional time to reflect upon experience as well as any applicable information from outside sources as to the probable value of that fouling resistance.

## G. Mechanical Design of Heat Exchangers

While the thermal–hydraulic design of heat exchangers is the central matter of this chapter, and in fact is the primary design problem, mechanical design is equally essential to the success of the total solution. In the normal range of operating conditions (say 0–800°F and up to 1000 psia), there are seldom serious problems in evolving a satisfactory mechanical design incorporating the essential thermal-hydraulic features. However, as temperatures and/or pressures go to extremes, mechanical problems begin to predominate and compromises may be required by the thermal–hydraulic designer.

Even in modest pressure–temperature ranges, there are often choices to be made that simplify mechanical design problems without seriously affecting (often even enhancing) the thermal–hydraulic performance. For example, long, small-diameter shells are generally to be preferred (compared with short, large-diameter shells) because a thinner shell wall and tube sheet can be used to accommodate the same pressure. Cost is correspondingly reduced both because less material is required and because manufacturing costs are lower: there are fewer (but longer) tubes for a given heat-transfer area and therefore fewer holes to be drilled in the tube sheets and fewer rolling and/or welding operations to fasten the tubes. And there are—up to a point—concomitant thermal–hydraulic advantages, such as higher velocities both tube side and shell side (thereby better utilization of pressure drop and reduced fouling), and better flow distribution. On the other hand, carrying the length-to-diameter ratio to extremes can lead to excessive pressure drops, to vibration and erosion problems, and to a long, skinny, springy tube bundle that requires an excessive clearway in the plant for pulling.

In summary, the thermal–hydraulic design and the mechanical design are not so neatly packaged and separated as the flow diagram of Figure 2.1 suggests, and the prudent designer knows this and takes it into account. But computer design programs do not know this, though one can introduce a certain amount of internal logic into the program to guide the design into the right general direction. So again the bottom line is that the computer-generated configuration needs to be examined by somebody who knows what the problem is and what the answer ought to look like, and who knows what to do when the computer printout is not a good answer.

## 2.3.  PROBLEM IDENTIFICATION

### A.  Process Design and Heat-Exchanger Design

The starting block in Figure 2.1 is "Problem Identification" and this is in many ways the key to the successful design of the heat exchangers or indeed anything else in a process plant. If the designer has not successfully identified the problem, it is not very likely that he will find a satisfactory solution to it. In this respect, it is important to remember that the successful operation of the process is essential and that the design of each individual component in the process is successful only to the extent that it contributes to the success of the process.

Hence it can be fairly concluded that the process designer is the key person in the total design function, and proper respect and due deference should be given to his needs and desires. It would be a very brave and naive heat-exchanger designer who presumed to challenge the process designer on matters relating to the process in general. At the same time, the process engineer cannot be presumed to understand all of the specifics of heat-exchanger design (even if he once designed them himself) and should not intrude himself into the work of the heat-exchanger designer (assuming competence on the part of both parties, of course).

Separating the functions of the process designer and the heat-exchanger designer is not as cleancut as may be implied by the foregoing statements. There is, in fact, a broad overlap particularly in the area of the amount of information that each needs from the other to do his job properly. As far as the heat-exchanger designer is concerned, there are many things about the process and about the plant environment that he really needs to know if he is to provide the best possible design for the purpose. I refer to this as the "context" of the design problem and give some specific examples of what I mean in the following section. Generally, however, the information given to the heat-exchanger designer is much more limited, frequently to what can be included upon a TEMA heat-exchanger specification sheet (TEMA, 1978, p.9). Numerous authorities in the heat-exchanger field have pointed out how inadequate this sheet is to convey even the most essential information to the designer. It is hopelessly lacking if multicomponent condensation or vaporization is involved. In those cases, condensing or vaporization curves (essentially temperature, mass flow rate, and composition profiles of the vapor and liquid phases as a function of the amount of heat added to or removed from the process stream) must be included.

It is fair to say that at this time there is no adequate, structured mechanism by which the heat-exchanger designer can be informed of what he needs to know about a problem to do a good design job. In practical terms, it is up to him to know what is needed and to attempt to seek it out by whatever communication channels he can utilize. If the heat-exchanger designer is not employed by the company doing the process design, he may find it almost impossible to obtain the desired information because of the practical difficulties of communication and perhaps company policy concerning proprietary information. From a purely economic point of view, the heat-exchanger designer working for an outside vendor probably cannot afford to spend the time necessary to obtain the desired information even if he knows that it exists. In that case, the process company (the customer) must be aware of this problem and provide all important information to the heat exchanger designer. The next section gives a few examples of the sort of information that may be required in these cases.


## B.  The "Context" of the Design Problem

Surrounding every heat-exchanger design problem there is a body of facts, circumstances, and traditions that play a vital role in what the design should be, how it will be used, and with what probable degree of success. A complete rendition of this information would occupy many pages of text, tables, and figures. As a practical matter, no one is going to write down all of this information, and the heat-exchanger designer would not have the time to read it in any case. How then is this information to be made available to the person who uses it? The best solution is for the essential design function to be carried out by someone in the same company who, through experience and previous opportunities for communication, knows something about the process, the plant in which the process is to be installed, and the people who are going to run the process. Until recently, this kind of background was fairly common in a number of the larger chemical and petroleum processors. But in recent years, there has been a definite attempt to reduce the amount of in-plant design expertise and the interchange of personnel between the plant and the engineering department. This is regrettable, but it is a fact. What then can be done to preserve at least the minimum essentials of communication?

The first responsibility lies with the process design engineer, who must identify clearly and as fully as reasonably possible those critical elements of the design without which the exchanger designer cannot intelligently proceed. Processors should take this responsiblity quite seriously, because they will make or lose money depending upon whether or not the plant operates successfully. But it is also true that the processor cannot anticipate all of the information requirements vital to the design. This is especially true for processors who do not maintain in-house design competence on specific components such as heat exchangers.

Therefore, it follows that the heat-exchanger designer should be generally aware of critical questions as they arise during his design calculations. The processor should provide to the designer a clear-cut and convenient communication channel through which he may enter his enquiries with a reasonable expectation of getting a prompt, accurate, and responsible answer.

To give these general and hopelessly idealistic ideas some concreteness, one can consider the problem of specifying allowable pressure drops in the heat exchanger. The allowable pressure drops for each fluid must be specified by the customer and one would like to believe that some thought has gone into those numbers. This is seldom true; a 10-psi allowable pressure drop is frequently specified for liquid and high-pressure gas streams, often because that was the number quoted last time and it worked well enough. But what does 10 psi actually mean? Does it mean that there is absolutely no more pressure drop available than 10.0 psi from nozzle to nozzle, and any heat exchanger that requires more than that pressure drop will fail to work satisfactorily? This is possible, but not very likely. Probably there is more than 10 psi available, but the process designer is assuming that he is going to use up some pressure drop in the piping and through a flow-control valve; and if the designer confines his pressure drop requirement to 10 psi, there will be enough to go around. Suppose that the designer puts in 10 psi as the maximum allowable pressure drop. The computer-design method will note this and will continue to modify the design of the heat exchanger until the predicted pressure drop is less than 10 psi. This may involve throwing out a very attractive design for which the computed pressure drop was 10.2 psi and possibly settling for a substantially larger heat exchanger for which the pressure drop is computed to be, say 6.7 psi. In all probability, there would be no difficulty in the plant in providing the 10.2 psi and the heat-exchanger designer should have the option to offer that design to the customer. But to do so, he must know that such a design does exist. So the computer program should provide at least an indication that it has found a design that is otherwise satisfactory except for the slightly larger-than-allowable pressure drop. Some design programs do this, while others do not. In the end, it is up to the designer to know that such a design might exist, and what to do with it.

But what if there *is* an absolute limit of 10.0 psi (or 5.25 or any other number)? Such a situation could arise if the flow is between two large flow manifolds whose pressures are set by conditions beyond the control of even the process designer. Now the design must observe a strict limit. This situation calls up another consideration that will be developed at greater length later in this chapter, namely the question of uncertainty in our design calculations. In fact, our ability to predict the single-phase shell-side pressure drop in a shell-and-tube heat exchanger is probably no better than about ±30%, using the most advanced and accurate design methods presently available. This means that, for a flow in which a computed pressure drop is 10 psi, the actual pressure drop might be as high as 13 psi. If only 10 psi were actually available, the shell-side flow would be reduced by about 15% with probably severe adverse effects upon the process. Therefore, if there is a real limitation to a 10-psi pressure drop, the heat-exchanger designer should have been designing a heat exchanger for 7 psi right from the beginning.

Another example using a more qualitative sort of information can also be cited. Assume that the problem is to design a reboiler to vaporize a fluid with some potential for fouling. The immediate choice that would occur to most heat-exchanger designers here would be a vertical thermosiphon reboiler. This configuration is known to have superior ability to resist fouling on the boiling surface because of the high fluid velocities past the surface. Also, the vertical thermosiphon permits relatively easy cleaning inside the tubes compared with shell-side boiling equipment. However, many years ago, thermosiphon reboilers received a widespread bad reputation because of a phenomenon known as flow instability, resulting in an extremely unstable and indeed violent oscillation of the flow. Many companies issued a rule that vertical thermosiphon reboilers were not to be considered for use in their plants, and these rules are often still in the engineering standards of these companies. Yet, in the fullness of time and a large-scale research program, the phenomenon of instability became generally well-understood, and criteria were developed for allowing the designer to design a reboiler that would be stable over the specified range of operation.

How shall a knowledgable heat-exchanger designer respond to a request for a quotation from a company that he knows has an inbred, if no longer justified, prejudice against vertical thermosiphon reboilers? Unless he communicates with the company, he has two choices: (1) he may specify a kettle or horizontal thermosiphon reboiler to the company, knowing full well that it might not be the optimal choice; or (2) he may specify a vertical thermosiphon reboiler to the company, possibly as an alternative to another design, and buttress his quotation with his best technical judgment and arguments in favor of the vertical thermosiphon.

There is a risk whichever way he goes. If he does specify a different type of reboiler, a competitor may convince the customer that in fact a vertical thermosiphon is the proper choice. If he recommends a vertical thermosiphon, even with convincing arguments, his bid may be dismissed as not being responsive to the engineering practices of the customer. How much better it would be to enter into a mutual exchange of technical views in which the needs, experience, and current technical competence of each of the parties are brought to bear on solving the problem!

These are only two examples in which the context dominates the decisions and actions of the parties involved, not to mention the probability of success of the outcome. More than further research on the technical details of various exotic heat, mass, and momentum transport processes, we need better ways to bring the context of the problem into the search for a solution.

## C.  Proper Goals for Optimization of Heat Recovery

A major driving force in the design of chemical processes since the energy crisis of the early 1970s has been the desire to optimize the recovery of heat in process plants. In the ''good old days,'' the thermal aspects of a process plant could be reasonably well visualized by imagining heat being supplied by steam and fired heaters to the process streams at the front end of the plant to bring them to the proper conditions for reaction and separation, and cooling water being supplied to the effluent end of the plant to condense and cool the products for storage. Prime fuel was used at the front end because of its low cost, and cooling water was used at the exit end because of its easy availability. These conditions are no longer true: high-equality energy sources such as oil and natural gas are now far too valuable to be used for low-temperature heating if there is any other alternative, and economic and environmental considerations preclude the use of surface water in unlimited quantities simply to remove unwanted heat.

We now find effluent streams that need to be cooled being used as heat sources for cold incoming streams that need to be heated. While the primary explanation for this has been the enormous cost of high-quality fuels and the decreasing availability of cooling water, it is also recognized that the earlier practices could result in a sharp increase in the cost of heat-transfer surface required. In recent years, great attention has been paid to the optimal design of plants using high-temperature effluent streams to heat incoming cold streams, resulting in two major lines of effort. One has been to minimize the amount of utilities, hot and cold, required in a given process, that is, primarily concentrating upon minimizing the operating cost. The other has been to minimize the number of heat exchangers required in the process, thereby minimizing the capital costs. The two efforts are not mutually exclusive, and the literature is beginning to show signs of accomplishing or at least accommodating both goals in a single attack to improve process efficiency and reduce capital cost. While there are many current literature references, perhaps the recent work of Linnhoff et al. (1982), for example, represents the best attempt at heat integration.

It is necessary to point out that, in any real plant situation, certain considerations do not always appear in the idealized models represented in the literature. One such consideration is that the equipment specified be equipment that can be manufactured and employed in the given service. For example, high-temperature heat exchangers, or those accommodating substantial temperature changes between the inlet and outlet streams, must generally make very specific provision for differential thermal expansion between tubes and shell. Floating-head heat exchangers are often used for this purpose, but a single, multiple tube-pass, floating-heat exchanger cannot work to the same temperature approaches between inlet and outlet streams that a true countercurrent heat exchanger can. By placing a sufficient number of such heat exchangers in a series arrangement, any temperature approach can be designed for; but this frequently results in a substantially larger investment in equipment and higher pressure drops than those that appear simply by drawing a process flow diagram with suitable heat balances on it.

Another aspect of the problem is that it is simply not economic to recover small amounts of heat (e.g., from streams having small flow rates) if such streams must be brought from one side of the plant to the other in order to meet the matching stream.

A final consideration is that the resulting plant must be operable; that is, there must be ways in which the operators can, in effect, reach in from outside to vary the conditions at any point in the process to make the process operate at the desired parameters for the production of the chemicals desired. There must, therefore, be some compromise between the amount of heat that is recovered, the capital cost of the equipment involved in heat transfer, and the operability and controllability of the plant during both normal and emergency procedures. The sophisticated computer optimization procedures suggested in the recent literature represent a first-cut approach to the problem, but both the process designer and the heat-exchanger designer must examine the idealized schemes to identify the real possibilities for the actual case.

## D. Pulling the Design Together

It must seem from the foregoing paragraphs that there is an infinite number of possibilities to be considered in each plant design and heat-exchanger design situation and indeed this is true on paper. However, there are almost always many feasible solutions to the problem whose actual economics lie within a very narrow range of the mathematical optimum. In fact, as we demonstrate in the next section, the introduction of uncertainty into the modeling means that it is literally impossible to define one and only one optimal configuration.

An experienced designer is well aware of several things, including the inherent uncertainties and the unpredictability of future operating conditions and economic parameters, which means that the optimal design is at best a moving target. While the designer can never know everything about the problem, he can make certain broad-based intuitive judgments which will leave him not far from a feasible as well as an economically and operationally optimal configuration. The computer solutions are fine as far as they go in identifying both the general direction in which one needs to work and regions of inherent infeasibility; but in the end, the human designer must step in to fix the final design in light of the nonquantifiable aspects of the problem as well as those that admit to mathematical manipulation.

## 2.4. UNCERTAINTY IN HEAT-EXCHANGER DESIGN

### A. Sources of Uncertainty

Uncertainties arise from almost every input into the heat-exchanger design process. Seldom does a heat exchanger operate at the conditions specified by the process design; flow rates, temperatures, composition, and even the products to be made usually change in the two or so years between design and operation. The physical properties of even pure substances are not very well known [the reported thermal conductivity of toluene decreased nearly 20% between 1920 and 1970, when improved experimental techniques seemed to stabilize the value (Mallan et al., 1972)], and mixture estimation procedures are a cross between crude fundamental modeling and pure empiricism. The basic heat-transfer and fluid-flow correlations have surprisingly large uncertainties—turbulent-flow heat-transfer coefficients in a plain round tube of $L/D > 50$ ($L$ is the length of the heated section and $D$ is the inside tube diameter) are predictable only within about $\pm 20\%$ for the typical case and much higher percentages for high Prandtl number or non-Newtonian fluids, for example.

The models incorporating these correlations into design procedures are necessarily approximate, and manufacturing variations in the nominal dimensions of the equipment cause further deviations between prediction and performance. Often, of more consequence than all of the foregoing, fouling of one or both surfaces—unpredictable, transient, and frequently dominating—may switch the control of the thermal performance of the equipment from the designer to the operator and the maintenance crew.

It is less surprising to me that heat exchangers often fail to operate as expected than that they usually operate well enough.

### B. Analyzing Uncertainty

It may seem from the foregoing that any attempt to quantify the uncertainties and even to do serious design is doomed to failure. This is not true, because through either engineering judgment or the principles and techniques of probability and statistics, we may estimate bounds and consequences of uncertainty. The latter approach works like this. From the experimental data going into the basic correlations and the weighted experiences of accomplished designers and operators, standard deviations of the major design variables can be estimated. Then the combined effects of these variables—through the complex and nonlinear exchanger design equations—can be estimated (in statistical terms) using a technique called Monte Carlo analysis (Al-Zakri and Bell, 1981). In effect, the input parameters are randomly selected over the range of possible values and the outcomes calculated. The calculations are repeated a sufficient number of times that a statistically meaningful distribution of outcomes can be obtained. From this, the designer can decide if the probability of an acceptable outcome is high enough. The procedure can be applied to a complex system of heat exchangers, and other statistical techniques can be used to identify the most critical uncertainties (Uddin and Bell, 1986). In spite of the large number of calculations to be carried out and interpreted, a large computer can handle the task in a few minutes, even for a system of 20 or 30 heat exchangers with feedback streams.

However, these techniques are still in their infancy in terms of development. In real-life terms, most heat exchangers and heat-exchanger systems are still analyzed and designed as if every number were precisely known. The designer may include some "fat" or safety factors in the input data, or he may add some conservatism in the final design; but these are based upon intuition and/or experience. In the hands of a competent designer, with a knowledge of the problem, this is still the best way.

But how does one know where to put the fat?

## C.  Designing Around Uncertainty

First, let the following fact be understood. The design specifications for a heat exchanger, as communicated to a designer, are usually a combination of nearly the worst conditions expected to be encountered in service (e.g., higher than expected flow rates, high air or cooling water temperatures, closer than required temperature approaches, and excessive fouling resistances). From this it follows, first, that the heat exchanger will probably never have to operate (or be given the chance to operate) under these conditions, and second, that there is a built-in bias, or margin of safety, already provided by the process designer. This is not entirely reassuring, because this margin of safety comes from a "generalist" who may not understand what constitutes a safety factor when it comes to heat exchangers.

Now the heat-exchanger designer must know how to provide the margin of safety. There are certain types of heat-transfer equipment that are almost self-fulfilling prophecies: they will work about as well or poorly as you expect them to work. As an example, kettle reboilers were traditionally designed for a high fouling resistance and a very conservative boiling heat flux, because "everyone knew" that kettles were prone to fouling. Just to make sure that they would work, a high-pressure (i.e., high-temperature) steam supply was provided to overcome the expected poor performance. The new clean kettles were started up with large $\Delta T$'s, resulting in transition and film boiling, until they were so badly fouled that the extra area and excessive $\Delta T$'s were needed to get them to work at design heat duties. Everyone stood around and congratulated themselves on their wisdom in "conservative design." Whereas, if the kettle is started up gently (low $\Delta T$'s) when clean and if the design heat fluxes are high, the kettle will tend to keep itself clean and work very well for long periods of time.

When conservative design does mean more area, it is generally best to provide it in the form of extra length (in a shell-and-tube exchanger) rather than extra diameter. This keeps velocities up without unduly disturbing the baffle spacing and cut and possibly reintroducing problems of vibration; of course, this assumes that the designer has a little spare pressure drop and enough space in the plant.

Sometimes conservative design means more flexibility in the piping arrangements, especially when more than one exchanger is involved. Or it may mean providing an extra exchanger (with additional piping, valving, and instrumentation) so that one shell can be taken out of service regularly for cleaning or other maintenance; this is a particularly common technique where severe fouling is expected, yet the plant as a whole must stay in service. This, of course, is the kind of decision the exchanger designer cannot make on his own, but he can call the problem—and possible solutions—to the attention of the process designer.

If it is objected that these alternatives cost too much, ask about the cost if the heat exchanger bottlenecks the process or is not available at all.


## 2.5. FLOW MALDISTRIBUTION

The inexperienced designer (of heat exchangers, packed columns, or piping systems) often assumes that when a flow is introduced into a device offering multiple parallel flow paths, the flow will split uniformly among the multiple paths and each stream may be analyzed as if exactly the same things—heat transfer, phase change, and pressure drop—are happening to it as to every other parallel stream. More often than not, small differences between the paths, or the relative opportunities of the fluid to get into them, result in large differences in flow rate and thermal–hydraulic performance. Even if the differences are anticipated by the designer, it may be impossible to eliminate or offset them, and so one may have to learn to live with them. There are many different forms of flow maldistribution, and the following sections describe and illustrate three general types.


## A. Bypassing

In general, there are many different routes by which fluid can flow from the inlet to the outlet nozzle of an exchanger, each one having different flow resistances and opportunities for the fluid to contact heat-transfer surface. For instance, in a shell-and-tube exchanger, there are clearances between tubes and baffle, and between baffles and shell, and between the tube field and the shell. Each of these permits part of the flow to short-circuit the prescribed path across the tubes and through the baffle windows.

The tube-to-baffle clearance is not too serious: the flow here is in intimate contact with heat-transfer surface (the tubes), which at least partially offsets the loss of heat-transfer coefficient in the tube bundle because of reduced cross-flow and the quite separate but even more serious distortion of the temperature profiles. On the other hand, the flow leaking between the baffle and the shell is not in contact with heat-transfer surface and has relatively poor mixing with the main cross-flow stream; this effect can reduce the overall heat-transfer performance by about one-third.

Why not seal the clearances by welding or gasketing? Because no one knows how to do this in a practical industrial unit that has to be assembled and regularly disassembled for cleaning and maintenance. So we incorporate these bypass flows into our models of the shell-side thermal–hydraulic phenomena, set manufacturing tolerances on the clearances, and learn to live with them—and watch for special cases where the bypassing problem may completely distort the assumed flow patterns. In that case, the designer may opt for an entirely different approach to the problem, possibly a double-pipe heat exchanger with only one parallel flow path, which is expensive, but predictable.

For just one more example of bypassing, consider an air-cooled heat exchanger in which the finned tubes are arranged in a square layout such that the fin tips and the small clearances between the fins in adjacent rows are in line with one another. The clearance path is a low-resistance path, compared with that through the fins close to the tube, so most of the flow goes through the clearance—where there is little heat-transfer surface to contact. The net effect is for inline high-fin tube banks to give an effective heat-transfer coefficient only about half that of the corresponding staggered tube arrangement (Bell and Kegler, 1978).

The point is that almost every type of heat exchanger can suffer from bypassing, and the designer needs to keep his antennae attuned to the possibility and the consequences.

## B. Nonuniform Stream Splitting

It is frequently necessary or desirable to split a stream into a few or many separate streams, and it is usually desirable that each of the resulting streams have the same flow rate. The tube side of a shell-and-tube exchanger is a common case. If the flow resistance through the tubes is identical for each tube, and if the pressure drop through the tubes is large compared to the kinetic head of the approaching stream (or if the approaching stream is diffused through a deep header, nested conical diffusers, or a series of perforated plates or screens), there is ordinarily no problem. If not, different tubes get different flows, thermal performance is adversely affected, and the exchanger may even tear itself apart because some tubes are running much hotter or colder than others.

A somewhat different example is manifolding several nominally identical exchangers in parallel to handle a large stream. This is often done in cryogenic services with plate-fin exchangers because of the limited individual unit size available. But these exchangers are designed for very close temperature approaches between the streams (high number of transfer units) and low pressure drop, to minimize compressor capital and operating costs. Any maldistribution (which may arise in the manifolding and/or in the exchanger flow characteristics) in the stream-splitting on either stream results in a thermal capacity mismatch in the exchanger and a decrease in the thermal effectiveness of the whole system. Flow instrumentation and control valves are minimized in such plants because of the difficulty of maintenance at low temperatures and because of the parasitic pressure loss, so the usual means of flow balancing are missing or limited. Thus the beautiful thermodynamic and economic optimization of such plants, based on ideal flow distribution, has to be tempered by the reality that we cannot guarantee perfect splitting.

## C.  Phase Separation

Two-phase (vapor/gas–liquid) flows frequently occur in process and energy plants and have to be accommodated in process equipment. These flows are particularly prone to nonuniform splitting: there is no way to insure that a two-phase flow will split uniformly with respect to either the total flow or the individual phase flows among multiple parallel channels. But even inside a single channel, the two phases may arrange themselves in a number of different patterns, depending upon the individual phase and total flow rates, the physical properties, and the geometry of the channel. In a horizontal tube, for example, at low flow rates, the liquid may be flowing on the bottom and the vapor on the top (stratified or wavy flow), while at high flow rates, the liquid may cover the entire surface with vapor in the core (annular flow). In the low-flow-rate case, the heat-transfer rate may be very high on the bottom and low on the top for a boiling flow, and exactly the opposite for a condensing flow. Failure to recognize these differences can lead to poor performance and even catastrophic failure (Bell, 1985).

## 2.6.  FOULING IN HEAT EXCHANGERS

As noted, fouling is simultaneously widespread and very detrimental, little understood, and poorly predicted. There is not much more that can be said, but we attempt here to put this phenomenon into context.

## A.  Types of Fouling

Epstein (1983) identifies the types of fouling as follows:

1. Crystallization fouling:
   a. Precipitation—crystallization from solution, sometimes called scaling.
   b. Solidification—freezing of a pure liquid or the higher-melting components of a liquid mixture (e.g., wax from oil).
2. Particulate fouling: deposition of suspended solids, "sedimentation" when gravity causes the settling.
3. Chemical-reaction fouling: deposit formation by chemical reaction not involving the surface material.
4. Corrosion fouling: accumulation of corrosion products.
5. Biological fouling:
   a. Macrobiofouling—deposition of large organisms (e.g., barnacles).
   b. Microbiofouling—deposition of slimes and microorganisms.

Chemical-reaction fouling further comprises a number of different cases, such as polymerization (e.g., of unsaturated hydrocarbons) and thermal degradation of heavy materials ("tar formation"). It is common for two or more mechanisms to occur together and even synergistically; for example, cooling-tower water often involves particulate, crystallization, corrosion, and microbiofouling simultaneously. Every specialist has his own bizarre favorites: mine include a special case of macrobiofouling in a cooling tower system (a population explosion of little green frogs) and frozen neon in a helium liquefaction plant.

## B. Design for Fouling

As shown in Equation (2.1), fouling resistances are included in the overall heat-transfer coefficient, the effect being to provide an additional amount of heat-transfer area in the exchanger. The problem with that approach is multifaceted: (1) the resistances are not usually related to actual process conditions; (2) the resistances are only guesses under the best of conditions; (3) the resistances are constants, whereas most fouling is a transient phenomenon; (4) if the designer is not paying attention, the extra exchanger area may result in a heat exchanger more prone to fouling than if the extra area had been left out.

Some explanation of the last point is required. If the extra area is provided by increasing the shell diameter, the velocities usually decrease on both sides, leading to lower fluid shear stress. But shear stress is one of the factors tending to limit or remove the fouling buildup. A generally better way is to design the exchanger without a fouling resistance and to a lower allowable pressure drop, and then add area by increasing the specified length. The final configuration may be very much the same, but the latter procedure involves the thought and attention of the designer more directly.

A comment about the transient behavior is in order. Some types of fouling (e.g., particulate fouling and some cases of corrosion) are inherently asymptotic; that is, as long as flow conditions remain constant, the fouling builds up to a maximum resistance and in effect stops. One can talk about providing sufficient extra area to allow the exchanger to operate ''forever'' without cleaning in these cases, and there are many real-life examples. Usually, however, the fouling builds up without a limit (though the rate of increase may slow down), and some action must be taken to restore the original clean or nearly clean surface. In these cases, the purpose of the fouling resistance and the excess area is to allow a reasonable operating time between corrective actions. Some examples of this situation are given in the next section.

## C. Operating Around Fouling

The most obvious—and most common—way to control fouling is to shut down the fouled exchanger and clean it. Cleaning techniques include blasting with high-velocity water jets or crushed walnut shells, brushing, scraping, drilling out (trying to distinguish between fouling and tube wall), or even burning the deposit off. But unless special provisions are made in the plant design, cleaning a heat exchanger usually requires shutting down the process. For a rapidly fouling exchanger, this alternative is unacceptable and other possibilities have to be considered.

On-line cleaning is a possibility. Power-plant condensers particularly have been fitted with Amertap® systems, in which slightly oversized sponge rubber balls are recirculated through the tubes for an hour or two a day. As the balls enter the tube with the normal flow of water, they squeeze down and scrub any deposits off the wall. For hard deposits, abrasive-coated balls can be used (carefully!). The system has high capital and operating costs and can only be justified in extremely valuable and critical services.

Biofouling deposits can sometimes be controlled by the very careful addition of biocides such as chlorine, but environmental restraints have to be observed. The relatively light microbiofouling observed with warm ocean water from far offshore can be controlled by the addition of 50–100 ppb of chlorine for an hour or two a day.

If all else fails, there is naught for it but to provide full standby—an extra heat exchanger waiting, ready to go, when the on-line heat exchanger is not transferring enough heat or requires too much pressure drop owing to fouling. One such example is the waste-water clean-up system in a coke chemical plant. The vertical thermosiphon reboilers on the column foul hopelessly (typical iron ferro- and ferricyanides, cemented in place by coal-tar pitch) about every 6 wk, and it takes about 2 wk to clean and recondition the exchanger. So at any given time, three reboilers are driving the column and one is off-line being cleaned. The multiple exchanger system is expensive, including complicated piping, instrumentation, and control systems. But the plant stays on-line, producing about $25 million worth of products every day.

In this business, that's success.

## 2.7.  SUMMARY

As I expected, this has turned out to be more a collection of anecdotes and examples than a carefully set-out system of principles. There may be some great and fundamental truth buried somewhere—a framework on which the examples can be hung like ornaments on a Christmas tree. But until that framework can be revealed and stated with mathematical precison amidst a holistic view of the heat-exchanger design process, my best advice has to be: Understand as best you can what is going on in mechanistic terms, try to accommodate this to the problem that needs to be solved, and figure out what your next move will be if that turns out to be wrong.

## REFERENCES

Al-Zakri, A. S. and K. J. Bell, "Estimating Performance when Uncertainties Exist," *Chem. Eng. Prog.*, **77**, (7), 39 (1981).

Bell, K. J., "Two-Phase Flow in Heat Exchangers," in *Two-Phase Flow and Heat Transfer*, X. Chen and N. Veziroglu (eds.), Hemisphere, Washington, D.C., 1984, pp. 341–361.

Bell, K. J. and W. H. Kegler, "Analysis of Bypass Flow Effects in Tube Banks and Heat Exchangers," in *Heat Transfer: Research and Application*, J. C. Chen (ed.), *AIChE Symp. Series*, Vol. 74, No. 174, American Institute of Chemical Engineers, New York, 1978, pp. 47–52.

Bergelin, O. P., G. A. Brown, and S. C. Doberstein, "Heat Transfer and Fluid Friction During Flow Across Banks of Tubes—IV: A Study of the Transition Zone Between Viscous and Turbulent Flow," *Trans. ASME*, **74,** 953 (1952).

Epstein, N., "Thinking about Heat Transfer Fouling: A 5 × 5 Matrix," *Heat Trans. Eng.*, **4** (1), 43 (1983).

Kakac, S., A. E. Bergles, and F. Mayinger (eds.), *Heat Exchangers: Thermal–Hydraulic Fundamentals and Design*, Hemisphere, Washington, D.C., 1981.

Kern, D. Q., *Process Heat Transfer*, McGraw-Hill, New York, 1950.

Linnhoff, B., D. W. Townsend, D. Boland, G. F. Hewitt, B. E. A. Thomas, A. R. Guy, and R. H. Marsland, *User Guide on Process Integration for the Efficient Use of Energy*, Institution of Chemical Engineers, London, 1982.

Mallan, G. M., M. S. Michaelian, and F. J. Lockhart, "Liquid Thermal Conductivities of Organic Compounds and Petroleum Fractions," *J. Chem. Eng. Data*, **17** (4), 412 (1972).

Marto, P. J. and R. H. Nunn (eds.), *Power Condenser Heat Transfer Technology*, Hemisphere, Washington, D.C., 1981.

Palen, J. W. and J. Taborek, "Solution of Shell-Side Flow Pressure Drop and Heat Transfer by Steam Analysis Method," *CEP Symp. Ser.*, No. 92, **65**, "*Heat Transfer—Philadelphia*," 53, American Institute of Chemical Engineers, New York, 1969.

Rohsenow, W. M., J. P. Hartnett, and E. N. Ganic (eds.), *Handbook of Heat Transfer*, 2nd ed., McGraw-Hill, New York, 1985.

Schlünder, E. U. (ed.), *Heat Exchanger Design Handbook*, Hemisphere, Washington, D.C., 1983.

Shah, R. K. and A. L. London, *Advances in Heat Transfer: Supplement 1: Laminar Flow Forced Convection in Ducts*, Academic, New York, 1978.

Singh, K. P. and A. I. Soler, *Mechanical Design of Heat Exchangers and Pressure Vessel Components*, Arcturus, Cherry Hill, NJ, 1984.

Taborek, J., T. Aoki, R. B. Ritter, J. W. Palen, and J. G. Knudsen, "Fouling—The Major Unresolved Problem in Heat Transfer," *Chem. Eng. Prog.*, **68** (2), 59; **68** (7), 69 (1972).

Taborek, J., G. F. Hewitt, and N. Afgan (eds.) *Heat Exchangers: Theory and Practice*, Hemisphere, Washington, D.C., 1983.

TEMA (Tubular Exchangers Manufacturers Association), *Standards*, 6th ed., New York, 1978.

Tinker, T., "Shell-Side Characteristics of Shell-and-Tube Heat Exchangers," *General Discussion on Heat Transfer*, Institution of Mechanical Engineers, London, England, 1951, p. 97.

Uddin, A. K. M. Mahbub and K. J. Bell, "Sensitivity Analysis Applied to Monte Carlo Simulation of Heat Exchangers," *Heat Transfer—1986*, Proceedings 8th International Heat Transfer, Vol. 2, San Francisco, Hemisphere, Washington, D.C., 1986, pp. 385–390.

# 3

## ENERGY-EFFICIENT SEPARATION PROCESS DESIGN

JAMES R. FAIR

## 3.1.  INTRODUCTION

In this chapter, we are concerned with methods for separating mixtures, with emphasis on the energy consumption of these techniques. The context is separation

71

of fluid mixtures, and the general area of application is in chemical processing. The context is also in the area of optimum design of chemical processes, approaches to which are many and varied, but all of which reduce to a common denominator: the result must be a design that produces one or more specified chemicals at the lowest cost. As the chemical and petroleum processing industries compete internationally during their enforced "restructuring" in the latter part of the twentieth century, it becomes very clear that the survivors will be the low-cost producers.

For processes that contain a significant complement of separation investment, energy is a key factor in bringing about the separation—and a key factor in the cost of production. The ever-present need is for *less energy-intensive separations*; hence the title and thrust of this chapter.

The need for separation steps in the chemical process is well known. It is necessary to isolate and purify products, recover valuable by-products, prepare raw material streams so that they are compatible with the processing equipment, and prevent toxic or otherwise undesirable materials from moving to the environment from the process. Examples are manifold: isolation of certain amino acids, separation of residual monomer from polymers, recovery of solvents from fiber-spinning operations, removal of sulfur-bearing catalyst inhibitors, and prevention of cooling-water system inhibitors from flowing into streams during the blowdown procedures. The place of separations in the conventional chemical process is shown in Figure 3.1.

## 3.2. ENERGY CONSUMPTION OF SEPARATION PROCESSES

Separations are "big business" in chemical processing. It has been variously estimated that the capital investment in separation equipment is 40–50% of the total for a conventional fluid processing unit. Of the total energy consumption of an average unit, the separation steps account for about 70%. And of the separation consumption, the distillation method accounts for about 95%.

From this type of information, several points emerge: improved capital investment economics for separations can have a significant impact on cash flow; relatively small-percentage improvements in energy consumption by separations can have a salutary effect on the cost of operation; and efforts are worthwhile if they are directed toward either replacing distillation with lower energy-consuming methods or finding ways to reduce the consumption of energy by distillation. An-

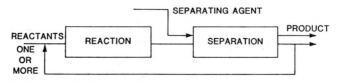

**FIGURE 3.1**  The place of separation in the typical chemical process.

other point is implicit in the points listed above: the impact of energy-reduction efforts will be greatest for the larger volume (commodity-type) processes. In fact, for many high-valued specialty chemicals, the cost of energy has such little impact that there is no incentive for improving on separation methods that have demonstrated satisfactory performance.

In a recent study sponsored by the United States Department of Energy (Bravo et al., 1986), this author and co-workers analyzed a number of separation methods with respect to (1) their likelihood for further development without government support, (2) whether they were generic in scope, and (3) whether they might hold promise for significant energy savings, if funded for further development (by government or other sources). For further study, a method required an affirmative response to all three considerations. This permitted a number of separation methods to be eliminated from further study, as shown in Table 3.1.

**TABLE 3.1   Separation Technologies Eliminated for Further Study**

| | | | |
|---|---|---|---|
| 1. Is the process of sufficiently high risk that industrial development is doubtful without governmental support? | | | |
| 2. Is the process generic in scope? | | | |
| 3. Does the process hold the promise of significant energy savings? | | | |
| Absorption (physical) | yes | yes | no |
| Bubble/foam fractionation (p. 17-64) | yes | no | yes |
| Distillation (conventional) | no | yes | no |
| Distillation (heat-pumping techniques) (p. 13-6) | yes | yes | no |
| Electrically induced permeation (p. 17-37) | yes | no | yes |
| Evaporation | no | yes | no |
| Extractive and azeotropic distillation (pp. 13-53 and 13-57) | no | no | yes |
| High-gravity distillation (Ramshaw, 1983) | no | yes | yes |
| Mass spectrometry | no | yes | yes |
| Molecular distillation (King, 1980, p. 25) | no | yes | yes |
| Pressure diffusion (p. 17-70) | yes | no | yes |
| Separation in an electric field (p. 17-34) | no | no | yes |
| Stripping | yes | yes | no |
| Sweep diffusion (King, 1980, p. 23) | yes | no | yes |
| Thermal diffusion (p. 17-68) | yes | no | yes |
| Ultracentrifuging (gas or liquid) (King, 1980, p. 24) | no | yes | yes |

*Source:* Bravo et al. (1986).

*Notes:* A ''yes'' answer was required if a method was to be considered further. Page number cited in the table (e.g., p. 17-64) refers to Perry and Green (1984).

The remaining methods were then graded by the several criteria shown in Table 3.2, and the final rankings of the methods are given in Table 3.3. The rankings form the basis of further discussion in this chapter, with the following methods indicating the greatest potential for study as well as for energy reduction: distillation (with high-efficiency mass-transfer devices), pressure-driven liquid permeation, pervaporation, adsorption, gas permeation, crystallization, and extraction. All of these methods are amenable to application in higher-volume processes.

**TABLE 3.2   How the Technologies Were Graded**

| Ranking Criterion | Weight Factor |
|---|---|
| Potential energy savings on a national scale | 3.0 |
| Potential for reducing required capital cost | 1.5 |
| Potential for reducing operating cost | 3.0 |
| Potential for increased productivity | 3.0 |
| Lack of potential mechanical/materials limitations | 1.5 |
| Lack of potential adverse environmental impacts | 2.0 |
| Ability to achieve the required level of separation | 3.0 |
| Potential for using/recovering waste energy | 1.5 |
| Degree of exposure of chemicals and products to damaging operating conditions | 1.0 |
| Lack of need for introduction of other separating agents that cause downstream problems | 1.0 |

*Source:* Bravo et al. (1986)

**TABLE 3.3   Scores of the Separation Technologies**

| Separation Technology | Final Score |
|---|---|
| 1. Distillation with high-efficiency mass-transfer devices | 277 |
| 2. Pressure-driven liquid permeation | 253 |
| 3. Pervaporation | 244 |
| 4. Adsorption | 233 |
| 5. Gas permeation | 232 |
| 6. Crystallization | 208 |
| 7. Supercritical extraction | 201 |
| 8. Absorption | 199 |
| 9. Liquid/liquid extraction | 197 |
| 10. Naturally driven permeation | 197 |
| 11. Chromatographic separations | 196 |
| 12. Clathration | 187 |
| 13. Imbibition | 166 |
| 14. Dual-temperature exchange reactions | 152 |

*Source:* Bravo et al. (1986)

## 3.3. SEPARATION-ENERGY RELATIONSHIPS

There is a minimum amount of thermodynamic work involved in separating a mixture into its constituents, regardless of the method of separation. Under isothermal conditions, this minimum work of separation is equal to the increase in Gibbs free energy of the products compared with the feed. The general expression for computing the minimum work, under isothermal and isobaric conditions, is (King, 1980, p. 662):

$$W_{min,T,P} = -\left( RT \left[ \sum_i x_{iF} \ln\left( \gamma_{iF} x_{iF} \right) - \sum_n \phi_{nF} \sum_i x_{in} \ln\left( \gamma_{in} x_{in} \right) \right] \right) \quad (3.1)$$

where

$W_{min,T,P}$ = minimum work per mole of feed (temperature $T$ and pressure $P$ being constant), J/(mole of feed)

$R$ = ideal gas constant

$i$ = $i$th component in feed

$x_{iF}$ = mole fraction of $i$th component in feed

$\gamma_{iF}$ = activity coefficient of $i$th component in feed

$n$ = number of product streams

$\phi_{nF}$ = mole fraction of total feed going to product stream $n$

$x_{in}$ = mole fraction of $i$th component in stream $n$

$\gamma_{in}$ = activity coefficient of $i$th component in stream $n$

If the feed is separated into two product streams $D$ and $B$, Equation (3.1) becomes

$$W_{min,T,P} = -\left( RT \left[ \sum_i x_{iF} \ln\left( \gamma_{iF} x_{iF} \right) - \phi_{DF}\left( \sum_i x_{iD} \ln\left( \gamma_{iD} x_{iD} \right) \right. \right. \right.$$
$$\left. \left. \left. - \phi_{BF}\left( \sum_i x_{iB} \ln\left( \gamma_{iB} x_{iB} \right) \right) \right] \right) \quad (3.2)$$

A simplified case is where there are only components 1 and 2 in the feed:

$$W_{min,T,P} = -RT\left[ x_{1F} \ln\left( \gamma_{1F} x_{1F} \right) + x_{2F} \ln\left( \gamma_{2F} x_{2F} \right) \right.$$
$$- \phi_{DF}\left[ x_{1D} \ln\left( \gamma_{1D} x_{1D} \right) + x_{2D} \ln\left( \gamma_{2D} x_{2D} \right) \right]$$
$$\left. - \phi_{BF}\left[ x_{1B} \ln\left( \gamma_{1B} x_{1B} \right) + x_{2B} \ln\left( \gamma_{2B} x_{2B} \right) \right] \right] \quad (3.3)$$

If the two product streams are essentially pure, $x_{2D}$ and $x_{1B} = 0$ and $\gamma_{1D}$ and $\gamma_{2B}$ = 1.0; also, $\phi_{DF} = x_{1F}$, $\phi_{BF} = x_{2F}$, and $\ln x_{1D}$ and $\ln x_{1B}$ are zero. Then

$$W_{\min, T, P} = -RT\left[x_{1F} \ln\left(\gamma_{1F} x_{1F}\right) + x_{2F} \ln\left(\gamma_{2F} x_{2F}\right)\right] \qquad (3.4)$$

Generalizing for the case of pure product streams gives (King, 1980, p. 661):

$$W_{\min, T, P} = -RT \sum_i x_{iF} \ln\left(\gamma_{iF} x_{iF}\right) \qquad (3.5)$$

As a simple example, consider the ethylbenzene/styrene separation at 100-mm Hg average pressure, and a 50–50 molar feed. This is an ideal system at all concentrations, so the activity coefficients are unity. Using Equation (3.3), the minimum work requirement for various levels of product purity is shown in Fig. 3.2. The significant influence of product purity on minimum work is apparent.

Although the minimum theoretical work requirement is far removed from the actual requirement for a given separation, it is a useful guideline for studying relative effects of variable changes.

Certain separations are usually heat-driven; that is, energy in the form of heat is supplied to the process, and some portion of the heat is later rejected from the process. Most distillations and extractions as well as conventional adsorptions are heat-driven. When such a separation process is under consideration, the inlet and outlet heat flows may be judged on the basis of their potential for performing work (as in a Carnot-type heat engine). The *net work potential* is based on the capability

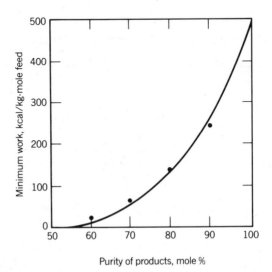

**FIGURE 3.2** Minimum work of separation: 50–50 molar mixture of ethylbenzene and styrene, and 100-mm Hg average pressure.

of the heat flows to drive heat engines (King, 1980, p. 665):

$$W_{\text{net}} = Q_H\left(\frac{T_H - T_0}{T_H}\right) - Q_L\left(\frac{T_L - T_0}{T_L}\right) \tag{3.6}$$

where

$W_{\text{net}}$ = net work potential for process per mole of feed, J/(mole of feed)

$Q_n$ = net heat into process at absolute temperature $T_H$, J

$Q_L$ = net heat out of process at absolute temperature $T_L$, J

$T_0$ = sink temperature, often taken to be the temperature of the ambient air or of a nearby stream of water where process heat is rejected, °K

The *thermodynamic efficiency* of the heat-driven separation method is based on minimum work calculated at the sink temperature:

$$\eta_I = \frac{W_{\text{min}, T_0, P}}{W_{\text{net}}} \tag{3.7}$$

Since the net work is computed on the basis of an ideal cycle, $\eta_I$ represents a maximum value (for a reversible process).

Traditionally, engineers have been concerned primarily with first-law thermodynamic efficiency, one form of which is defined above. A *second-law efficiency* can also be important:

$$\eta_{II} = \frac{\text{minimum exergy}}{\text{actual exergy}} = \frac{\Delta B_{\text{min}}}{\Delta B_{\text{act}}} = \frac{\left(\begin{array}{c}\text{least amount of availability}\\ \text{that could have been used}\end{array}\right)}{\left(\begin{array}{c}\text{actual amount of availability}\\ \text{used}\end{array}\right)} \tag{3.8}$$

where the exergy represents the maximum potential work that a stream can deliver when reaching equilibrium with the surroundings (at the sink condition). The availability function $B$ is equal to $H - T_0 S$ (where $H$ is the enthalpy and $S$ the entropy), and thus $\Delta B = (H - H_0) - T_0(S - S_0)$. It can be shown that the net work potential and the exergy are related functions (King, 1980, p. 664).

Another measure of energy efficiency is the *coefficient of performance*, COP. A first-law definition of this function is:

$$\text{COP} = \frac{\text{heat delivered}}{\text{work done on system}} = \frac{Q}{W} \tag{3.9}$$

From a second-law point of view, availability assumes importance and an alternate coefficient is defined:

$$COP' = \frac{Q}{W}\left(\frac{T - T_0}{T}\right) \tag{3.10}$$

These latter relationships are presented for information, but are not normally used in separation-energy comparisons.

## 3.4. DISTILLATION SEPARATIONS

As noted earlier, distillation is the large consumer of energy among the separation methods used in the processing industry. There is a general axiom that *if distillation appears feasible for a separation need* (relative volatility not too low, materials amenable to the temperatures involved), *then it quite likely will be the method of choice*. It has been called the "enemy" of the development of other separation methods, traditional and new, by Keller (1982), and the ubiquitousness of distillation in chemical plants and refineries appears to be proof that it has economic advantages. However, distillation is also known to have energy disadvantages; in the traditional situation, energy at a high level is fed to the base of the distillation column and the same approximate quantity is rejected at the top of the column, but at a level so low that additional use of the rejected heat is unlikely.

When energy costs spiraled upward in the late 1970s, attention was given to distillation in two ways:

1. Modifications to the system in order to retain the advantages of distillation but with lower energy consumption.
2. Comparison of distillation with other methods to uncover areas where such methods might indeed have advantages, especially for new designs and developments.

A simple distillation process is diagrammed in Figure 3.3. For the purposes of this discussion, a binary mixture is considered. Heat is supplied at temperature $T_H$ and is removed at temperature $T_L$, which for practical purposes is taken as the sink temperature (e.g., an air-cooled condenser might be used). The condenser duty is

$$Q_c = \Delta H_c(L + D) = D\Delta H_c(R' + 1) \tag{3.11}$$

where

$Q_c$ = condenser duty (a rate), W

$L$ = reflux flow rate, kg mole/hr

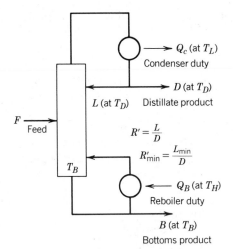

$Q_c$ (at $T_L$)
Condenser duty

$D$ (at $T_D$)
Distillate product

$L$ (at $T_D$)

$R' = \dfrac{L}{D}$

$R'_{min} = \dfrac{L_{min}}{D}$

$Q_B$ (at $T_H$)
Reboiler duty

$F$ Feed

$T_B$

$B$ (at $T_B$)
Bottoms product

**FIGURE 3.3**  Simple distillation process.

$D$ = distillate product rate, kg mole/hr

$R'$ = reflux ratio

$\Delta H_c$ = latent heat of condensation, kJ/mole

Since the sink temperature has been taken to be $T_L$, and for the usual case (with saturated liquid feed) of condenser duty being approximately the same as the reboiler duty, the net work potential from Equation (3.6) becomes, for $Q = Q_B = Q_c$,

$$W_{net} = Q\left(\frac{1 - T}{T_0}\right) \tag{3.12}$$

or, for a flow process,

$$W'_{net} = W_{net}F = Q\left(\frac{1 - T_0}{T_H}\right)F = Q'\left(\frac{1 - T_0}{T_H}\right) \tag{3.13}$$

where $F$ is the molar flow rate of the feed mixture. Thus, for a given separation, where $W_{min}$ is constant, the thermodynamic efficiency depends directly on the net work potential.

For the special (but quite important) case of separating a close-boiling binary mixture into high-purity product streams, and where the relative volatility is low (say less than 1.5), the required reflux ratio will be high and indicative of high energy input. An approximation of the thermodynamic efficiency of this case, for total reflux conditions, has been presented by Steinmeyer (1984):

$$\eta_I = \frac{x_{1F} \ln{(\gamma_{1F})} + x_{2F} \ln{(\gamma_{2F}x_{2F})}}{1 - (\alpha - 1)x_{1F}} \tag{3.14}$$

For the styrene/ethylbenzene example used earlier, a 50–50 molar feed and a relative volatility of 1.40 gives an efficiency of 0.52. However, this value neglects temperature differences at the condenser and reboiler, pressure drops in the column and piping, and operation at a reflux above the minimum. When these factors are considered, the efficiency can drop to a value of 0.20 or less. The reader may refer to Section 4.3 for a related analysis of investment and energy costs for an ethylbenzene/styrene column.

Recognizing that the net work potential is a guide toward optimizing distillation conditions, let us consider the separation of propylene from propane under several alternate processing arrangements. Diagrams for these alternates are shown in Figure 3.4. Conditions for the analysis are as follows:

| | |
|---|---|
| Light key | Propylene |
| Feed composition | 50 mole % propylene |
| Feed condition | Saturated liquid |
| Overhead product | 99.0 mole % propylene |
| Bottoms product | 95.0 mole % propylene |
| Temperature difference in condenser and in reboiler | 10°C |
| Sink temperature | 30°C (303 K) |
| Cooling water temperature | 30°C (303 K) |

*Scheme A.*   This is the base case. The column is operated at a pressure high enough for the overhead product to be condensed by cooling water. For the bottom temperature driving force of 10°C, a heating medium temperature of 59°C is indicated. The net work potential for the operation is, from Equation (3.12),

$$W_{net} = Q_A\left(1 - \frac{303}{332}\right) = 0.087Q_A \tag{3.15}$$

Here, $Q_A$ is the heat duty of condenser (or reboiler) for Scheme A.

By Equation (3.13), the indicated thermodynamic efficiency is 52%, but without the simplifying assumptions for the equation, the efficiency drops to 16%.

Scheme A calls for a heating medium of 59°C (138°F). For the case of a non-condensing heating medium, a practical plant condition would call for a higher temperature for the medium flowing to the reboiler, and the availability of such a stream might be unlikely. In such a case, the net work potential would become approximately $W_{net} = 0.21Q_A$. This would lower the efficiency to 6.6%.

This brings up an important point, however. Low-pressure steam tends to be surplus at many chemical plants and if steam is used for distillation, it is usually the low-pressure variety. Thus the cost of energy for many distillation separations is relatively low.

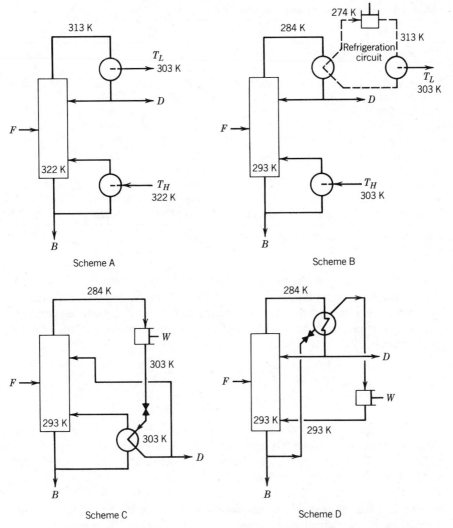

**FIGURE 3.4** Possible operating schemes for propylene/propane fractionation.

Referring back to Equation (3.15), if the heat-exchanger temperature differences could be eliminated,

$$W_{\text{net}} = Q_A\left(1 - \frac{303}{312}\right) = 0.029Q_A \tag{3.16}$$

The indicated thermodynamic efficiency would then be about 48%, and a reversible process would be approached in a reasonable way. More details on Scheme A and the other schemes are shown in Table 3.4.

TABLE 3.4   Propylene-Propane Fractionation

| Scheme | A | B | C | D |
|---|---|---|---|---|
| $C_3H_6$ in dist., mole % | 99.0 | 99.0 | 99.0 | 99.0 |
| $C_3H_6$ in btms., mole % | 5.0 | 5.0 | 5.0 | 5.0 |
| Top pressure, atm | 16.0 | 8.2 | 8.2 | 8.2 |
| Relative volatility | 1.15 | 1.20 | 1.20 | 1.20 |
| Min. reflux ratio, $R'$ | 13.1 | 9.78 | 9.78 | 9.78 |
| Operating reflux ratio, $R'$ | 15.7 | 11.7 | 11.7 | 11.7 |
| Minimum stages | 54 | 41 | 41 | 41 |
| Theoretical stages | 100 | 78 | 78 | 78 |
| Heat required, kcal $(10^{-3})$ | 24.5 | 18.7 | 18.7 | 18.7 |
| Minimum work, kcal | 339 | 339 | 339 | 339 |
| Net work potential, kcal | 2135 | 2330 | 1180 | 1215 |
| Thermodynamic efficiency | 0.159 | 0.145 | 0.288 | 0.279 |

Basis: 1.0 kg-mole feed of 50-50 molar mixture.

*Scheme B.*   In this case, the process is operated with refrigeration, at a level such that the heating medium is cooling water at 30°C. Now the net work potential for the overall operation is zero, since the heating medium is at the sink temperature. However, it is necessary to provide mechanical work of compression which, for an ideal cycle, is

$$W_{net} = Q_B \left( \frac{T_0 - T_c}{T_0} \right) = Q_B \left( \frac{313 - 274}{313} \right) = 0.125 Q_B \qquad (3.17)$$

Since this scheme operates at a lower pressure than Scheme A, the relative volatility is increased to 1.20 and this permits a 25% reduction in reflux ratio, as shown in Table 3.4. Thus, on a comparable basis with Scheme A, the net work potential is $W_{net} = 0.75(0.125)Q_B = 0.094Q_A$.

*Scheme C.*   This process utilizes vapor-recompression distillation, with only mechanical work being added to the system. This is the familiar "heat pump" system that has been well known for many years. While the need for a condenser is eliminated, the cost of the compressor must be added. For this scheme, the net work potential is:

$$W_{net} = Q_c \left( \frac{T_H - T_c}{T_H} \right) = Q_c \left( \frac{303 - 284}{303} \right) = 0.063 Q_c$$

$$= 0.75(0.063)Q_A = 0.047 Q_A \qquad (3.18)$$

*Scheme D.*   This is an alternate form of vapor-recompression distillation, in which the bottom liquid is expanded for refrigerant value, to condense the overhead stream, and is then recompressed as reboiler vapor. The net work potential is

$$W_{net} = Q_D \left( \frac{293 - 274}{293} \right) = 0.065 Q_D = 0.049 Q_A \qquad (3.19)$$

The analysis presented above is intended to be descriptive and introductory only. It deals with ideal, not actual, cycles. It does not take into account the many operating limitations and economic factors. It does suggest, however, that a net work potential analysis is indicative of the thermodynamic efficiency that might be expected from a simple, single-column distillation.

Much recent study has gone into ways and means for reducing the energy intensiveness of distillation. A comprehensive analysis of the energy requirements of U.S. distillation columns has been reported by Mix and associates (1978); the full report has been published by the U.S. Department of Energy (1981). A report by the Electric Power Research Institute deals exclusively with vapor-recompression distillation (1984). A detailed analysis of the applicability of vapor-recompression distillation has been published by Null (1976). There are numerous publications dealing with thermodynamic analysis of distillation systems; examples are Kaiser and Gourlia (1985), Steinmeyer (1984), Kenney (1984), Mostafa (1981), Mah et al. (1977), and Fitzmorris and Mah (1980). Elaahi and Luyben (1983) looked at both simple and complex configurations of distillation column arrangements to separate a four-component mixture and included complete process simulations for selected configurations.

From a practical point of view, there are three approaches to reducing the cost of energy for distillation systems:

1. Utilize the heat in the overhead vapor instead of rejecting it to the sink.
    a. Cascade the column with another column or heat user (e.g., use the vapor to reboil another column or to preheat a process stream).
    b. Pump the heat to a level where it can be usefully transferred to another process stream.
    c. Recover a portion of the heat in a Rankine cycle arrangement whereby a working fluid is vaporized.
2. Decrease the amount of heat needed.
    a. Decrease the amount of reflux used, by more efficient contacting devices (e.g., structured packings), adding more contacting stages, eliminating excess separation, or using intermediate reboilers and condensers.
    b. Use feed/product heat exchange.
    c. Decrease heat losses from equipment.
    d. Change the separation requirement and/or sequencing of columns in a distillation train.
3. Use a lower-cost energy source.
    a. Use exhaust steam of low value.
    b. Use process fluids that are available at elevated temperatures but which

are currently rejecting their heat to the sink. In some cases, the available temperature level might justify the use of an intermediate reboiler.

As noted, many existing distillation systems utilize steam as a heating medium and this steam is often of a low-pressure (low-availability) variety that because of the total-plant steam balance would not have other uses. This is a point to consider in retrofit cases, but may not be germane for new designs.

Another point to consider is that although distillation is a large user of energy, because of the latent heat of vaporization inherent in the process, it is not as inefficient thermodynamically as one might suspect. King (1980, p. 699) discusses several approaches to decreasing the irreversibility of distillation; unfortunately such approaches have large capital costs and operating difficulties.

One may conclude that distillation will continue to be the separation method of choice when it can be applied in a reasonable way. To reiterate statements made earlier, the reasonableness of application hinges on relative volatility and whether the materials being processed can be subjected to the temperature levels required for vaporization. Still, there are specialized cases where distillation can, and perhaps should, lose out to other separating techniques. These are discussed in the following sections.

## 3.5.  EXTRACTION SEPARATIONS

Extraction involves the use of a selective solvent to separate two or more components of a liquid mixture. The solvent may be in a subcritical or supercritical state. The process may be operated at pressures from atmospheric to those above the critical pressure of the solvent. In general practice, little or no energy is required for the extraction step itself; however, separation of the extracted material from the solvent may require a significant amount of energy. In addition, there may be a separation required for the raffinate stream. General characteristics of solvent extraction have been reviewed recently by Humphrey et al. (1984), and the books by Treybal (1963) and Laddha and Degaleesan (1978) provide details on analysis and design of conventional extraction systems.

The solvent recovery step in the process usually involves vaporization of the extracted material (i.e., for a heavy solvent) and this step may be a simple one- or two-stage flash, or it may be a distillation system with reflux. For cases where the solvent is a light component, as in carbon-dioxide extraction under supercritical conditions, the solvent is flashed or fractionated as distillate product. In either case, principles related to distillation energy consumption apply to the process.

Null (1980) compared extraction and distillation energy requirements, taking into account general economic factors and energy requirements and providing for alternate methods for solvent separation from extract and raffinate streams. He found that extraction is approximately more energy-efficient than distillation when the reflux ratio $R'$ for the base-case distillation step is

$$R' > \left[ 1 + R_E + (1 + R_E') \left( \frac{B}{D} \right) \right] \left( \frac{T_{H,D}}{T_{H,E}} \right) \left( \frac{T_{H,E} - T_0}{T_{H,D} - T_0} \right) - 1 \quad (3.20)$$

where

$R'$ = reflux ratio for the base-case distillation

$R_E$ = reflux ratio for the extract stripper

$R_E'$ = reflux ratio for the raffinate stripper

$B$ = molar flow rate of bottoms product (from distillation or for equivalent extraction), kg mole/hr

$D$ = molar flow rate of distillate product (for distillation or for equivalent extraction), kg mole/hr

$T_{H,D}$ = temperature of steam (or other heating medium) used for distillation process, K

$T_{H,E}$ = temperature of steam (or other heating medium) used for the extract and raffinate strippers, K

$T_0$ = sink temperature, K

If the same temperature level is represented by both media for heating the extract stripper and the distillation column (i.e., $T_{H,D} = T_{H,E}$), Equation (3.20) becomes

$$R' > R_E + (1 + R_E') \frac{B}{D} \tag{3.21}$$

Figure 3.5 is a graphical representation of Equation (3.21), with results for the identical parameters.

For the special case of an extraction solvent that needs no fractionation for separation from the extracted material (e.g., a nonvolatile solvent), a simple flash step suffices for the stripping process. In this case, Equation (3.20) becomes

$$R' > \frac{\left(1 + \dfrac{B}{D}\right)(T_{H,D}/T_{H,E})(T_{H,E} - T_0)}{(T_{H,D} - T_0) - 1} \tag{3.22}$$

and if $T_{H,E} = T_{H,D}$, a very simple relationship results:

$$R' > \frac{B}{D} \tag{3.23}$$

Graphical representations of Equations (3.22) and (3.23) are shown in Figure 3.6.

Figures 3.5 and 3.6 permit a quick but approximate estimate of conditions under which extraction might be favored over distillation. It should be borne in mind that the comparison is on the basis of energy use only. The figures show that for a typical case with a large amount of the distillation feed going to distillate product,

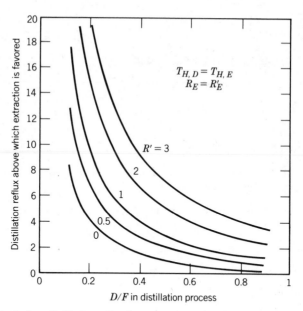

**FIGURE 3.5** Equivalent distillation reflux for extraction with refluxed extract and raffinate strippers based on Equation (3.21) (Null, 1980). Reprinted with permission from the American Institute of Chemical Engineers.

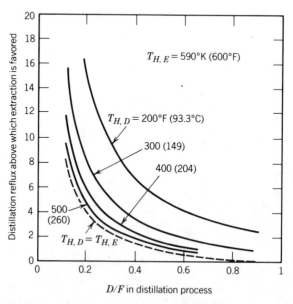

**FIGURE 3.6** Equivalent distillation reflux for extraction with simple flash separation for extract and raffinate streams based on Equations (3.22) and (3.23) (Null, 1980). Reprinted with permission from the American Institute of Chemical Engineers.

an extraction of the minority component of the feed would then require relatively less energy for the separation of extraction from solvent. In general, this is a situation that can be favorable to extraction. At the same time, one will find that most extractions are based on an improved selectivity of key components, when compared with distillation.

## 3.6. CRYSTALLIZATION SEPARATIONS

There are two general categories of crystallization separations: solution and melt. We are concerned here only with melt crystallization, since it is the more likely of the two categories to compete with distillation.

Melt crystallization involves the separation of solid crystals from a melt, and for typical mixtures involving eutectics, it is possible to obtain a product with very high purity. A representative binary system-phase diagram is shown in Figure 3.7 (King, 1980, p. 743). This diagram is for the system of *m*-xylene (melting point 225 K) and *p*-xylene (melting point 287 K), for which the mixture is separated

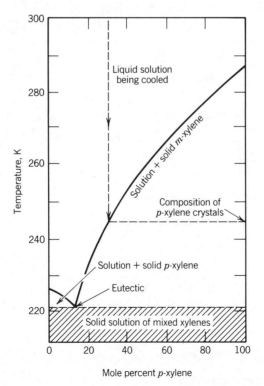

**FIGURE 3.7** Liquid–solid phase diagram for *m*-xylene/*p*-xylene system (King, 1980, p. 743). Reprinted with permission from McGraw-Hill Book Company.

commercially by melt crystallization. With reference to the diagram, if a mixture containing 30 mole % p-xylene is cooled to 245 K, a phase boundary is reached and crystals of p-xylene begin to separate from the melt. If occluded material can be removed from the crystals, it will be found that they are essentially pure para-isomer.

In a continuous crystallization system, refluxing of equilibrium melt is needed to achieve removal of occluded impurities. This reflux is provided by melting a portion of the solid product and passing it in countercurrent fashion to the forward flow of liquid–solid mixture. Examples of countercurrent crystallizers are the Brodie purifier and the Phillips pulsed column, shown in Figure 3.8 (Atwood, 1972). As shown for both crystallizers, heat is removed in one part of the system and is added in another. For the xylene isomer example, refrigeration is needed to form a solid phase, and this is typical. Thus the energy input for refrigeration must be considered. In addition, a heating medium is required for melting the reflux stream as well as the "distillate" product, representing an additional energy input. The parallel with a distillation column requiring a refrigerated overhead condensation is apparent.

The 1980 Null paper includes a comparison between base-case distillation and melt crystallization. On the basis of a number of simplifying assumptions, Null found that the equivalent distillation reflux ratio $R'$ may be based on the following relationship:

$$R' > (R_c + 1)\left(\frac{\Delta H_F}{\Delta H_v}\right)\left(\frac{B_c}{D}\right)$$

$$\cdot \left[\left(\frac{T_{H,C} - T_0}{T_H - T_0}\right)\left(\frac{T_H}{T_{H,C}}\right) + \frac{1}{\epsilon_P \epsilon_R}\left(\frac{T_0 - T_{RC}}{T_H - T_0}\right)\left(\frac{T_H}{T_0}\right)\right] - 1 \quad (3.24)$$

where

$R_c$ = melt-crystallization reflux ratio

$\Delta H_F$ = latent heat of fusion of the component at the melting end of the crystallizer, kJ/kg mole

$\Delta H_V$ = latent heat of vaporization for distillation, kJ/kg mole

$B_C$ = molar flow of crystallization "bottoms," kg mole/hr

$D$ = molar flow of distillate for the equivalent distillation case, kg mole/hr

$T_{H,C}$ = temperature of heating medium for melting crystals, K

$T_H$ = temperature of heating medium for distillation, K

$T_0$ = sink temperature, K

$\epsilon_P$ = Carnot cycle efficiency for heating medium

$\epsilon_R$ = Carnot cycle efficiency for refrigerant

$T_{RC}$ = refrigeration temperature for melt chilling, K

Null prepared the chart shown as Figure 3.9 by making the following simplifying assumptions for Equation 3.24: $T_H = T_{H,C}$; $\Delta H_F/\Delta H_v = 0.2$; $B_C = D$; $T_0 = 311K$; and $\epsilon_p\epsilon_R = 0.36$. The chart shows values of the distillation reflux ratio

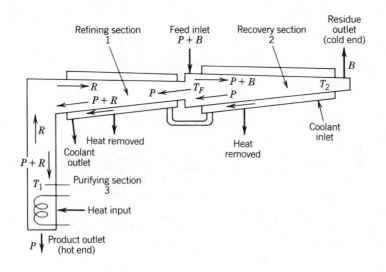

Brodie purifier

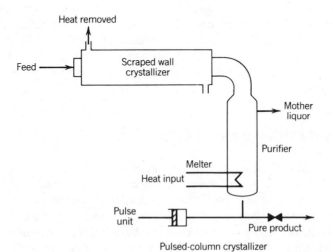

Pulsed-column crystallizer

**FIGURE 3.8** Examples of continuous melt crystallizers (Atwood, 1972). Reprinted with permission from CRC Press, Boca Raton, FL.

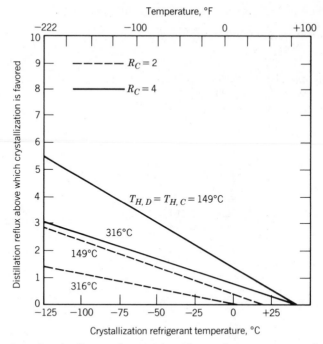

**FIGURE 3.9** Equivalent distillation reflux ratio for melt crystallization separations based on Equation (3.24) with equal heating-medium temperatures (Null, 1980). Reprinted with permission from the American Institute of Chemical Engineers.

above which crystallization would be favored from an energy standpoint. Crystallization operation incurs many operating problems not present with distillation. The commercial success of the crystallization method for separating the xylene isomers has been due to the extremely high reflux ratio that would be required for distillation, owing to the very close boiling points of the isomers. The specific energy advantages of crystallization have not been a factor in its selection for commercial use.

## 3.7.  ADSORPTION SEPARATIONS

Essentially all adsorption separations are carried out in fixed beds of solid adsorbents, and are cyclic in nature. When a bed becomes loaded with adsorbate, it is switched to a regeneration mode, where a significant portion of the separation energy is consumed. (When very large volumes of gas are handled, the pressure drop through the bed may also represent a significant energy requirement.) In general, adsorption may be considered a heat-driven separation. Adsorption applications may normally be divided into two categories, lean and bulk mixture.

Traditionally, fixed-bed adsorption has been applied to lean mixtures, for example, to recover a solvent from an exhaust air stream; and in such an application, it is not considered a competitor of distillation. Many installations for recovering organic solvents from air employ steam regeneration of solvent (enabling recovery by condensing the stripping vapors), and typically 3–5 kg of steam are used per kg of solvent recovered. This is a much higher ratio of steam to product than would be typical for distillation.

Bulk-mixture adsorptive separations that are commercially significant include m-xylene/p-xylene, oxygen/nitrogen, and n-paraffins/isoparaffins. Regeneration may be by temperature swing (as for the solvent recovery example) or by pressure swing. For comparison with distillation, only the temperature-swing regeneration need be considered, since the pressure-swing method normally takes advantage of the available pressure level of the mixture to be separated and thus is integrated with the upstream process in ways not easily reducible to generalities.

The net work potential, and thus the thermodynamic efficiency of the bulk-mixture separation (with temperature-swing regeneration) may be obtained from Equations (3.1), (3.6), and (3.7). Heat is fed to the bed at heating medium temperature $T_H$ and is rejected to the sink or cooling water at temperature $T_L$. Conversion to an equivalent feed basis is made by considering time cycles for adsorption and regeneration.

In his 1980 paper, Null compared energy requirements for temperature-swing adsorption with those for distillation. He found that the equivalent reflux ratio for distillation could be obtained from the following relationship:

$$R' > \left(\frac{N_{\text{ADS}}}{D}\right)\left(\frac{\Delta H_{\text{REG}}}{\Delta H_v}\right)\left(\frac{T_{H,D}}{T_{RG}}\right)\left(\frac{T_{RG} - T_0}{T_{H,D} - T_0}\right) - 1 \qquad (3.25)$$

where

$R' = $ distillation reflux ratio

$N_{\text{ADS}} = $ average rate of adsorption, kg mole/hr

$D = $ distillate rate for distillation, kg mole/hr

$\Delta H_{\text{REG}} = $ total regeneration heat, kJ/kg mole adsorbed

$\Delta H_v = $ latent heat of vaporization for distillation, kJ/kg mole

$T_{H,D} = $ heating-medium temperature for distillation, K

$T_{RG} = $ supply temperature of heating medium for adsorption, K

$T_0 = $ sink temperature, K

Equation (3.25) is shown graphically in Figure 3.10, taken from the paper by Null, and with the restrictions noted. Normally the preferential adsorption is for the heavier component(s) in the feed mixture and thus $N_{\text{ADS}}/D$ corresponds to the

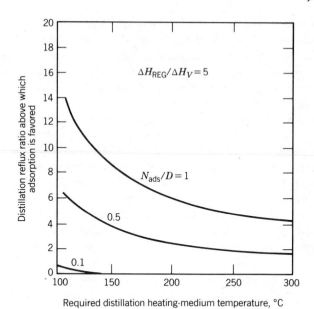

**FIGURE 3.10**  Equivalent distillation reflux ratio for temperature-swing adsorption based on Equation (3.25) with regeneration temperature for adsorption at 316°C (Null, 1980). Reprinted with permission from the American Institute of Chemical Engineers.

bottoms/distillate ratio for the equivalent distillation. A ratio of total regeneration heat to latent heat of vaporization for distillation, $\Delta H_{REG}/\Delta H_v$, of five is typical for organics adsorption systems.

General characteristics of adsorption and adsorption systems are covered in the recent treatise by Ruthven (1984). An interesting application of large-scale adsorption is for the separation of the meta- and para-xylene isomers, and a novel system varies feed location to a fixed bed to simulate a continuous, moving-bed system (Broughton et al., 1970). An earlier concept of moving-bed adsorption (Berg, 1946) has been restudied recently (U.S. Department of Energy, 1982) and an alternate version of a continuous adsorber, in which there is both moving-bed and fluidized-bed operation (Figure 3.11), is now in commercial use (Keller, 1984).

Lean-mixture separations by adsorption appear to have special attractiveness, especially when the adsorbate content of the feed is in the parts-per-million range, and energy consumption is not normally a significant factor in choosing the separation method. For bulk mixtures, however, adsorption may be making some inroads on other and more established techniques. An example is in the pressure-swing adsorption separation of air into oxygen and nitrogen, where cryogenic distillation has been the traditional method for separation. Cases are now being studied where the compression energy for distillation, to operate the columns at the needed pressure level, is not an adequate trade-off for the lower recovery and lack of versatility of the adsorption method.

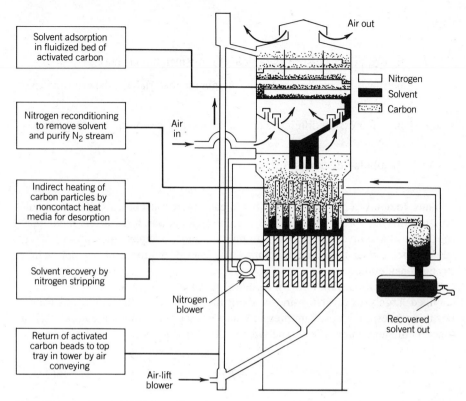

**FIGURE 3.11**  Purasiv HR® process for solvent recovery: a continuous adsorption process. Courtesy of Union Carbide Corporation.

## 3.8.  MEMBRANE SEPARATIONS

Separation of liquid or gaseous mixtures by semipermeable membranes has been studied for decades, but only in recent years have larger-scale commercial processes been designed to use the products of earlier research. For the vast majority of the applications, a phase change does not accompany the separation and thus the heat-driven aspects of energy requirements are not present. (The exception to this is pervaporation, where the feed mixture is a liquid and the permeate is withdrawn as a vapor.) Accordingly, compression requirements constitute the major energy consideration.

The flux equation for separating a gas mixture by means of a membrane is (Perry and Green, 1984, p. 17–15)

$$J = \frac{\bar{P}\Delta p_i}{\Delta z} \tag{3.26}$$

where

$J$ = flux of permeate species $i$, mole/hr·m$^2$(membrane area)

$\overline{P}$ = permeability coefficient for the membrane material and the diffusing species $i$, mole/hr·Pa

$\Delta p_i$ = partial pressure difference across the membrane for the diffusing species $i$, Pa

$\Delta z$ = membrane thickness, m

Thus, to obtain a reasonable flux of permeate, there must be a significant partial-pressure driving force. Many of the present applications for gaseous membrane separations have involved an upstream mixture under adequate pressure for extraneous process reasons; hence a compression energy requirement has not been a strong consideration.

Liquid-mixture separations by membranes, in a large-scale sense, have been confined largely to reverse osmosis applications for removing salts from water. For this process, the upstream pressure must exceed the osmotic pressure, itself a function of salt content (Perry and Green, 1984, p. 17–23):

$$P_{os} = \rho_m RT \ln (\gamma_i x_i) \qquad (3.27)$$

where

$P_{os}$ = osmotic pressure, Pa

$\rho_m$ = molar density of water (solvent), kg mole/m$^3$

$\gamma_i$ = activity coefficient of water (solvent)

$x_i$ = mole fraction of salt (solute) in the feed solution

Even for very low salt concentrations, the osmotic pressure is quite large. Even so, research is in progress to find nonwater applications for reverse osmosis, and the cost of compression (pumping) energy will always be a key consideration in proving out the economic feasibility of the method.

The published literature on membrane separations is voluminous; only four recent books that can provide the reader with any needed introduction to the subject are mentioned here. The treatise by Belfort (1984) deals with processes, and those by Kesting (1985), Sourirajan and Matsuura (1985), and Lloyd (1985) deal specifically with reverse osmosis and ultrafiltration.

## 3.9. ALTERNATE METHODS

The preceding sections dealt with comparisons between the several different separation methods and the basic method of distillation. It was implied that for a number of mixtures, more than one separating method would be possible and perhaps feasible. In making comparisons of alternates, the basic criteria are cost and operability. The lower-cost alternate may not be feasible because of inadequate scale-up knowledge or potential operating problems such as corrosivity, equipment maintenance difficulties, or general inexperience with the techniques for running the process. Thus the low-energy alternate may lose out both on total economic factors and on unpredictable success with day-to-day operations, especially on a large scale.

An example of a study of alternates has been given by Bravo et al. (1986), relating to the important commercial separation of ethylbenzene from styrene. This separation is practiced by vacuum distillation, with higher temperature restrictions, because of the tendency for the styrene to polymerize. Table 3.5 shows the results reported by the authors, and it would appear that methods involving membranes show distinct advantages over the vacuum distillation method currently practiced. Present plants have single-train units with styrene annual capacities up to 500,000 metric tons; deviation from distillation would indeed involve insurmountable design and operating problems, based on the current state of the art for membrane separations. This should not imply that tomorrow's knowledge might permit replacing the traditional methods.

Pursuing further the ethylbenzene/styrene example, Table 3.5 shows that meth-

**TABLE 3.5 Separation of Ethylbenzene/Styrene: Summary of Comparison of Alternative Technologies**

| Process | Energy Consumption Btu/lb Styrene (KJ/kg) | Expected Capital Cost Compared to Currently Used Process | Remarks |
|---|---|---|---|
| Refluxed membrane | 30 (69.8) | Higher | Low permeation rates |
| Crystallization/refluxed membrane | 35–60 (81.4–139.6) | Slightly higher | Membrane used to enhance recovery |
| Pervaporation | 20–74 (46.5–172.1) | Higher | Low permeation rates |
| Vacuum distillation with heat pump | 100–330 (232.6–1213.1) | Slightly higher | Low pressure-drop devices reduce energy significantly |
| Vacuum distillation | 1,180 (2744.7) | Same | Currently used process |

*Source:* Bravo, et al. (1986)

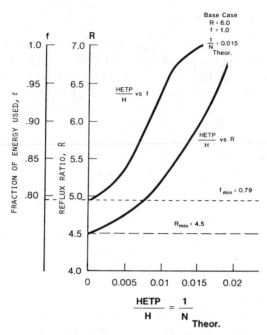

**FIGURE 3.12** Required reflux ratio and energy consumption for an ethylbenzene/styrene distillation column with retrofitted higher-efficiency contacting devices (Bravo et al., 1986). HETP is height equivalent to a theoretical stage, m. H is column height, m. $N_{Theor}$ is number of theoretical stages.

ods such as those discussed in Section 3.4 could lead to economies without operating difficulties. The energy requirement for vapor-recompression distillation is significantly lower than that for straight vacuum distillation, but the question is how increased capital can be traded off against lower energy costs. This matter is discussed in some detail by Steinmeyer (1982).

For either vapor recompression or straight distillation, a vacuum fractionation such as ethylbenzene/styrene can profit from reduced pressure drop in the column. Figure 3.12, from Bravo et al. (1986), shows the effect of lowering the column pressure drop by replacing cross-flow trays with high-efficiency, low-pressure-drop packing. The base case is for a reflux ratio of 7.0. Depending on the packing used, the energy consumption can be lowered to 85% or so of its base-case value. Such retrofit studies have been widespread and have led to positive economies.

Another well-known separation process that has been studied extensively is that of recovering anhydrous ethanol from dilute fermentate. The standard method is by distillation in three columns, breaking the water–ethanol azeotrope with a third component such as benzene. Current work indicates that the capital investment and the cost of energy can be reduced greatly by using adsorption or membrane permeation to break the azeotrope. Liquid–liquid extraction has been considered for this purpose also. Representative studies have been reported by Teo and Ruthven (1986) for adsorption, Lee and Pahl (1985) for extraction and Garg and Ausikaitis (1983) for adsorption and Mehta (1982) for membranes.

## 3.10. CONCLUSIONS

In the foregoing sections, several important separation methods were discussed from the standpoint of energy requirements. Throughout the discussion, it was emphasized that distillation is the base-case method against which other methods must be compared. If the separation is feasible by distillation, it is likely that the other methods will not be able to compete with it easily. This means that the mission should continue that will lead to lower energy requirements for distillation.

But there are cases where distillation is not feasible or is at the least quite clumsy and expensive. If the materials to be separated have a very close volatility that would lead to an excessive number of contacting stages or an exorbitant amount of reflux, if the materials are heat-sensitive, if the mixture involves a solid phase, or if an exceedingly large amount of energy might be needed—then there is the opportunity for other methods. This is why the meta/para-xylene isomer mixture is separated by crystallization or by adsorption; why benzene and other aromatics are recovered from reformate by extraction; or why more and more carbon dioxide is being separated from methane by the use of membranes.

The chemical engineer must be global in his or her thinking about the optimum means for separating a mixture at hand. There must be sufficient knowledge available for a proper choice to be made and a realistic assessment of costs to be carried out. Much work remains in areas of nondistillation separation before these goals can be achieved.

## REFERENCES

Atwood, G. Z., "Developments in Melt Crystallization," in *Recent Developments in Separation Science*, Vol. I, N. N. Li (ed.), CRC Press, Cleveland, OH, 1972.

Belfort, G., *Synthetic Membrane Processes*, Academic Press, Orlando, FL, 1984.

Berg, C., "Hypersorption Process for Separation of Light Gases," *Trans. AIChE*, **42**, 665 (1946).

Bravo, J. L., J. R. Fair, J. L. Humphrey, C. L. Martin, A. F. Seibert, and S. Joshi, *Fluid Mixture Separation Technologies for Cost Reduction and Process Improvement*, Noyes Publications, Park Ridge, NJ, 1986.

Broughton, D. B., R. W. Neuzil, J. M. Pharis, and C. S. Brearley, "The Parex Process for Recovering Paraxylene," *Chem. Eng. Prog.*, **66** (9), 70 (1970).

Elaahi, A. and W. L. Luyben, "Alternative Distillation Configurations for Energy Conservation in Four-Component Separations," *Ind. Eng. Chem. Process Des. Dev.*, **22**, 80 (1983).

Electric Power Research Institute, "Heat Pumps in Distillation Processes," Final Report, Project 1201-23, by Radian Corporation, August 1984.

Fitzmorris, R. E. and R. S. H. Mah, "Improving Distillation Column Design Using Thermodynamic Availability Analysis," *AIChE J.*, **26**, 265 (1980).

Garg, D. R. and J. P. Ausikaitis, "Molecular Sieve Dehydration Cycle for High Water-Content Streams," *Chem. Eng. Prog.*, **74** (5), 60 (1983).

Humphrey, J. L., J. A. Rocha, and J. R. Fair, "The Essentials of Extraction," *Chem. Eng.*, **91** (19), 76 (Sept. 17, 1984).

Kaiser, V. and J. P. Gourlia, "The Ideal Column Concept: Applying Energy to Distillation," *Chem. Eng.*, **92** (17), 45 (Aug. 19, 1985).

Keller, G. E., "On Knowing the Enemy," *Proceedings of the 2nd GRI Gas Separations Workshop*, Gas Research Institute, Chicago, IL, 1982.

Keller, G. E., "Gas Adsorption Processes—An Update," *Proceedings of the 6th Annual Industrial Energy Conservation Technology Conference*, Houston, TX, April 1984.

Kenney, W. F., *Energy Conservation in the Process Industries*, Academic Press, Orlando, FL, 1984.

Kesting, R. E., *Synthetic Polymeric Membranes*, 2nd. ed., Wiley, New York, 1985.

King, C. J., *Separation Processes*, 2nd ed., McGraw-Hill, New York, 1980.

Laddha, G. S. and T. E. Degaleesan, *Transport Phenomena in Liquid Extraction*, McGraw-Hill, New York, 1978.

Lee, F-M. and R. H. Pahl, "Use of Gasoline to Extract Ethanol from Aqueous Solution for Producing Gasohol," *Ind. Eng. Chem. Process Des. Dev.*, **24**, 250 (1985).

Lloyd, D. R. (ed.), *Materials Science of Synthetic Membranes*, American Chemical Society, Washington, DC, 1985.

Mah, R. S. H., J. J. Nicholas, and R. B. Wodnik, "Distillation with Secondary Reflux and Vaporization: A Comparative Evaluation," *AIChE J.*, **23**, 651 (1977).

Mehta, G. D., "Comparison of Membrane Processes with Distillation for Alcohol/Water Separation," *J. Membrane Sci.*, **12**, 1 (1982).

Mix, T. W., J. S. Dweck, M. Weinberg, and R. C. Armstrong, "Energy Conservation in Distillation," *Chem. Eng. Prog.*, **24** (4), 49 (1978).

Mostafa, H. A., "Thermodynamic Availability Analysis of Fractional Distillation with Vapour Compression," *Can. J. Chem. Eng.*, **59**, 487 (1981).

Null, H. R., "Heat Pumps in Distillation," *Chem. Eng. Prog.*, **72** (7), 58 (1976).

Null, H. R., "Energy Economy in Separation Processes," *Chem. Eng. Prog.*, **76** (8), 42 (1980).

Perry, R. H. and D. Green (eds.), *Perry's Chemical Engineers' Handbook*, 6th ed., McGraw-Hill, New York, 1984.

Ramshaw, C., " 'Higee' Distillation—an Example of Process Intensification," *Chem. Eng. (London)*, **389**, 13 (1983).

Ruthven, D. M., *Principles of Adsorption and Adsorption Processes*, Wiley, New York, 1984.

Sourirajan, S. and T. Matsuura, *Reverse Osmosis/Ultrafiltration Process Principles*, National Research Council of Canada, Ottawa, Canada, 1985.

Steinmeyer, D. E., "Take Your Pick: Capital or Energy," *Chemtech*, **12** (3), 188 (1982).

Steinmeyer, D. E., "Process Energy Conservation," in *Encyclopedia of Chemical Technology*, Suppl. Vol., 3rd ed., M. Grayson (ed.), Wiley, New York, 1984.

Teo, W. K. and D. M. Ruthven, "Adsorption of Water from Aqueous Ethanol Using 3-A Molecular Sieves," *Ind. Eng. Chem. Process Des. Dev.*, **25**, 17 (1986).

Treybal, R. E., *Liquid Extraction*, 2nd ed., McGraw-Hill, New York, 1963.

U.S. Department of Energy, "Energy Conservation in Distillation," Final Report DOE/ CS/40259 by Merix Corporation, July 1981.

U.S. Department of Energy, "Hypersorption Process for Separation of Components of a Medium-BTU Gas," Final Report DOE/MC/16447-1239 by Dravo Engineers and Constructors, July 1982.

# 4

# PROCESS INTEGRATION

FUN-GAU HO

GEORGE E. KELLER II

## 4.1. INTRODUCTION

Process integration has been a very popular topic during the last few years, especially in regard to the issue of reducing the amount of energy usage. Early work (see, for example, Linnhoff and Flower, 1978a,b; Linnhoff and Vredeveld, 1984) dealt primarily with optimal integration of heat exchangers. It is now possible to specify the minimum practical number of heat exchangers in a process and the minimum hot and cold utility usages, subject to real-process constraints. More recently, there has been a growing interest in heat integration of separation processes (Null, 1980; Umeda et al., 1979; Morari and Faith, 1980; Linnhoff et al., 1983; Hindmarsh and Townsend, 1984; Andrecovich and Westerberg, 1985). This is only natural, because utility usage in most chemical and petroleum-refining processes generally occurs in the separation parts of these processes.

Distillation and related vapor-liquid-based separation processes are by far the most widely used methods in the petroleum-refining, organic–chemical, and allied industries. At the same time, distillation is clearly a high-energy-usage process, since it involves supplying the latent heat of vaporization to much if not all of the feed, and in many cases much more than the latent heat. A recent estimate (Mix et al., 1978, 1981) places the energy usage by distillation in the United States at about two quadrillion Btu's in 1976, or about 3% of the total national energy usage. This level is greater than that of the entire commercial aviation industry in the same year (U.S. Department of Commerce, 1983).

In the following sections, we investigate the energy usage of distillation, its economics, and how energy usage can be economically reduced through process integration. A real-life example is given of heat-integrating a multicolumn process. Other recent developments on process integration can be found in references suggested in the section on Further Reading.

## 4.2. ENERGY USAGE AND EFFICIENCY IN SEPARATION

Energy usage in simple distillation per unit of material separated is typically 1–2 orders of magnitude greater than the minimum work, or free energy, of separation (Keller, 1982). A simple distillation column is defined here as having one feed stream, one reboiler, and one condenser. It is possible in a straightforward manner to calculate the minimum reboiler duty for such a column making a complete separation of a binary mixture. (Such a separation requires an infinite number of stages, but that problem is sidestepped in this analysis.) Figure 4.1 shows the effect of relative volatility on what is termed the maximum efficiency for complete separation of an equal-molar binary mixture. This value is calculated by dividing the minimum *work* of separation by the minimum amount of *heat* required in the distillation column. The minimum work of separation is the minimum (reversible) mechanical work required for separation of a homogeneous mixture into pure products at constant pressure and constant temperature (Abrams, 1978; Krishna, 1978;

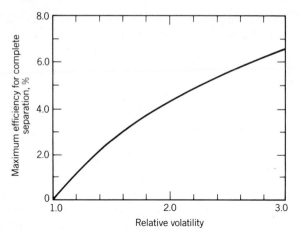

**FIGURE 4.1** Effect of relative volatility on maximum efficiency for complete separation of a binary mixture by simple distillation. Maximum efficiency = (minimum work of separation x 100)/(minimum heat for separation). Bases: boiling liquid feed, mole fraction of lower boiling component = 0.5, heat of vaporization given by Trouton's Rule.

King, 1980):

$$W_{\min, T, P} = -RT \sum_j x_{jF} \ln \left( \gamma_{jF} \, x_{jF} \right) \qquad (4.1)$$

where $W_{\min, T, P}$ is the minimum work, $R$ is the gas constant, $T$ is the absolute temperature, $P$ is the pressure, $x_{jF}$ is the mole fraction of component $j$ in the feed, and $\gamma_{jF}$ is the activity coefficient of component $j$ in the feed mixture.

Most distillation columns do not use mechanical work but rather heat; hence, because of the inevitable problem of Carnot efficiency, the energy requirement is substantially larger. The minimum heat usage for a simple distillation column separating a binary mixture occurs at the minimum reflux ratio; and this value, per mole of feed, is equal to (Keller, 1982)

$$\frac{Q_R}{F_M} = \lambda_M \left[ \frac{1}{\alpha - 1} + x_F \right] \qquad (4.2)$$

where $Q_R$ is the heat supply, $F_M$ is the molar feed rate, $\lambda_M$ is the molar heat of vaporization, $\alpha$ is the relative volatility, and $x_F$ is the mole fraction of the low-boiling component in the feed.

Trouton's rule, which states that the molar heats of vaporization of liquids at their atmospheric-pressure boiling points divided by their absolute boiling points is a constant, can be used to estimate the heat of vaporization in Equation (4.2) to give

$$\frac{Q_R}{F_M} = CT \left[ \frac{1}{\alpha - 1} + x_F \right] \qquad (4.3)$$

where $C$ is the Trouton's-rule constant (approximately $88 \times 10^3$ J/kg-mole-K).

The maximum efficiency $E$ for complete separation is defined as the ratio of Equation (4.1) to Equation (4.2) times 100, or

$$E = \frac{100\ W_{min,T,P}}{Q_R/F_M} = \frac{-100\ R\left[x_F \ln x_F + (1 - x_F) \ln (1 - x_F)\right]}{C\left[\dfrac{1}{\alpha - 1} + x_F\right]} \quad (4.4)$$

There are some assumptions and approximations in Equation (4.4). First, Equation (4.1) holds strictly for an isothermal process, while the column represented by Equation (4.2) is clearly not isothermal. Also, it is assumed in Equation (4.2) that the "pinch," or the point at which the operating line intersects the equilibrium curve, occurs at the feed point, and the relative volatility is that at the feed composition. (This assumption infers that the system does not have an azeotrope or a near-azeotrope which could cause a "tangent pinch" instead of a pinch at the feed point.) The operating lines are assumed to be straight, and the feed is assumed to be liquid at its boiling point. Trouton's rule is only an approximation for some liquids, with variations in the estimated heats of vaporization of up to about $\pm 20\%$, and it can be less accurate at pressures far from 1 atm. But in spite of these caveats, Figures 4.1 and 4.2 (which shows the effect of feed composition of the maximum efficiency) portray a reasonably accurate picture of the tremendous gap between what can almost be achieved in a typical distillation column and what could be achieved with a reversible separation process driven by mechanical energy.

Actually, practical efficiencies are somewhat lower than the values given in Figures 4.1 and 4.2, because real columns require more reflux than the minimum value. Practical efficiencies in well-run columns making nearly complete separations are usually 5–20% (of the value) lower.

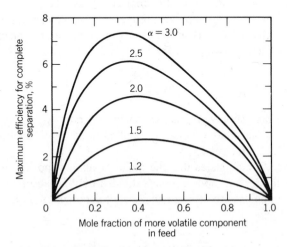

**FIGURE 4.2** Effect of relative volatility and feed composition on the maximum energy efficiency for complete separation of a binary mixture by simple distillation. Bases: boiling liquid feed, heat of vaporization given by Trouton's Rule.

These maximum efficiencies, which are well below 10%, seem shockingly low, and this realization has spawned numerous research projects to raise these values by various means or to develop alternative processes that will compete favorably for large portions of distillation's present uses. These efforts have in general been less than spectacularly successful, and in the following section we explore some of the economics of distillation that make it so dominant. We then look at the question of how, in light of the economic constraints so often encountered, we can reduce energy usage in distillation through process integration.

## 4.3. SEPARATION PROCESS ECONOMICS

We investigate the economics of separation processes by means of two examples. The first deals with the separation of ethylbenzene and styrene—a large-scale, high-energy-usage distillation. The flowsheet (Frank et al., 1969) is shown in Figure 4.3. Here we see that the column is large (70 trays) and the separation is difficult. The relative volatility varies between 1.32 and 1.43 from one end of the column to the other. Steam usage is about 1.3 kg per kilogram of styrene, which would definitely classify this distillation as a high-energy-usage process. The story is not complete with just a discussion of the energy usage, however, as can be seen in Table 4.1. Here we see capital and energy costs summarized. Capital costs consist of the return-on-investment income, depreciation, and other small, investment-related charges. Typically this yearly cost is about 30% of the investment in the process. Table 4.1 (Keller, 1983) shows that, despite the use of a substantial quantity of fairly expensive steam, the steam cost (including the capital cost of the steam plant) is only slightly larger than the capital cost of the column and auxiliaries. The steam operating cost is substantially smaller than the capital cost. Thus we begin to see that (1) capital cost may be an extremely important factor in process selection, and (2) *reduction of energy costs cannot be investigated without simultaneous reference to the implications for investment.* Conclusion (2) has not been properly stressed in the recent literature on process integration, most of which emphasizes either minimizing the loss of available energy (see, for example, Naka et al., 1982; Umeda et al., 1979) or bounding the utility consumption (see, for example, Andrecovich and Westerberg, 1985; Westerberg and Andrecovich, 1985; Glinos et al., 1985).

In the second example, we investigate the effect of relative volatility, column pressure, and other variables on capital and utility costs for a simple distillation column separating an ideal, equimolar binary mixture. The relative volatility range covered is between 1.2 and 10—a range that describes well over 90% of all commercial key-component relative volatilities. The cases studied are listed in Table 4.2 with the economic criteria used in the evaluation. Calculations were carried out using part of the Union Carbide's ADVENT process optimization system (Gautam and Smith, 1985). This package makes use of computer power and "pinch technology" (Linnhoff and Vredeveld, 1984) to minimize utility and capital costs and to improve raw material efficiency. The ADVENT system has been extended

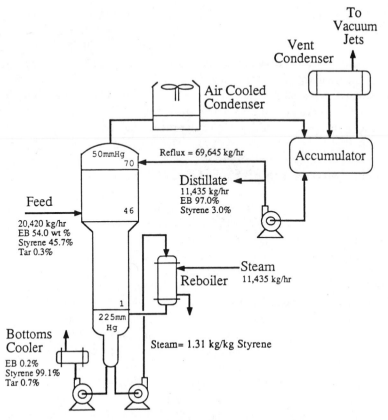

**FIGURE 4.3**  Flowsheet for styrene–ethylbenzene splitter (Frank et al., 1969). Reprinted with permission from the American Institute of Chemical Engineers.

beyond the heat-exchanger network into the separation processes and reaction systems. Figure 4.4 shows the capital costs, set at 30% of the investment, for these cases; Figure 4.5 shows the utility operating costs, which include all costs except utility–plant capital costs. The steam operating cost makes up about 97% of the utility operating cost.

As expected, both of these costs rise as the relative volatility decreases. The important point is that *the rises in both investment and utility costs become quite precipitous below a relative volatility of 2.0 and especially so below about 1.5.* The effect on investment caused by operating below atmospheric pressure is quite high, while the effect of moderate super-atmospheric pressure is quite low. The use of 316 stainless steel also causes a substantial increase in investment.

The ratios of the capital to utility direct costs can be depicted as shown in Figure 4.6. For all but the very-high-investment systems, this ratio is nearly constant over the range of relative volatility between 2.0 and 10. For the very-high-investment systems, there is a slow rise in the ratio as the relative volatility decreases over this range. Below 2.0, the ratio rises in all cases. This means that, perhaps contrary

**TABLE 4.1  Rough Investment and Energy Cost Analysis of an Ethylbenzene–Styrene Column[a]**

| *Capital Charges[b]* | | |
|---|---|---|
| System installed cost = $10,600,000 | | |
| Depreciation (8% of IC) | | $ 848,000 |
| Interest on borrowed money (2% of IC) | | 212,000 |
| Before-tax return on investment (20% of IC) | | 2,120,000 |
| | TOTAL | $3,180,000 |
| *Energy Charges* | | |
| Yearly steam usage = 99,900,000 kg/yr | | |
| Steam operating cost at $13.22/1000 kg | | $1,320,000 |
| Steam capital charges (30%) for investment of $73.80/1000 kg | | 2,212,000 |
| | TOTAL | $3,532,000 |
| Energy charges as percent of capital plus energy charges | | 53 |
| Steam operating cost as percent of capital plus steam operating cost | | 29 |

[a]Basis: Column and associated equipment shown in Figure 4.3, 1986 economics, 8400 hr/yr operation. Investment accuracy: ±25%.

[b]Does not include capital charges on allocated working capital and other investment-related charges, for example, maintenance and certain overheads.

**TABLE 4.2  Base-Case Specifications and Economic Assumptions for a Simple Distillation Column**

*Base Case (1986 dollars)*

Investment includes sieve-tray column, reboiler, condenser, associated piping, foundations, erection costs, instrumentation, control room, engineering design costs, and so on.

Reflux ratio = 1.1 times minimum reflux ratio.
Feed rate = 350 lb-mole/hr (binary) = 159.1 kg-mole/hr.
Feed composition = 0.5 (mole fraction of the more volatile component).
Distillate composition = 0.99.
Bottoms composition = 0.01.
Pressure = 14.7 psia (101.3 kPa).
Approach temperature in reboiler and condenser = 10 K.
Heat of vaporization = 14,000 Btu/lb-mole (32, 564 kJ/kg-mole).
Steam cost = $6.00/$10^6$ Btu (0.57¢/$10^6$J).
Cooling water cost = $0.20/$10^6$ Btu (0.019¢/$10^6$J).
Material of construction for equipment = carbon steel.
Capital cost/yr = 0.3 times investment.

to intuition, capital costs rise even more rapidly than utility costs as the relative volatility decreases.

The ratios of capital to utility cost given in Figure 4.6 range from a low of about 0.55 to a high of 3.5 over the full range of relative volatility investigated. Thus in all cases the capital cost becomes a significant economic factor in process

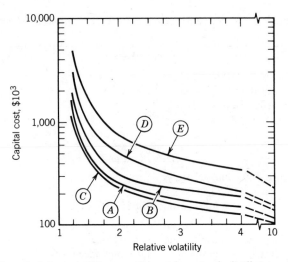

**FIGURE 4.4** Effect of relative volatility on capital cost for a simple distillation column. Deviation from base case: (A) base case (see Table 5.2); (B) increasing feed rate to 550 lb-moles/hr (220 kg-moles/hr); (C) increasing column pressure to 100 psia (700.7 kPa); (D) decreasing column pressure to 2 psia (13.8 kPa); (E) material of construction for equipment = 316 stainless steel (SS); (F) 316 SS, 2 psia (case D + case E).

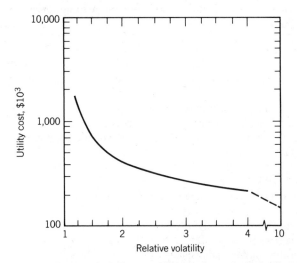

**FIGURE 4.5** Effect of relative volatility on utility cost for a simple distillation column.

selection, and in many cases is a more important factor than the utility cost, even though distillation is the epitome of a simple process. Thus it might be argued that, for the same degree of separability (relative volatility, selectivity, etc.), distillation is almost always the process of choice, unless there are other mitigating factors. Beyond that, distillation can often perform high degrees of separation between key components—many separation processes cannot—and it does not require the use

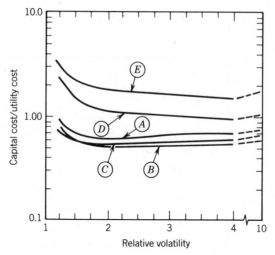

**FIGURE 4.6** Effect of relative volatility on capital cost/utility cost ratio for a simple distillation column. Deviation from base case: (A) base case (see Table 5.2); (B) increasing feed rate to 550 lb-moles/hr (220 kg-moles/hr); (C) increasing column pressure to 100 psia (700.7 kPa); (D) decreasing column pressure to 2 psia (13.8 kPa); (E) material of construction for equipment = 316 stainless steel (SS); (F) 316 SS, 2 psia (case D + case E).

of a mass-separation agent that must be regenerated, adding to the overall process complexity and cost.

Some quantification of the effect of mechanical complexity on process economics was given by Souders (1964), who compared distillation, extractive distillation, and solvent extraction, three processes that increase in mechanical complexity in the order given. This comparison was done by determining the relative volatility for extractive distillation and the selectivity for extraction required to give costs equal to that of the distillation of a given (constant) relative volatility. The results are given in Figure 4.7. For example, an extractive distillation process requires a relative volatility of 2.0, and an extraction process a selectivity of 6.0, to be cost-competitive with a distillation with a relative volatility of 1.5. For a distillation with a relative volatility of 2.5, the figure shows that the chances of extractive distillation or extraction being competitive are almost nil: the separation factors have to be unreasonably large.

The curves in this figure were developed before the rapid rise in energy prices and, to a somewhat lesser extent, investment costs. Therefore, this figure should be taken as indicative, not quantitative; but most certainly the general conclusion that extractive distillation and solvent extraction require larger separation factors than distillation to be cost-competitive remains unaffected.

## 4.4.  PRINCIPLES OF HEAT INTEGRATION

Given that we must live with distillation in many cases, we now turn to the question of how energy usage can be reduced. Many heat-integration schemes have

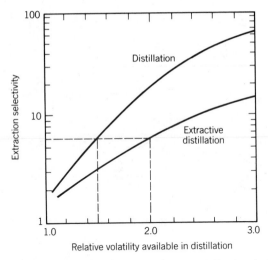

**FIGURE 4.7**  Selectivities required for equal costs (Sauders, 1964). Reprinted with permission from the American Institute of Chemical Engineers.

been successfully developed, such as column sequencing, heat pumping, sloppy cuts, feed preheat or subcool, and heat integration with other process streams. In this chapter, we only illustrate how to use a temperature–enthalpy (T–H) diagram to analyze a given distillation system and to identify the heat-integration opportunities of multiple effecting, side reboilers, and side condensers.

## A.  The Temperature–Enthalpy (*T–H*) Diagram

A conventional distillation column, as shown in Figure 4.8, contains a feed stream $F$ at temperature $T_F$ and two products: the distillate $D$ at $T_D$ and the bottoms $B$ at $T_B$. A heat flow of $Q_R$ units at temperature $T_R$ is introduced into the column through the reboiler. About the same amount of heat, $Q_c$, is recovered from the condenser at a lower temperature, $T_c$. The separation task is accomplished at the price of consuming the "temperature," or degrading the ability to do work, of a heat flow. This heat-degrading phenomenon of distillation processes can easily be visualized on a *T–H* diagram, which can be considered as an approximate heat-availability diagram, as shown in Figure 4.8.

The area $Q(T_R - T_C)$ is proportional to the net work consumption of separation, $W_{\text{net}}$ (King, 1980; Umeda et al., 1979):

$$
\begin{aligned}
W_{\text{net}} &= Q_R\left[1 - \frac{T_0}{T_R}\right] - Q_C\left[1 - \frac{T_0}{T_C}\right] \\
&= \frac{[Q(T_R - T_C)]T_0}{\tilde{T}^2}
\end{aligned}
\tag{4.5}
$$

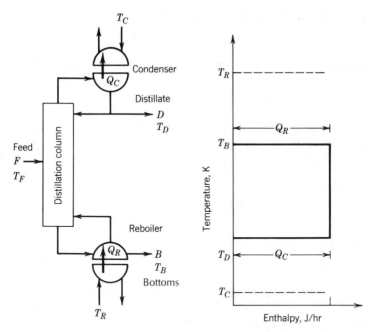

**FIGURE 4.8** A temperature–enthalpy depiction of a distillation column.

where $T_0$ is the absolute temperature of the surroundings, $\tilde{T}$ is the geometric mean of the absolute temperature of heat supply and removal streams, $(T_R T_C)^{1/2}$, $Q = Q_R - Q_C$, and $[1 - T_0/T]$ is the Carnot efficiency.

The area $Q(T_R - T_C)$ consists of two subareas: the heat supply and removal subarea, $Q[(T_R - T_B) + (T_D - T_C)]$, and the distillation column subarea, $Q[T_B - T_D]$. The net work consumption in the heat supply and removal subsystem can be reduced by applying the multiple-effect principle (see Section 4.4.B). The net work consumption in the distillation-column subsystem can be reduced by applying side reboilers and side condensers. The combination of applying side reboilers and side condensers, and the multiple-effect principle, can provide a host of heat-integration opportunities.

## B. Multiple-Effect Principle

A distillation column can be represented by a rectangle on the $T$–$H$ diagram. The width indicates the amount of heat required to run the column, assuming that $Q_R$ equals to $Q_c$. The height indicates the temperature drop through the column. Since the width on the enthalpy axis only shows the required heat duty, the rectangle can be moved horizontally without changing column operating conditions. The rectangle can be moved up or down between the upper and lower temperature bounds by increasing or decreasing the column pressure, respectively. Generally, both the heat duty and the temperature drop through the column increase as the column operating pressure increases (Westerberg and Andrecovich, 1985; Glinos et al., 1985). A rectangle may also be split vertically into two or more parts and be moved independently. The addition of a new column is needed for each split and vertical move. Figure 4.9 shows a vertical move, a horizontal move, and a column split and vertical move.

The temperature bounds are usually determined by one of the following considerations: the available and economical utility levels, the critical temperatures of all the components, the product decomposition temperature, the undesired reaction kick-off temperature, or equipment limitation.

The application of the multiple-effect principle may be described as how to effectively pack the rectangles between the upper and lower temperature bounds

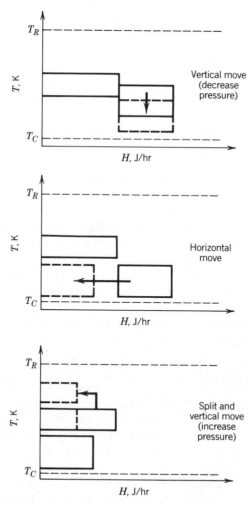

**FIGURE 4.9** Column moves on a *T–H* diagram.

to have a minimum total energy use. This may generate several combinations of column configurations for engineering concerns, such as operability and safety, as well as business concerns, such as payback time.

## C. Side Reboilers and Side Condensers

It is quite common to have situations in which the rectangles do not fit well between the temperature bounds. As shown in Figure 4.10a, the temperature range

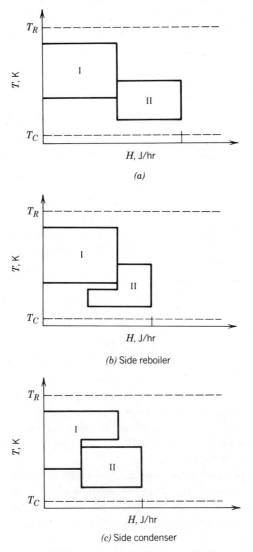

FIGURE 4.10 Application of side reboilers and side condensers.

$T_R - T_C$ is too small for a straight condenser–reboiler match between column I and column II. If we truncate an upper corner out of column II, by supplying a portion of the required heat through a side reboiler at a temperature lower than that of the main reboiler, it is possible to heat-integrate column II's side reboiler with column I's overhead vapor, as shown in Figure 4.10b. We may truncate a lower corner out of column I and heat-integrate column I's side condenser with column II's reboiler to reduce total energy usage, as shown in Figure 4.10c.

The application of side reboilers and side condensers, which is depicted by truncating corners out of the original rectangle, creates more heat-integration opportunities, especially in a crowded $T–H$ diagram situation. The maximum heat loads that can be shifted to side reboilers and side condensers at various temperature levels can be estimated from the heat-distribution curve of the system (Flower and Jackson, 1964; Timmer, 1969; Naka et al., 1980).

## D. Heat-Distribution Curve

Figure 4.11a shows a typical heat-distribution curve of an ideal binary system with saturated liquid feed. The curve BM is constructed from a series of minimum-boil-up calculations for various points in the stripping section according to the equation

$$H'_P = \frac{B\lambda}{S'_P - 1} \tag{4.6}$$

In the equation, $H'_P$ is the minimum heat flow corresponding to point $P'$ in the stripping section; $S'_P$ is the slope of the operating line, assuming a straight line, corresponding to point $P'$, as shown in the McCabe–Thiele diagram, Figure 4.11b, and $\lambda$ is the heat of vaporization, which is assumed to be constant. The curve DM is constructed from a series of minimum-reflux calculations for various points in the rectifying section according to the equation

$$H_P = \frac{D\lambda}{1 - S_P} \tag{4.7}$$

where $H_P$ and $S_P$ are the minimum heat flow and the slope of the operating line corresponding to point $P$ in the rectifying section, respectively.

The enclosed area BDM is proportional to the minimum work of separation. The ratio of area BDM to the area $Q(T_R - T_C)$ is proportional to the thermodynamic efficiency of separation.

Suppose that a portion of the required heat is added to the column through a side reboiler at temperature $T_{IR}$. Then the maximum heat load that can be shifted from the main reboiler to the side reboiler is $H_M - H_R$, as shown in Figure 4.11a. The heat distribution curve provides a boundary for truncating the rectangles on the $T–H$ diagram.

The use of a saturated-vapor or subcooled-liquid feed can be treated as a special

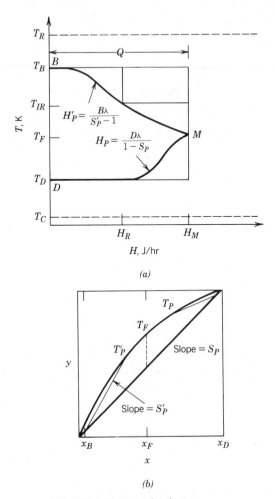

**FIGURE 4.11** Heat-distribution curve.

type of side reboiler or side condenser, respectively. The saturated-vapor-feed case is illustrated in Figure 4.12. The heat-distribution curve $DM$ is extended until it intersects the saturated-vapor-feed temperature $T_{FV}$. The new truncated rectangle indicates that a lower-level energy may be used to vaporize the feed, but overall more energy, $H_{MV} - H_M$, is required.

The shape of the heat-distribution curve is a unique characteristic of a distillation system; it depends on the nature of the mixture, the feed concentration and thermal quality, and product purities. Figure 4.13 illustrates the feed-concentration effect. The calculations for the heat-distribution curve are based on a benzene–toluene binary mixture at atmospheric pressure with saturated liquid feed, $X_D = 0.99$ (mole fraction of benzene), and $X_B = 0.01$. Figure 4.13a is the case of $X_F = 0.2$ and Figure 4.13b for $X_F = 0.8$. Two heuristic rules can be abstracted:

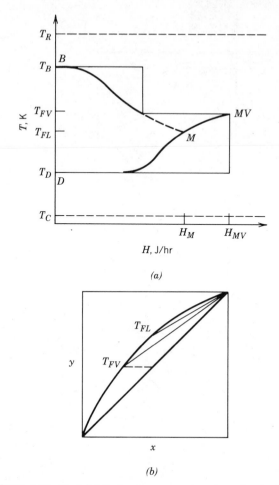

*(a)*

*(b)*

**FIGURE 4.12**   Heat-distribution curve: use of a saturated vapor feed.

1. Side condensers or feed subcooling should be considered when the feed contains mostly high-boiling components.
2. Side reboilers or vapor feed should be considered when the feed contains mostly low-boiling components.

The heat-distribution curve thus not only provides a boundary for truncating corners but also gives insights into a distillation system.

## 4.5.  APPLICATIONS OF HEAT INTEGRATION

Figure 4.14 shows the flow schematic and its corresponding $T$–$H$ diagram of a four-column separation process in a chemical plant. Two duplicate extractive-distillation trains—where column A's are the extractive distillation columns and col-

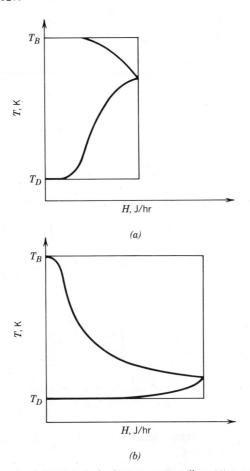

(a)

(b)

**FIGURE 4.13** Heat-distribution curve: feed-concentration effect. (A) $x_F$ = 0.2; (B) $x_F$ = 0.8.

umn B's are the strippers—are used to handle the huge feed stream. This combination of separations was considered to be a state-of-the-art design. On this $T–H$
diagram we can easily see how the energy has been cascaded from 200-psig (1480-
kPa) steam to columns IA and IIA, from IA and IIA to IB and IIB, respectively,
and then down to cooling water. [The extra amounts of heat required by IB and
IIB are supplied by 200-psig (1480-kPa) steam.] In addition, we can identify some
important factors for heat integration:

1. The temperature ranges:
   a. The total available temperature range, 146°C, that is, the range between
      200-psig (1480-kPa) steam, 196°C, and cooling water, 50°C.
   b. The temperature range consumed by columns IA and IIA, 9°C, and that
      consumed by columns IB and IIB, 28°C.

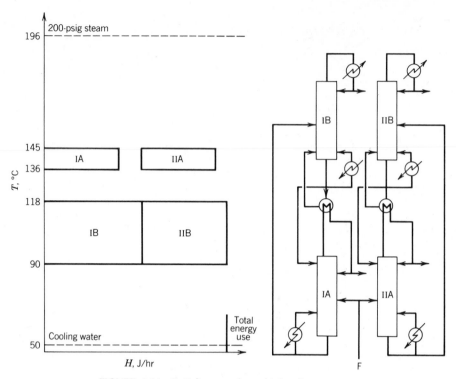

**FIGURE 4.14**   *T–H* diagram of a multiple-column system.

    **c.** The temperature range consumed by the reboilers and condensers, 146
    $- 9 - 28 = 109°C$.

**2.** The total energy use being determined by columns IB and IIB, which are
    the energy-use bottlenecks of this system.

If a 10°C temperature approach ($\Delta T_{min}$) for the reboilers and condensers is
assumed, a 48°C($10 + 28 + 10$) temperature range is needed to operate column
IIB. From Figure 4.14 we can quickly identify that we may run column IIB either
at high pressure, in the temperature range of 196 and 145°C, or at vacuum conditions in the temperature range of 90°C and cooling water temperature, thus using
a triple-effect operation. The *T–H* diagram of the high-pressure scheme is shown
in Figure 4.15. As column IIB's operating pressure increases, its reboiler duty
increases, but the temperature span changes little. Because the recycle solvent
stream brings more sensible heat to column IIA, the reboiler duty on column IIA
decreases, even though its column pressure does not change.

    Both the high- and low-pressure schemes require the purchase of new columns.
The high-pressure scheme saves $5.5 \times 10^6$/yr of steam operating costs, with an
associated capital investment of about $3.0 \times 10^6$. The low-pressure scheme saves

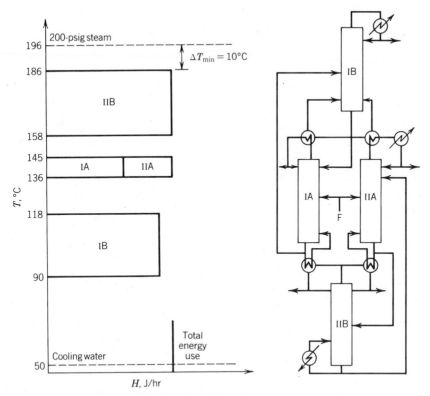

**FIGURE 4.15** Triple-effect scheme: high-pressure alternative.

$4.8 \times 10^6$/yr steam operating cost, with about a $4.0 \times 10^6$ capital investment. The high-pressure scheme has higher savings and a shorter payback time, about 7 months, but the large capital investments for both schemes could be a major drawback.

To retrofit the existing columns, and thus reduce investment requirements, we are constrained by equipment limitations. New upper- and lower-temperature bounds for heat integration are set according to the column pressure ratings. After moving columns IA and IB up and column IIB down, as shown in Figure 4.16, we can see that the new temperature range is too small for a straight triple effect. If we truncate an upper corner out of column IIB, by supplying a portion of the required heat through a side reboiler at a temperature lower than that of the main reboiler, it is possible to heat-integrate column IIB's side reboiler with column IB's overhead vapor. The $T$–$H$ diagram is shown in Figure 4.17. The side-reboiler scheme saves $1.90 \times 10^6$/yr in steam operating cost. Only a side reboiler and some piping rework are needed, and the capital investment is less than $1.0 \times 10^6$/yr.

The $T$–$H$ diagram and the heat-distribution curve provides a simple and powerful tool for heat integration of distillation processes.

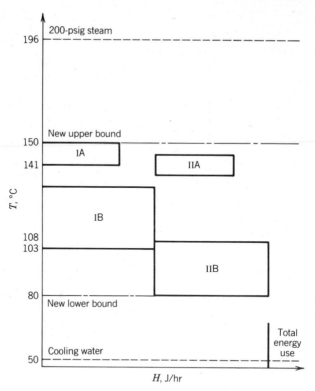

**FIGURE 4.16**   New bounds set by the equipment limitations.

## 4.6.  OTHER HEAT-INTEGRATION ADVANTAGES

In this chapter we emphasized heat-integrating distillation systems, and showed that substantial savings in utility costs are possible at acceptable levels of investment. In a number of cases, it appears that other savings can also result. For example, Terrill (1985) and Terrill and Douglas (1987) showed that, if large savings are possible in retrofitting a distillation train, a heat-exchanger network, or both, the process flows should almost always be reoptimized. This is because cost reductions in these systems decrease recycle costs and allow for a shift in reaction variables—conversion, feed and purge compositions, and so on—which in turn can result in raw-material savings. In some cases these savings can be even larger than the energy savings.

## 4.7.  CONCLUSION

Distillation has achieved its place of prominence in the petroleum-refining, organic-chemical, and allied industries by a combination of tolerable energy costs plus investments which are lower in most cases than alternative processes. Several

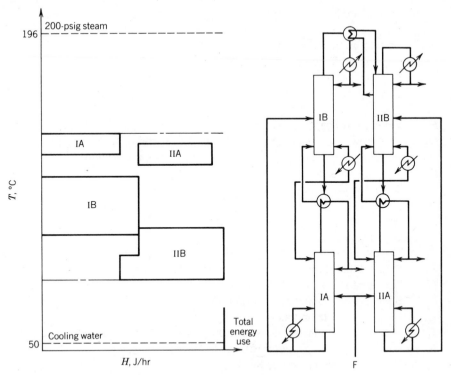

**FIGURE 4.17** Heat-integration with side reboilers.

possibilities exist for reducing energy costs, and one of the more powerful is the reuse of heat liberated from one column as the heat source in another. Simple but powerful methods for analysis of the heat-reuse problem are now available.

Not only can heat be added and removed at the ends of columns, but by the use of side reboilers and condensers, it can also be added and removed at intermediate temperatures. These strategies present additional opportunities for heat integration. This chapter illustrates that such strategies can be particularly effective in retrofit situations in which the latitude for changing column pressures is often constrained. New computer tools, such as Union Carbide's ADVENT system, are now available to analyze and optimize these opportunities.

## ACKNOWLEDGMENTS

The authors thank Union Carbide Corporation for permission to publish this chapter, and we especially recognize the many members of the Research and Development and Engineering Departments in South Charleston, West Virginia, for their encouragement in this effort.

## NOMENCLATURE

| | |
|---|---|
| $B$ | Bottom products, kg/hr |
| $C$ | Constant in Trouton's rule |
| $D$ | Distillate, kg/hr |
| $F$ | Feed, kg/hr |
| $F_M$ | Feed, kg-mole/hr |
| $H$ | Enthalpy, J/hr |
| $H_P$ | Heat flow corresponding to point $P$ in the rectifying section, J/hr |
| $H'_P$ | Heat flow corresponding to point $P'$ in the stripping section, J/hr |
| $P$ | Pressure, kPa |
| $Q$ | Total energy use, J/hr |
| $Q_C$ | Amount of heat removal, J/hr |
| $Q_R$ | Amount of heat supply, J/hr |
| $R$ | Gas constant |
| $S_P$ | Slope of the operating line corresponding to point $P$ in the rectifying section |
| $S'_P$ | Slope of the operating line corresponding to point $P'$ in the stripping section |
| $T$ | Temperature, K |
| $T_B$ | Bubble-point temperature of bottoms, K |
| $T_C$ | Heat-removal temperature, K |
| $T_D$ | Bubble-point temperature of distillate, K |
| $T_F$ | Bubble-point temperature of feed, K |
| $T_R$ | Heat-supply temperature, K |
| $T_O$ | Surroundings temperature, K |
| $x_B$ | Mole fraction of low-boiling component in bottoms |
| $x_D$ | Mole fraction of low-boiling component in distillate |
| $x_F$ | Mole fraction of low-boiling component in feed |
| $x_{jF}$ | Mole fraction of component $j$ in feed mixture |
| $\alpha$ | Relative volatility |
| $\gamma_{jF}$ | Activity coefficient of component $j$ |
| $\Delta T_{\min}$ | Minimum temperature difference in heat exchanger, K |
| $\lambda$ | Heat of vaporization, J/kg |
| $\lambda_M$ | Heat of vaporization, J/kg-mole |

## REFERENCES

Abrams, H., "Energy Reduction in Distillation," in *Alternatives to Distillation, I. Chemical Engineers Symposium Series*, No. 54, Institution of Chemical Engineers, London, UK, 1978, pp. 295–306.

Andrecovich, M. J. and A. W. Westerberg, "A Simple Synthesis Method Based on Utility Bounding for Heat-Integrated Distillation Sequences," *AIChE J.*, **31**, 363 (1985).

Flower, J. R. and R. Jackson, "Energy Requirements in the Separation of Mixtures by Distillation," *Trans. Inst. Chem. Eng.*, **42**, T249 (1964).

Frank, J. C., G. R. Geyer, and H. Kehde, "Styrene–Ethylbenzene Separation with Sieve Trays," *Chem. Eng. Prog.*, **65** (2), 79 (1969).

Gautam, R. and J. A. Smith, "A Computer-Aided System for Process Synthesis and Optimization," Paper No. 146, AIChE Meeting, Houston, TX, March 1985.

Glinos, K., M. F. Malone, and J. M. Douglas, "Shortcut Evaluation of $\Delta T$ and $Q\Delta T$ for the Synthesis of Heat-Integrated Distillation Sequences," *AIChE J.*, **31**, 1039 (1985).

Hindmarsh, E. and D. W. Townsend, "Heat Integration of Distillation Systems into Total Flowsheets: A Complete Approach," Paper 88B, AIChE Meeting, San Francisco, CA (November, 1984).

Keller, G. E., *Adsorption, Gas Absorption, and Liquid–Liquid Extraction: Selecting a Process and Conserving Energy*, The MIT Press, Cambridge, MA, 1982.

Keller, G. E., "Economic and Energy Considerations in Separation Process Selection," in *Proceedings of the Symposium on Separation Technology*, National Taiwan Institute of Technology, Taipei, Taiwan, May 16–18, 1983.

King, C. J., "Energy Requirements of Separation Processes," in *Separation Processes*, 2nd ed. McGraw-Hill, New York, 1980, Ch. 13.

Krishna, R., "A Thermodynamic Approach to the Choice of Alternatives to Distillation," in *Alternatives to Distillation, I. Chemical Engineers Symposium Series*, No. 54, Institution of Chemical Engineers, London, UK, 1978, pp. 185–214.

Linnhoff, B. and J. R. Flower, "Synthesis of Heat Exchanger Networks: I. Systematic Generation of Energy Optimal Networks," *AIChE J.*, **24**, 633 (1978a).

Linnhoff, B. and J. R. Flower, "Synthesis of Heat Exchanger Networks: II. Evolutionary Generation of Networks with Various Criteria of Optimality," *AIChE J.*, **24**, 642 (1978b).

Linnhoff, B., H. A. Dunford, and R. Smith, "Heat Integration of Distillation Columns into Overall Processes," *Chem. Eng. Sci.*, **38**, 1175 (1983).

Linnhoff, B. and D. R. Vredeveld, "Pinch Technology Has Come of Age," *Chem. Eng. Prog.*, **80** (7), 33 (1984).

Mix, T. W., J. S. Dweck, M. Weinberg, and R. C. Armstrong, "Energy Conservation in Distillation," *Chem. Eng. Prog.*, **74** (4), 49 (1978).

Mix, T. W., J. S. Dweck, M. Weinberg, and R. C. Armstrong, *Energy Conservation in Distillation*, Report No. DOE/CS/40259-1, National Technical Information Services, Springfield, VA, July, 1981.

Morari, M. and D. C. Faith, "The Synthesis of Distillation Trains with Heat Integration," *AIChE J.*, **26**, 916 (1980).

Naka, Y., M. Terashita, S. Hayashijuchi, and T. Takamatsu, "An Intermediate Heating and Cooling Method for a Distillation Column," *J. Chem. Eng. Jpn.*, **13**, 123 (1980).

Naka, Y., M. Tasayuki, and T. Takamatsu, "A Thermodynamic Approach to Multicomponent Distillation System Synthesis," *AIChE J.*, **28**, 812 (1982).

Null, H. R., "Energy Economy in Separation Processes," *Chem. Eng. Prog.*, **76** (8), 42 (1980).

Souders, M., "The Countercurrent Separation Processes," *Chem. Eng. Prog.*, **60** (2), 75 (1964).

Terrill, D. L., "Heat Exchanger Network Design: Operability Evaluation," Ph.D. Dissertation, Chemical Engineering, University of Massachusetts, Amherst, MA, May 1985.

Terrill, D. L. and J. M. Douglas, "Heat Exchanger Network Analysis. 1. Optimization and 2. Steady-State Operability Evaluation," *Ind. Eng. Chem. Res.*, **27**, 685 and 691 (1987).

Timmer, Jr., A. C., "Use of Cascade Theory and the Concept of Energy in Distillation Unit Design," Distillation Symposium Proceeding, Brighton, UK, 1969.

Umeda, T., K. Niida, and K. Shiroko, "A Thermodynamic Approach to Heat Integration in Distillation Systems," *AIChE J.*, **25**, 423 (1979).

U. S. Department of Commerce, Bureau of the Census, *Statistical Abstract of the United States, 1982–83*, 103rd ed., Washington, D. C., 1983.

Westerberg, A. W. and M. J. Andrecovich, "Utility Bounds for Nonconstant $Q\Delta T$ for Heat-Integrated Distillation Sequence Synthesis," *AIChE J.*, **31**, 1475 (1985).

## FURTHER READING

### A. General References

Henley, E. J. and J. D. Seader, "Energy Conservation and Thermodynamic Efficiency," in *Equilibrium-Stage Separation Operations in Chemical Engineering*, Wiley, New York, 1981, Ch. 17.

King (1980) cited in References.

I. Chem. E., *Understanding Process Integration*, I. Chem. E. Symp. Series No. 74, The Institution of Chemical Engineers, London, UK, 1982.

Linnhoff, B., D. W. Townsend, B. Boland, G. F. Hewitt, B. E. A. Thomas, A. R. Guy, and R. H. Marsland, *A User Guide on Process Integration for the Efficient Use of Energy*, Institute of Chemical Engineers, London, UK, 1982.

Steinmetz, F. J. and M. O. Chaney, "Total Plant Process Energy Integration," *Chem. Eng. Prog.*, **81** (7), 27 (1985).

Tjoe, T. N. and B. Linnhoff, "Use Pinch Technology for Process Retrofit," *Chem. Eng.*, **93** (8), 47 (April 28, 1986).

### B. Multieffect Distillation and Evaporation Systems

Andrecovich, M. J. and A. W. Westerberg, "An MILP Formulation for Heat-Integrated Distillation Sequences," *AIChE J.*, **31**, 1461 (1985).

Hillenbrand, J. B., "Studies in the Synthesis of Energy-Efficient Evaporation Systems," Report No. DRC-02-21-84, Design Research Center, Carnegie-Mellon University, Pittsburgh, PA, December 1984.

Nishitani, H. and E. Kunugita, "The Optimal Flow-Pattern of Multiple-Effect Evaporation Systems," *Comput. Chem. Eng.*, **3**, 261 (1979).

Nishitani, H. and E. Kunugita, "Multiobjective Analysis for Energy and Resource Conservation in an Evaporator System," Paper presented at the Second World Congress of Chemical Engineers, Montreal, Canada, 1981.

Tyreus, B. D. and W. L. Luyben, "Two Towers Cheaper Than One?" *Hydrocarbon Proc.*, 93 (July 1975).


## C.  Thermodynamic Second-Law Analysis and Synthesis in Process Integration

Fitsmorris, R. E. and Mah, R. S. H., "Improving Distillation Column Design Using Thermodynamic Availability Analysis," *AIChE J.*, **26,** 265 (1980).

Itoh, J., K. Niida, K. Shiroko, and T. Umeda, "Analysis of the Available Energy of a Distillation System," *Intern. Chem. Eng.*, **20,** 379 (1980).

Krishna, R., "A Thermodynamic Approach to the Choice of Alternatives to Distillation," in *Alternatives to Distillation*, Institute of Chemical Engineers Symposium Series No. 54, London, UK, 1978, pp. 185–214.

Linnhoff, B., "Entropy in Practical Process Design," in *Foundations of Computer-Aided Chemical Process Design*, Vol. II, R. S. H. Mah and W. D. Seider (eds.), Engineering Foundation, New York, 1981, pp. 537–572.

Liu, Y. A. and W. J. Wepfer, "Second Law Analysis of Processes: A Bibliography," *ACS Symp. Series*, No. 235, *Efficiency and Costing*: *Second Law Analysis of Processes*, R. A. Gaggioli (ed.) American Chemical Society, Washington, D.C., 1983, pp. 415–446.

Seader, J. D., *Thermodynamic Efficiency of Chemical Processes*, MIT Press, Cambridge, MA, 1982.

Townsend, D. W., "Second Law Analysis in Practice," *Chem. Eng. (London)*, **361,** 628 (1980).


## D.  Heat Pumps in Process Integration

Freshwater, D. C., "The Heat Pump in Multicomponent Distillation," *Brit. Chem. Eng.*, **6,** 388 (1961).

Townsend, D. W. and B. Linnhoff, "Heat and Power Networks in Process Design: I. Criteria for Placement of Heat Engines and Heat Pumps in Process Networks," *AIChE J.*, **29,** 742 (1983).


## E.  Side (Intermediate) Reboilers and Condensers in Process Integration

Kayihan, F., "Optimum Distribution of Heat Load in Distillation Columns Using Intermediate Condensers and Reboilers," in *Recent Advances in Separation Techniques-II*, Norman N. Li (ed.), *AIChE Symposium Series*, Vol. 76, No. 192, 1980, pp. 1–5.

Naka, Y., M. Terashita, S. Hayashiguchi, and T. Takamatsu, "An Intermediate Heating and Cooling Method for a Distillation Column," *J. Chem. Eng. J.*, **13,** 123 (1980).

## F.  Heat Pumps and Side Reboilers/Condensers in Process Integration

Flower, J. R. and R. Jackson, "Energy Requirements in the Separation of Mixtures by Distillation," *Trans. Inst. Chem. Eng.*, **42**, T249 (1964).

Freshwater, D. C. "The Heat Pump in Multicomponent Distillation," *Brit. Chem. Eng.*, **6**, 388, (1961).

Lynd, L. R. and H. E. Grethlein, "Distillation with Intermediate Heat Pumps and Side-stream Return," *AIChE J.*, **32**, 1347 (1986).

## G.  Heat-Exchange Integration in Separation Processes

Freshwater, D. C. and E. Ziogou, "Reducing Energy Requirements in Unit Operation," *Chem. Eng. J.*, **11**, 215 (1976).

Morari and Faith (1980) cited in References.

Rathore, R. N. S., K. A. Van Wormer, and G. J. Powers, "Synthesis Strategies for Multicomponent Separation Systems with Energy Integration," *AIChE J.*, **20**, 491 (1974a).

Rathore, R. N. S., K. A. Van Wormer, and G. J. Powers, "Synthesis Strategies for Multicomponent Separation Systems with Energy Integration," *AIChE J.*, **20**, 491 (1974b).

Sophos, A., G. Stephanopoulos, and B. Linnhoff, "A Weak Decomposition and the Synthesis of Heat-Integrated Distillation Sequences," Paper presented at AIChE Meeting, New Orleans, LA (November, 1981).

Sophos, A., G. Stephanopoulos, and M. Morari, "Synthesis of Optimum Distillation Sequences with Heat Integration," Paper presented at AIChE Meeting, Miami, FL, November 1978.

## H.  Heat and Power Integration

Nath, R., D. J. Liddy, and H. J. Duhom, "Joint Optimization of Process Units and Utility System," *Chem. Eng. Prog.*, **82** (5), 31 (1986).

Nishio, M., J. Itoh, K. Shiroko, and T. Umeda, "Thermodynamic Approach to Steam-Power System Design," *Ind. Eng. Chem. Proc. Des. Dev.*, **9**, 306 (1980).

Nishio, M., I. Koshijima, K. Shiroko, and T. Umeda, "Synthesis of Optimal Heat and Power Supply Systems," Paper presented at *AIChE* Meeting, Orlando, FL (February–March, 1982).

Papoulias, S. A. and I. E. Grossmann, "Structural Optimization Approach in Process Synthesis: III. Total Processing Systems," *Comput. Chem. Eng.*, **7**, 723 (1983).

Petroulas, T. and G. V. Reklaitis, "Computer-Aided Synthesis and Design of Plant Utility Systems," *AIChE J.*, **30**, 69 (1984).

Townsend, D. W. and B. Linnhoff, "Heat and Power Networks in Process Design: II. Design Procedure for Equipment Selection and Process Matching," *AIChE J.*, **29**, 748 (1983).

# 5

# PROCESS SYNTHESIS: A MORPHOLOGICAL VIEW

ARTHUR W. WESTERBERG

## 5.1.  INTRODUCTION

In this chapter, we first examine the activity of process synthesis in an abstact manner, noting activities currently identified as synthesis. We expose several interesting extensions that might be so identified in the future, suggesting that many tough questions still exist for design research.

After reviewing some of the process synthesis literature in general terms, this chapter examines in more detail the literature that deals with the flow of heat in processes. An understanding of heat flows has been one of the major contributions to date in this area and should be part of the training of future chemical engineers.

## 5.2.  A VIEW OF PROCESS SYNTHESIS

### A.  The Conventional View

Most people would readily accept the following definition for process synthesis. Process synthesis is the discrete decision-making activities of conjecturing (1) which of the many available component parts one should use, and (2) how they should be interconnected to structure the optimal solution to a given design problem. The challenge may be to create a new process (grass-roots design) or to improve an existing one (retrofit design), as Figure 5.1 illustrates. And as the figure also indicates, the activity is aided by the engineer's insights into the problem.

Associated with the synthesis step is always an analysis step, which determines if the conjectured solution will perform as desired (see Figure 5.2). This step usually exposes deficiencies in the original ideas, requiring that they then be modified to create either a feasible or an improved solution. This loop of conjecture followed by analysis is iterated until a satisfactory solution is found or until the design problem itself is abandoned. The analysis can involve the use of very simple to quite complex models to assess the performance of the resulting process.

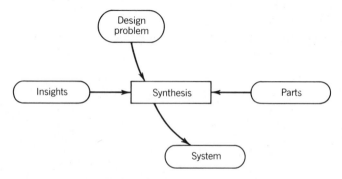

**FIGURE 5.1**  Conventional view of the synthesis step.

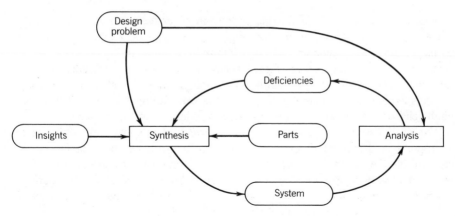

**FIGURE 5.2**   The analysis step associated with synthesis.

## B.  Innovation

Virtually all the work to date attempting to automate process synthesis limits itself to this view. In Figure 5.3 we add a new dimension to the problem, one that has been recognized, but whose solution has yet to be discovered. When designing, the deficiencies discovered in a design can also inspire the *invention of building blocks which are not in the catalogue of given parts.*

If one attends a conference about "the activity of design," one frequently witnesses an outbreak of disagreement among the participants that would seem to take direct aim at this invention step. There are those who argue that little can be done to automate invention and those who argue that, if not, there is little to be contributed to design by synthesis. Both camps, of course, have interesting points to make.

We might note that invention as described here is distinctly different from the

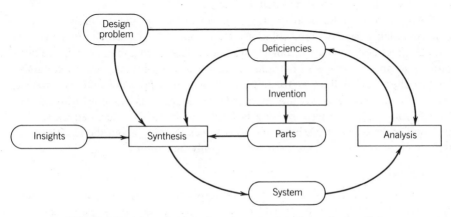

**FIGURE 5.3**   A loop that introduces invention into the synthesis activity.

act of creating a design from existing parts. It involves the use of lateral thinking, where one brings ideas that are seemingly unrelated to bear on the problem—perhaps from one's personal experience with fixing a car or frying potatoes. There is an interesting question being asked: "Is automatic innovation precluded because innovation comes through the use of a so-called garbage-can-effect of having all of one's experiences available to help invent novel solutions? If true, what would be the chances of encoding and using these experiences on a computer?"

For all this argument, it would seem the correct view is that both parties have valid points and both have perhaps not been willing to listen to the other side as carefully as they should. Clearly, automatic innovation is an intriguing goal. It should be worth investigating what is required to accomplish it, if only to prove that it cannot be done. Somehow I suspect we all do believe that someday there will be a computer out there called Hal, or at least Hal, Jr.

Although automatic innovation may be an elusive goal, one can well imagine the existence of computer aids that help the engineer to innovate. The simple existence of automatic synthesis tools would in fact serve this purpose, but here I am talking about aids that take direct aim at this problem of guiding innovation. The aid may not be able to carry out the invention step, but it could certainly point out where invention would significantly change the structure and perhaps the economics (or safety or reliability) of the solution. Johns and Romero (1979) presented a synthesis algorithm for separation processes in which the computer identified separation tasks that could not be done within the given technology and which, if free, would do the most to change the economics for the design.

To state that learning how to aid in innovation should be the only concern of those doing synthesis is equally invalid. Without innovation being involved, the number of alternative designs that can be configured from existing components is staggering. Numbers such as one trillion that most of us would accept as being equivalent to infinity are easy to generate for what seem to be fairly simple design problems. Thus there is an important need for the creation of efficient means to search among these alternatives without creating each of them, and this activity has been the thrust of most research to date on synthesis.

We must work on this problem. If we can develop automatic synthesis programs that can quickly develop good solutions to the combinatorial search problems that are the direct consequence of our innovative decisions, imagine the help we will provide to an engineer who is trying to design a new process. Almost certainly another alternative will not be tried by a tired engineer who is not anxious to repeat a lengthy analysis to assess the consequences. It really is not fun to invent the fiftieth heat-exchanger network for a problem involving 20 process streams. The first two or three may have been fun, as one was learning about the ideas behind the synthesis techniques, but the fiftieth can only be tedious.

## C. Learning

We can add another dimension to the synthesis activity, as illustrated by Figure 5.4. Here we see a loop being added, which requires solving several synthesis problems of a given type. By analyzing the solutions obtained, one can add to the insights available for the design activity. If this loop is automated, one has created

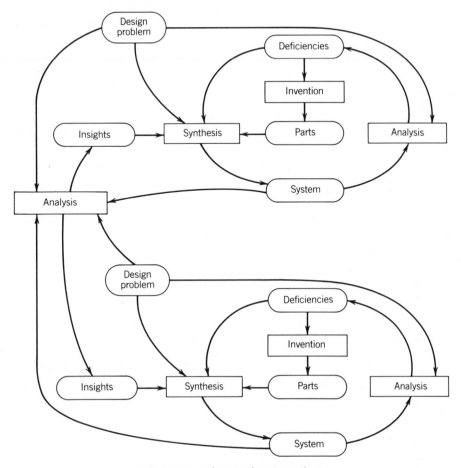

**FIGURE 5.4**   A learning loop in synthesis.

a design aid that becomes smarter with time, that is, it learns. To gain insight from the solution of problems is to spot patterns in those solutions. To what extent we can hope to do this step automatically and with credible results is an open question. I have held the suspicion that pattern matching really involves knowing or guessing the patterns to be found, solving a variety of problems, and then simply adjusting some parameters. That does not seem too exciting. This view could in part, however, turn into adding this learning capability into a synthesis problem itself, where one is attempting to create the patterns from catalogued building blocks whose combinations will create alternative patterns to be tested on the problem at hand.

## D.   Decomposition

Figure 5.5 illustrates another dimension to the synthesis activity. We must almost certainly decompose large problems if we ever hope to solve them. We will get wiped out on ''debugging'' the final solution to a problem if we do not decompose

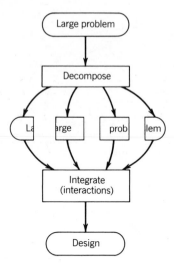

**FIGURE 5.5** Decomposition and integration steps for solving large problems.

it and debug the component parts first. However, if we decompose, we must then integrate the decomposed problem, and this step will lead to the creation of new and unexpected ''bugs'' that are a function only of the integrated system and not of the parts.

Many decompositions can often be proposed for solving a problem. Several may be used alternately in the solution of a given problem to gain different views of the problem. We can identify a few different decompositions here.

First we could decompose the problem functionally, as done for example by the system AIDES (Siirola et al., 1971), by selecting the reactors, allocating the species, solving the separation problems, and finally designing the heat-recovery system.

We could also decompose the problem spatially by synthesizing all parts before the reaction step, the reaction step, and all parts following.

We can see the distinction between these two decompositions perhaps a bit more clearly in the synthesis of high-rise buildings, where for the functional decomposition, one first designs the building superstructure, then the floor plans, the heating systems, and so on. The spatial decomposition suggests designing all aspects of each floor, proceeding perhaps from the ground floor and moving to each of the upper floors in sequence.

For these two decompositions to be performed on a rigorous basis, there must be a feedback to the earlier steps of the results obtained by the later ones. Based on this feedback, the earlier ones should then modify their solutions and pass their new solutions again to the steps that follow. These decompositions are really ''projections'' (Geoffrion, 1969), where projection is a strategy often proposed to solve mathematical programming problems.

The computations can also be decomposed when solving the large models for the system being synthesized. They can often be partitioned and solved in a precedence order. Even if they cannot be partitioned, computational decompositions are possible which deal with the equations for a single unit at a time, both for solving the equations and for optimizing the problem represented by the equations. The decompositions could involve converging the individual parts in an innermost loop, converging the total system in the next level of loop, converging the continuous variable optimization of each alternative structure in the next level, and finally iterating through the alternatives at the outermost level. In contrast, one could attempt to create schemes where the whole problem is simply constructed modularly, but converged all at once in a single loop.

Many other decompositions have been and will be discovered to solve synthesis problems. Each will almost certainly offer a different insight to the problem. They are not easy to discover nor are their implications readily noted. Thus entire PhD projects may be devoted to the proposal and assessment of a decomposition.

In all cases, for the decompositions to be valid, one must realize that there will be an integration step that has to be handled correctly.

## E.   Trade-offs Between Insights and Search

Given a design problem, we can visualize different outcomes of the approach to search among the alternatives. All synthesis problems involve the use of insights or heuristics. Solving the complete problem is out of the question. Thus heuristics (rules of thumb) must be used to reduce the problem to a manageable size. It is of course hoped that these heuristics, based on past experience or on approximate analyses, will allow us to trim away the bulk of the alternatives without losing the actual best solution in the process.

The heuristics can lead to (1) a small (hopefully it will be small) set of options to be searched by enumeration, (2) the construction of a superstructure that can be optimized by standard optimization codes such as mixed-integer linear or nonlinear programming codes, and (3) a unique answer. The best of all worlds occurs when the heuristics are actually able to isolate the best solution directly. This happened to us for the problem of multieffect evaporation-system synthesis (Hillenbrand and Westerberg, 1984), when we discovered that we could write down directly the one alternative for the liquid-flow pattern among the $N!$ options (where $N = 10!$ leads to 3,628,800 alternatives) that results in the minimum consumption of energy for a fully heat-integrated process. What was thought to be a large search problem was reduced to a nonproblem.

There is an opportunity to work toward many different outcomes in synthesis research. One can work on the insights to reduce the search problem as far as possible. This may be risky, as the insights may not exist or will be too elusive. One may also work on the creation of simple models that can be searched quickly, on mixed-integer (non)linear programming [MI(N)LP] codes that are efficient, or on posing problems in a superstructure form so they can be solved using existing or improved MI(N)LP codes. Much of the synthesis work by Grossmann falls into this latter category. All of these methods constitute real contributions to synthesis research.

## F.  Multiobjectives

In designing a process, we want it to be as profitable, safe, reliable, flexible, and controllable as possible. It must also be easy to start up. Obviously these are competing objectives. How do we synthesize in the face of all these objectives? One approach is to create first a process we can show to be profitable. If no profitable solution is possible, there is no need to worry about its safety. (We could also assess first the safety, if we think that it would be easy to ascertain, and then assess the economics if the process is deemed safe enough.) Given that we can discover a design that is profitable, we can then worry about its other attributes, altering the design to improve these attributes as we discover difficiencies. However, we must check after each alteration that the process remains acceptable in terms of the earlier objectives.

We really are in poor shape when it comes to these other objectives. To place a quantitative measure on the objectives, we need models of the process that are likely quite different from those used to determine profitability. For example, profitability is usually assessed by solving the heat and material balances and equipment design equations for the process being invented. Safety on the other hand is usually assessed by constructing a fault tree for the process and analyzing this tree to see that no potential problem has a high probability of occurring. (See Chapter 9 for a discussion of safety in process and plant design.) Reliability requires quite a different model again, one based on historical data for failures on the equipment being considered. Analysis techniques for flexibility are only now being developed by Grossmann (on our faculty) and Morari (at Cal Tech) and their students.

Almost all the work on these objectives has been to allow us to analyze a process to see how it measures up quantitatively to other proposed designs. For those objectives that we are able to quantify, we are still only at the beginning when it comes to creating process synthesis techniques related to them. At present we are forced to propose a solution and analyze it to see how well we did—a probe-and-test approach. We really need to discover the principles that allow us to propose directly the solutions that will optimize these measures.

If these objectives seem elusive, how do we quantitatively assess the appearance of the process to assure ourselves that it will look acceptable within its environment. That is certainly a significant concern of an architect. Many objectives are used which are not yet well-stated. An assessment based on them may only be intuitive, albeit one that is easy to make. Skilled designers can often look at a design and reject it as being unsuitable and not be able to explain why. They may simply be unwilling to accept a process design where the pipework does not run down the middle of the layout. One senses they are anticipating the end result of the design, much like a skilled chess player makes moves that can only be explained by their long-term potential to make the opponent's position untenable.

As stated earlier, if we choose to decompose the problem to produce first a profitable process, then a safe one, and so on, we must concern ourselves with the integration step of feeding the consequences of the later solutions back to the earlier ones. We must take care to perform this decomposition correctly. We need also to learn how to alter the earlier decisions to take the later ones into account.

We could of course try to account for all attributes in a single enormous analysis, much like we have been doing when trying to remove the loops within loops used in the recent past to optimize the models set up for processes. It will likely be a long time before we see how to do this type of loop-breaking for the broader problem, since the outer loops involve very different structure from the models used for cost estimation, and are replete with discrete decisions which we are not well able to handle yet, if we ever shall be.

## G.   Expert Systems in Design

There has been considerable publicity about the use of expert systems for solving engineering problems, not all of which has been positive. I would like to argue that there are several extremely important roles for expert-system concepts in design. The ideas that follow summarize some of those in a recent work of ours (Lien et al., 1986).

All of us appreciate the importance of using models based on first principles to characterize the phenomena occurring in processes. Examples include modeling the detailed behavior of combustion processes, the complexities of reaction-injection molding, and so forth. These computations can involve large systems of ordinary and partial differential, integral, and "algebraic" equations. The models may also be subjected to equality and inequality constraints that can involve minimization operators, discrete variables, and so on. Their complexity can be extreme by today's standards.

Obviously designers today do not solve design problems using what we might term "ab initio" calculations, NOR IS IT LIKELY THAT THEY EVER WILL. As noted earlier, a design problem can give rise to an enormous number of alternative configurations. Detailed calculations may be possible in the future to investigate the behavior of the one or two processes that are finally selected, but certainly they will not be used in the enormous searches that are needed for design.

Also, detailed computations for modeling a process are full of local behavior which makes them very poorly behaved. Their detailed behavior may preclude us from getting in the vicinity of the "globally" best solution, while simpler computations may not, a case of not being able to see the forest for all the trees. It may not be a good strategy, therefore, to use detailed computations too early in the design process.

The solution of large, complex problems is itself an interesting problem. Humans frequently do this activity with a remarkable degree of success. The contributions of expert systems are and will be at least these: (1) to understand how to model the process by which complex problems can be solved; and (2) to provide computer environments that help to implement and test these models. This type of modeling is obviously very different from the quantitative modeling with which we are familiar. If we can successfully model this process, then we can automate it and study alternative approaches for different classes of complex problems that we would like to solve. Our understanding of the methodology of problem solving will be brought to a new level.

Let us follow very briefly the approach an experienced designer might take to solve a process design problem. He will first classify it as either one that he expects to find in the available literature, a complex one that he will have to work on from scratch, or a very easy one that has an obvious solution. For example, he knows that the separation of propane from propylene is difficult, but well-studied from a process economics point of view. The literature will be directly helpful. On the other hand, he also knows that separating benzene from toluene is easy; there will not be a useful literature.

If a design problem is in the literature, there will be suggested designs which can form the basis of his work. He can use these available insights to help him judge which will be the better alternatives. If he is starting from scratch on a difficult problem, he will know that he has to be more careful about ruling out alternatives.

He will frequently sketch out over half of the structure for a process without doing a single computation, and his decisions will likely be very difficult to over-turn. What is he doing? He is clearly using internal representations that allow him to see the total problem. He has a good idea of where the trouble spots will be for the design, what will be the key decisions, where to check his assumptions, which constraints he should impose, and which technologies are likely going to be the better ones.

In an opportunistic fashion, he will alternately use complex and simplified models—the simplified ones to appreciate the global concerns and the complex ones to check that the details do not ruin his approach. Some computations will be qualitative and will involve a few sketches on a piece of paper, comparing "large" to "medium" to "small," or will involve simply knowing that adding a more volatile component to the mixture will decrease its bubble point significantly. Some will be to learn about the problem and will not, in the end, be directly used in solving the problem. He will constantly be replanning his next moves based on what he has just learned.

Buried in the expert-system literature are several important ideas on how humans solve complex problems. It was our goal to describe many of these in Lien et al. (1986), generally using process-design examples as illustrations. For example, we can look briefly at the concept of search, distinguishing between "chronological search" and "domain-dependent backtracking." In a design problem, one is often required to guess in order to proceed with the design. Later in the activity, a check should be made to verify that the guess was valid. For example, we may guess that we should solve the problem using distillation technology for the separation process. The later check is that we succeed with the design and that it has led to a fairly straightforward solution.

Guesses can be embedded. One guess leads us to another and then to another. If we find the design is failing, we can (1) back up to the last guess made and alter it, or (2) go back to an earlier guess that we believe should be reversed given the failure found. Always deciding to reverse the last guess made is chronological search. "Knowing" that an earlier guess than the last is a better one to reverse is domain-dependent backtracking.

Suppose that we later find the earlier decision should not have been reversed. We return it to the value originally guessed and find that we will be tracking down the same path we traversed before, which could involve a significant amount of computation. If we threw out the earlier work, we are doomed to repeat it. If we did not, we can simply change a few flags, suggesting that we now believe the work again to be valid. This notion of retaining earlier work that we now think invalid, but which we later think might become valid, forms the basis of the ideas behind "belief revision systems."

Expert systems have a role in the area of tool integration. Engineers often spend considerable time setting up and solving a complex computation. From the results, they make qualitative decisions about which computations to do next (planning). Often these qualitative decisions are substantial, but fairly routine, and could be automated if we could handle fairly arbitrarily structured, qualitative information. Expert-system technology is giving us this capability.

A last comment is in order. Expert systems today seem to be very slow to execute. Indeed we have had exactly this experience. However, if one can compile the code for the system, considerable increases in speed (like factors of 40) result. Also one can use what are at present relatively expensive computers that execute LISP much faster. These programs do not have to be as slow as they might seem at present.

## 5.3. SOME KEY PROCESS SYNTHESIS RESULTS

### A.  An Overview

In this section, we summarize some of the key synthesis results presented in the literature. To get a more detailed look at the literature one can consult the several review articles on synthesis (general: Hendry et al., 1973; Hlavacek, 1978; Westerberg, 1980; Stephanopoulos, 1981; Nishida et al., 1981; Umeda, 1982; distillation-based separation systems in particular: Westerberg, 1985). Also, the Institution of Chemical Engineers, United Kingdom, publishes a book that discusses the principles behind the synthesis of heat-exchanger networks and power systems (Linnhoff et al., 1982).

Process synthesis is the invention of the structure of processes. The term was first defined by Rudd (1968), although some earlier work has been identified (e.g., Lockhart, 1947). Since 1968, well over 200 articles on the topic can be listed. The synthesis literature blossomed during the 1970s, with the major contribution being a much improved understanding of the energy flows in processes. We can now much more clearly see how to design heat-recovery systems and to worry about integrating power-generation systems with existing processes.

Much of the literature (perhaps 20%) also concerned itself with the synthesis of separation systems, but almost all that work dealt with distillation-based separation systems where "well-behaving" mixtures are being separated into pure-component products. Here the insights are not as powerful. Useful but far from infallible heuristics based on plausible arguments are a major contribution. The work shows clearly how rich the solution space is for even this apparently highly restricted problem class (Westerberg, 1985).

Around 1980, the publication of research seemed to slow substantially. It was suggested that the only really significant "theoretical" result would be the understanding of the so-called "pinch point." (See further discussion in Section 5.3.B and Figure 5.6 below.) As with many predictions, this one is proving to be premature. Lately much emphasis has been on the synthesis of processes that heat-integrate well.

Other problems in synthesis have also been the subject of intense research and development. In chemistry, work first appearing in 1969 by Corey, Wipke, Hendrickson, Ugi, and others (references are reviewed in Nishida et al., 1981) has led to major computer programs to aid in the synthesis of complex organic molecules. Govind and Powers (1981) (concentrating on commodity chemicals), May and Rudd (1976) (considering chemical cycles to perform reactions that are not directly thermodynamically feasible), and Agnihotri and Motard (1980) (mixing some chemicals in a pot and wondering what might appear as a result of reactions) have contributed to this class of problem in chemical engineering.

Govind and Powers (1978, 1982) performed some of the earliest synthesis work in process control. They provided ideas on how to generate automatically from minimal information the structure of feasible single-input, single-output control systems for complete processes. Work by Morari et al. (1980) and Morari and Stephanopoules (1980) provided a broader definition of the synthesis problem for process-control systems and gave some ideas that could be used to attack the problem. There is continuing interest in this synthesis research.

As mentioned earlier, very interesting work is under way to develop models and synthesis procedures for processes that can assess measures other than cost. Flexibility or resiliency, depending on whether one is reading the work by Grossmann (Grossmann and Floudas, 1985; Grossmann et al., 1983; Grossmann and Morari, 1984; Swaney and Grossmann, 1985) or Morari (Marselle et al., 1982; Morari, 1982; Morari and Skogestad, 1985; Saboo and Morari, 1984; Saboo et al., 1985, 1987), is a measure receiving attention at this time.

## B.  Understanding the Heat Flows in Processes

We return to the synthesis work that has been the most complete and that is having the most impact in industry—the design of energy-efficient processes. It is probably safe to say that an understanding of the concepts in this area can give a novice designer the ability to out-perform an experienced design engineer who has not mastered them. The most important first problem to study is the synthesis of heat-exchanger networks for fixed processes.

The problem statement is typically given as follows.
Given:

1.  A set of hot streams to be cooled and a set of cold streams to be heated.

2.  The flow rates and the inlet and outlet temperatures for all these streams.

3. The heat capacity versus temperature of all streams as they pass through the heat-exchange process to be invented.

4. The available utilities and their costs per unit of heat provided or removed.

Determine the heat-exchanger network for energy recovery that will cost the least in terms of the total of the annual cost of utilities and annualized costs for the heat-exchanger equipment.

This problem should be studied first as it provides the understanding needed to appreciate later work. It is not that this is the most important problem in design. It is rather like learning all about linear programming to improve one's appreciation of the results in nonlinear programming.

The earliest work was by Hohmann (1971). He showed that one could (1) predict precisely the minimum amount of utilities and the type of utilities needed to solve the problem, and (2) estimate the fewest number of exchanges needed for the solution WITHOUT INVENTING A HEAT-EXCHANGER NETWORK to achieve the solution. The first result is readily obtained by plotting, on a figure of temperature versus heat, a composite heating curve for all the cold streams to be heated in the process against a composite cooling curve for all the hot streams to be cooled. These curves are strategically placed to reflect countercurrent heat exchange. Where the curves do not overlap, utilities are necessary (see the text by Linnhoff et al., 1982).

Linnhoff and Flower (1978) rediscovered Hohmann's results (which were never published except in his thesis with Lockhart—his now obvious major contribution was twice rejected by peer review) and extended them. Linnhoff, in several later publications, began to expose the real power of understanding the so-called "pinch point" in a process. This is the point that appears on the Hohmann/Lockhart plot that precludes further heat integration. A pinch point decomposes a process into two parts: a high-temperature heat-sink portion and a low-temperature heat-source part. Figure 5.6 illustrates this concept. No heat should be passed from streams above the pinch into streams below it, or both the consumption of hot and cold utilities must increase by exactly this amount.

Umeda et al. (1979) and later Linnhoff and co-workers noted that one should look at the streams causing the pinch point to see where to modify the process if one wishes to improve its energy efficiency.

Cerda et al. (1983) and later Papoulias and Grossmann (1983) showed that the minimum utility-use problem can be modeled as a linear program, allowing the calculation to be extended to situations where there are matches between certain streams that are a priori forbidden (Cerda and Westerberg, 1983).

Townsend and Linnhoff (1983) show very clearly that for minimum utility-use designs, one should never place a heat engine into a process such that the source of any process heat into the engine is above the pinch, and the sink for the heat back into the process from the engine is below the pinch. They also showed that a heat pump must transfer heat from below the pinch to above it to be of any use. With this insight, it becomes easy to look at earlier designs to decide if they are violating these principles. Apparently many have, since studies to improve the

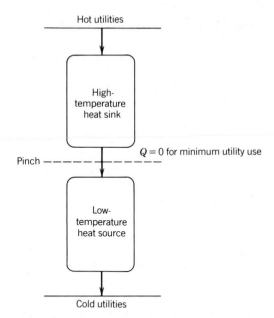

**FIGURE 5.6** The pinch point for heat flow for most processes.

energy efficiencies of existing processes can frequently claim 30% plus reductions in utility use.

The "pinch-design" method (Linnhoff and Hindmarsh, 1983) for inventing heat-exchanger networks that feature the minimum use of utilities argues that the most important design decisions are made where the process pinches. At this point, the temperature-driving forces are the smallest, allowing the minimum room for error in which streams should be matched against which in the final network.

More recent work has attempted to design processes that are energy-efficient. They expose what is now rather obvious—that well-integrated processes tend to have pinch points everywhere. To create a process with this feature renders the earlier analysis tools that look for pinch points less useful. The temperatures at which to operate units in the process are not fixed; the flows of streams in the process are adjustable; the pressure and thus the temperature at which a stream can change phase are not a priori fixed. Thus the heating and cooling curves cannot be plotted.

In Linnhoff et al. (1982), this class of problem underlies their discussion of "profile matching," a topic they cover when trying to decide the appropriate placement of heat pumps and engines in processes where the other parts of the process are fixed.

We have looked at the problem of synthesizing distillation-based separation processes where heat integration is taken into consideration (Andrecovich and Westerberg, 1985). A better illustration of the heat flow in such processes is a heat-cascade representation that ties the heat into the column in the reboiler to the heat removed from the column in the condenser (see Figure 5.7). Now columns

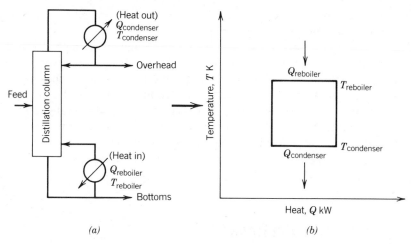

**FIGURE 5.7**  (a) Heat flow in a distillation column, and (b) a heat-cascade representation for the column on a T-Q diagram.

can be moved around on the plot of temperature versus heat and the occurrence of multiple pinch points poses no difficulty. Our work also allowed us to account for the use of multieffect distillation columns, that is, using more than a single column to accomplish a single separation task. (A single separation task can be accomplished in one column.) This diagram is also used in Chapter 4 by Ho and Keller.

Shelton and Grossmann (1985, 1986) did a similar study for the design of refrigeration systems, where the heat flows, temperature levels, and so forth are selected to allow for optimal heat integration. Two significant outcomes of this work are the approximate method to rank-order refrigerants for use over given temperature ranges and a very fast automatic algorithm that can select the number of stages of refrigeration and the heat-recovery network simultaneously. As mentioned earlier, we (Hillenbrand, 1984; Hillenbrand and Westerberg, 1984) completed a study of how to design multieffect evaporator systems.

As mentioned above, mixed integer (non)linear programming [MI(N)LP] techniques are proving very powerful in the selection of better process alternatives. This approach has been used for years by the petroleum industry to aid in the selection of subprocesses to be included when designing new refineries—yes, at one time new refineries were designed here. The integer (usually binary) variables indicate the existence (equals to 1) or nonexistence (equals to 0) of various parts of the process. Grossmann and co-workers presented many interesting MILP models for synthesis.

In some recent studies (Grossmann, 1985; Duran and Grossmann, 1984, 1986; Dong et al., 1986), an MI(N)LP model is used to select among alternatives for the design of an entire flowsheet. The model includes a cost for utilities that corresponds to the process being heat-integrated to consume the minimum utilities possible. Note that this model has to be pretty clever, as the location of the pinch point cannot be a priori determined; in fact, it will move through discrete jumps as the optimization progresses.

One important observation from this study is that a multiple-step procedure that first selects the process configuration with fixed utility charges, then adds the optimal heat integration, and finally optimizes the resulting structure, does not give the optimal answer, as indeed it should not. By assuming an optimal heat integration, the apparent cost of utilities per unit mass of product will typically decrease. Thus one can invest more in the process and try to convert a larger percentage of the raw materials to final desired products. The utility consumption can actually increase as a result, because more product is produced from the same raw-material consumption.

## 5.4.  CONCLUDING REMARKS

This chapter attempted to show that the process-synthesis problem is far from being fully explored, even though there has been an explosion in the literature on the subject. We teach our students how to look at processes from the point of view of material flow; now we can give them a much improved understanding of the flow of energy through processes too. However, from looking at the problem of synthesis in a broader context, as we did in the first part of this chapter, we see that there is plenty of room for significant future contributions in the areas of (1) automatic innovation (if it can be done), (2) automatic learning, (3) problem decompositions, (4) better search techniques, (5) problem insights, (6) alternative objectives, and (7) understanding the role of artificial intelligence, expert-system techniques.

## REFERENCES

Agnihotri, R. B. and R. L. Motard, "Reaction Path Synthesis in Industrial Chemistry," in *Computer Applications to Chemical Process Design and Simulation*, R. G. Squires and G. V. Reklaitis (eds.), *ACS Symposium Series*, Vol. 124, American Chemical Society, Washington, D.C., 1980, pp. 193–206.

Andrecovich, M. J. and A. W. Westerberg, "A Simple Synthesis Method Based on Utility Bounding for Heat-Integrated Distillation Sequences," *AIChE J.*, **31,** 363 (1985).

Cerda, J. and A. W. Westerberg, "Synthesizing Heat Exchanger Networks Having Restricted Stream/Stream Matches Using Transportation Problem Formulations," *Chem. Eng. Sci.*, **38,** 1723 (1983).

Cerda, J., A. W. Westerberg, D. Mason, and B. Linnhoff, "Minimum Utility Usage in Heat Exchanger Network Synthesis," *Chem. Eng. Sci.*, **38,** 373 (1983).

Dong, L. Y., L. T. Biegler, and I. E. Grossmann, "Simultaneous Optimization and Heat Integration with Process Simulators," Paper presented at AIChE Annual Meeting, Miami, FL (November, 1986).

Duran, M. A. and I. E. Grossmann, "A Mixed-integer Programming Algorithm for Process Synthesis," paper 104e, AIChE Meeting, San Francisco, CA (November 1984).

Duran, M. A. and I. E. Grossmann, "Simultaneous Optimization and Heat Integration in Chemical Processes," *AIChE J.*, **32,** 592 (1986).

Geoffrion, A. M., "Elements of Large-Scale Mathematical Programming." Research Report No. R-481-PR, RAND Corporation, Santa Monica, CA, 1969.

Govind, R. and G. J. Powers, "Synthesis of Process Control Systems," *IEEE Trans. Syst. Man. Cyber.*, **FMC-8,** 792 (1978).

Govind, R. and G. J. Powers, "Studies in Reaction Path Synthesis," *AIChE J.*, **27,** 429 (1981).

Govind, R. and G. J. Powers, "Control System Synthesis Strategies," *AIChE J.*, **28,** 60 (1982).

Grossmann, I. E., "Mixed-Integer Programming Approach for the Synthesis of Integrated Process Flowsheets," *Comput. Chem. Eng.*, **9,** 463 (1985).

Grossmann, I. E. and C. A. Floudas, "Active Constraint Strategy for Flexibility Analysis in Chemical Process Design," Paper 1f, AIChE Meeting, Chicago, IL (November, 1985).

Grossmann, I. E., K. P. Halemane, and R. E. Swaney, "Optimization Strategies for Flexible Chemical Processes," *Comput. Chem. Eng.*, **7,** 439 (1983).

Grossmann, I. E. and M. Morari, "Operability, Resilience and Flexibility: Process Design Objectives for a Changing World," in *Proceedings 2nd International Conference on Foundations of Computer-Aided Process Design*," A. W. Westerberg and H. H. Chien (eds.), CACHE Corporation, Austin, TX, 1984, pp. 931–1010.

Hendry, J. E., D. F. Rudd and J. D. Seader, "Synthesis in the Design of Chemical Processes," *AIChE J.*, **19,** 1 (1973).

Hillenbrand, J. B., Jr., "Studies in the Synthesis of Energy-Efficient Evaporation Systems," PhD Thesis, Carnegie-Mellon University, Pittsburgh, PA, 1984.

Hillenbrand, J. B., Jr., and A. W. Westerberg, "Synthesis of Evaporation Systems Using Minimum Utility Insights," AIChE Meeting, San Francisco, CA (November, 1984).

Hlavacek, V., "Journal Review: Synthesis in the Design of Chemical Processes," *Comput. Chem. Eng.*, **2,** 67 (1978).

Hohmann, E. C., "Optimum Networks for Heat Exchange," PhD Thesis, Chemical Engineering, University of Southern California, Los Angeles, CA, 1971.

Johns, W. R., and D. Romero, "The Automated Generation and Evaluation of Process Flowsheets," *Proc. 12th Symposium on Computer Applications in Chemical Engineering*, Montreaux, Switzerland, 1979, p. 435.

Lien, K., G. Suzuki and A. W. Westerberg, "The Role of Expert System Technology in Process Design," paper presented at ISCRE9, Philadelphia, PA, May 18-21 (1986). Appearing in *Chem. Eng. Sci.*, **42,** 1049 (1987).

Linnhoff, B. and J. R. Flower, "Synthesis of Heat Exchanger Networks. I: Systematic Generation of Energy Optimal Networks," *AIChE J.*, **24,** 633 (1978).

Linnhoff, B. and E. Hindmarsh, "The Pinch Design Method of Heat Exchanger Networks," *Chem. Eng. Sci.*, **38,** 745 (1983).

Linnhoff, B., D. W. Townsend, D. Boland, G. F. Hewitt, B. E. A. Thomas, A. R. Guy and R. H. Marsland, *A User Guide on Process Integration for the Efficient Use of Energy*, The Institution of Chemical Engineers, London, United Kingdom (1982).

Lockhart, F. J., "Multi-column Distillation of Natural Gasoline, *Petrol Refiner*, **26,** 104 (1947).

May, D. and D. F. Rudd, "Development of Solvay Clusters of Chemical Reactions," *Chem. Eng. Sci.*, **31,** 59 (1976).

Marselle, D. F., M. Morari, and D. F. Rudd, "Design of Resilient Processing Plants. II: Design and Control of Energy Management Systems," *Chem. Eng. Sci.*, **37,** 259 (1982).

Morari, M., "Flexibility and Resiliency of Process Systems," *Proceedings of Process Systems Engineering Symposium*, Kyoto, Japan, 1982, p. 223.

Morari, M., Y. Arkun, and G. Stephanopoulos, "Studies in the Synthesis of Control Structures for Chemical Processes: Part I. Formulation of the Problem. Process Decomposition and the Classification of the Control Tasks. Analysis of the Optimizing Control Structures," *AIChE J.*, **26,** 220 (1980).

Morari, M. and S. Skogestad, "Effect of Model Uncertainty on Dynamic Resilience," *IChE Symp. Series*, No. 92, Institution of Chemical Engineers, London, UK, 1985, p. 493.

Morari, M. and G. Stephanopoulos, "Studies in the Synthesis of Control Structures for Chemical Processes: Part II. Structural Aspects and the Synthesis of Alternative Feasible Control Structures." "Part III. Optimal Selection of Secondary Measurements within the Framework of State Estimation in the Presence of Persistent Unknown Disturbances," *AIChE J.*, **26,** 222, 247 (1980).

Nishida, N., G. Stephanopoulos, and A. W. Westerberg, "A Review of Process Synthesis," *AIChE J.* **27,** 321 (1981).

Papoulias, S. A. and I. E. Grossmann, "A Structural Optimization Approach in Process Synthesis. I: Utility Systems. II: Heat Recovery Networks. III: Total Processing Systems," *Comput. Chem. Eng.*, **7,** 695 (1983).

Rudd, D. F. "The Synthesis of System Design, I. Elementary Decomposition Theory," *AIChE J.*, **14,** 343 (1968).

Saboo, A. K. and M. Morari, "Design of Resilient Processing Plants—Some New Results on Heat Exchanger Network Synthesis," *Chem. Eng. Sci.*, **39,** 579 (1984).

Saboo, A. K., M. Morari, and D. C. Woodcock, "Design of Resilient Processing Plants— A Resilience Index for Heat Exchanger Networks," *Chem. Eng. Sci.*, **40,** 1553 (1985).

Saboo, A. K., M. Morari, and R. D. Colberg, "Resilience Analysis of Heat Exchanger Networks. I: Temperature-Dependent Heat Capacities. II: Stream Splits and Flowrate Variations," *Comput. Chem. Eng.*, in press (1987).

Shelton, M. R. and I. E. Grossmann, "A Shortcut Procedure for Refrigeration Systems," *Comput. Chem. Eng.*, **9,** 615 (1985).

Shelton, M. R. and I. E. Grossmann, "Optimal Synthesis of Integrated Refrigeration Systems. I: Mixed-Integer Programming Model. II: Implicit Enumeration Algorithm," *Comput. Chem. Eng.*, **10,** 445 (1986).

Siirola, J., G. Powers, and D. F. Rudd, "Synthesis of System Designs, III: Toward a Process Concept Generator," *AIChE J.*, **17,** 677 (1971).

Stephanopoulos, G., "Synthesis of Process Flowsheets: An Adventure in Heuristic Design or a Utopia of Mathematical Programming?" in *Foundations of Computer Aided Chemical Process Design*, R. S. H. Mah and W. D. Seider (eds.), Engineering Foundation, New York, 1981, p. 439.

Swaney, R. E. and I. E. Grossmann, "An Index for Operational Flexibility in Chemical Process Design. I: Formulation and Theory. II: Computational Algorithms," *AIChE J.*, **31,** 621 (1985).

Townsend, D. W. and B. Linnhoff, "Heat and Power Networks in Process Design. I: Criteria for the Placement of Heat Engines and Heat Pumps in Process Networks," *AIChE J.*, **29,** 742 (1983).

Umeda, T., "Computer Aided Process Synthesis," *Proceedings of Process Systems Engineering Symposium*, Kyoto, Japan, 1982, p. 79.

Umeda, T., K. Niida, and K. Shiroko, "A Thermodynamic Approach to Heat Integration in Distillation Systems," *AIChE J.*, **25,** 423 (1979).

Westerberg, A. W., "A Review of Process Synthesis," in *Computer Applications to Chemical Engineering*, R. G. Squires, and G. V. Reklaitis (eds.), *ACS Symposium Series*, No. 124, American Chemical Society, Washington, D.C., 1980, pp. 53–87.

Westerberg, A. W., "The Synthesis of Distillation-Based Separation Systems," *Comput. Chem. Eng.*, **9,** 421 (1985).

# 6

# PROCESS SYNTHESIS: SOME SIMPLE AND PRACTICAL DEVELOPMENTS

Y. A. LIU

Acknowledgments
Nomenclature
References

## 6.1. INTRODUCTION

In the preceding chapter, Westerberg gives an excellent overview of the field of process synthesis and also an enlightening discussion of recent developments and future directions. Despite the significant progress made in the field, however, the usual emphasis in chemical engineering-design teaching has been directed more toward process analysis rather than process synthesis, and more toward the components and short-cut methods rather than the systems and computer-aided approaches. Currently, the most popular chemical engineering-design text, *Plant Design and Economics for Chemical Engineers* by Peters and Timmerhaus (1980), contains no discussion of either process synthesis or computer-aided design; and it emphasizes mainly process analysis and equipment design. An early text on *Process Synthesis* by Rudd et al. (1973) is already 14 years old. For the benefit of engineering-design teaching, there apparently exists a need for an organized presentation and illustration of some simple and practical developments in the field of process synthesis.

This chapter describes in some detail two specific topics of process synthesis, namely, the systematic synthesis of (1) multicomponent separation sequences and (2) heat-exchanger networks. These topics are chosen because *they are industrially significant and can be readily taught to undergraduate students and applied by practicing engineers*. Particular emphasis is placed on selected synthesis techniques *that do not require special mathematical background and computational skill from the user*. To do so, we present mainly the heuristic methods based on rules of thumb, and evolutionary techniques wherein improvements are systematically made to an initial design or flowsheet.

## 6.2. MULTICOMPONENT SEPARATION SEQUENCES

### A. Problem Statement and Background

Multicomponent separation systems are widely used in the chemical and petroleum industries. An important process-design problem in multicomponent separations is separation sequencing, which is concerned with the selection of the best method and sequence for the separation. This problem is often solved by first arranging the components in a mixture to be separated in some ranked lists of appropriate physical and/or chemical properties such as relative volatility and solubility in water. The resulting ranked list for each property gives the component name and its property ranking relative to other components in the mixture. This list allows a separation step to be considered as a list-splitting operation, in which components above a certain property value are separated from components below that value.

The generation of a separation sequence then involves the selection of an appropriate property list and the choice of separation keys within the list. For example, sequences for separating a three-component mixture of A, B, and C into pure components by two ordinary distillation columns can be found by first arranging the components in a ranked list of relative volatility from the most (A) to the least volatile (C), and then examining the different splits in the two columns. One possible sequence involves making the split A/BC in the first column followed by the split B/C in the second column. Another possible sequence is AB/C in the first column and A/B in the second. These separation sequences may be represented schematically as follows:

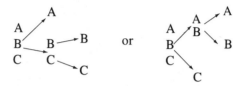

In the preceding example, each component to be separated appears in one and only one product stream and the type of separation that yields such products with nonoverlapping components is called *the high-recovery or sharp separation.* In industrial practice, however, it is sometimes useful to permit certain components to be separated to appear in two or more product streams. For example, a scheme for separating the same three-component mixture considered above may correspond to the following:

This type of separation, resulting in products with overlapping components, is called *the sloppy or nonsharp separation.* In this chapter, we are primarily concerned with the synthesis of sharp separation sequences. For methods applicable to synthesizing sloppy separation sequences, the reader may refer to the work of Cheng (1987).

To appreciate the need for some simple and practical methods for synthesizing multicomponent separation sequences, let us consider the number of theoretically possible sequences for separating a multicomponent mixture into pure-component product streams (Henley and Seader, 1981, pp. 530–533).

1. The number of theoretically possible sequences $S_N$ for separating an $N$-component mixture into $N$ pure-component products by one separation method only:

$$S_N = \frac{[2(N - 1)]!}{N!(N - 1)!} \tag{6.1}$$

2. The number of theoretically possible sequences $S$ for separating an $N$-component mixture into $N$ pure-component products by $T$ separation methods:

$$S = T^{N-1} \cdot S_N = T^{N-1} \frac{[2(N - 1)]!}{N!(N - 1)!} \tag{6.2}$$

Table 6.1 compares the numbers of theoretically possible sequences for $T = 1$ and $T = 2$. We see that the number of such sequences for separating a 10-component mixture into pure-component products approaches 10 million! In view of the enormous number of solution alternatives available, there is obviously a need for some "creative" methods to synthesize good separation sequences for multicomponent mixtures.

Some excellent discussion of the background information related to the multicomponent separation-sequencing problem can be found in the texts by Rudd et al. (1973, pp. 155–208 and 288–295), King (1980, pp. 710–720), and Henley and Seader (1981, pp. 527–555). The general techniques developed for solving the separation-sequencing problem have included algorithmic approaches involving some established optimization principles (e.g., Hendry and Hughes, 1972), heuristic methods based on the use of rules of thumb (e.g., Rudd et al., 1973, pp. 155–208), evolutionary strategies wherein improvements are systematically made to an initially created separation sequence (e.g., Stephanopoulos and Westerberg, 1976), and thermodynamic methods involving applications of thermodynamic principles (e.g., Gomez-Munoz and Seader, 1985). In some situations, two or

**TABLE 6.1 Number of Theoretically Possible Sequences for Separating an N-Component Mixture into Pure-Component Products with One ($T = 1$) and Two ($T = 2$) Separation Methods**

| Number of Components, $N$ | Theoretically Possible Sequences | |
|:---:|:---:|:---:|
| | $T = 1$ | $T = 2$ |
| 3 | 2 | 8 |
| 5 | 14 | 224 |
| 7 | 132 | 33,792 |
| 10 | 4,862 | 9,957,376 |

more of these techniques have been used together in the synthesis (e.g., Seader and Westerberg, 1977). Reviews of previous studies on the multicomponent separation sequencing can be found in Nishida et al. (1981) and Westerberg (1985).

## B.  Heuristic Synthesis of Multicomponent Separation Sequences

### 1.  A Simple Heuristic Method

Heuristic rules to guide the order of separation sequencing have long been available. Appendix 6.1 gives a survey of heuristics for the synthesis of multicomponent separation sequences published since 1947. It should be noted that in most of the heuristics reported thus far, no indication has been given of the conditions under which a specific heuristic is favored in selecting the separation sequence other than the general statement that "all other things being equal, favor the sequence which. . . ." Further, many of the heuristics apparently contradict or overlap others. For instance, for a feed mixture containing a component in excess and with this component being also a key component of a difficult separation, the heuristic of removing the most plentiful component first suggests its early removal. This is contradictory to the heuristic of performing difficult separations last, which favors the late removal of this plentiful component in the sequence.

The recent development of ordered heuristic methods (Seader and Westerberg, 1977; Nath and Motard, 1981; Nadgir and Liu, 1983) has effectively resolved the apparent conflicts among heuristics and enhanced the applicability of heuristic methods for multicomponent separation sequencing. In these methods, certain heuristics are selected and ranked in a specific order. In other words, the chosen heuristics are to be applied one by one in the order specified by the method. If one heuristic is not important or not applicable to a given synthesis problem, the next one in the method is considered.

Perhaps the most simple ordered heuristic technique is that proposed by Nadgir and Liu (1983). In this method, heuristic rules for separation sequencing are broadly classified into four categories: (1) *method heuristics* (designated as M heuristics), which favor the use of certain separation methods under given problem specifications; (2) *design heuristics* (designated as D heuristics), which favor specific separation sequences with certain desirable properties; (3) *species heuristics* (designated as S heuristics), which are based on the property differences between the species to be separated; and (4) *composition heuristics* (designated as C heuristics), which are related to the effects of feed and product compositions on separation costs. The method by Nadgir and Liu involves the systematic application of the following seven ordered heuristics. It is straightforward to apply by hand and does not require any mathematical background or computational skill.

1. Heuristic M1 (favor ordinary distillation and remove mass separating agent first). (a) All other things being equal, favor separation methods using only energy separating agents (e.g., ordinary distillation), and avoid using separation methods

(e.g., extractive distillation) that require the use of species not normally present in the processing, that is, the mass separating agent (MSA) (Rudd et al., 1973, pp. 174–181). However, if the separation factor or relative volatility of the key components $\alpha_{LK,HK} < 1.05$ (Van Winkle, 1967, p. 381; Seader and Westerberg, 1977) to 1.10 (Nath and Motard, 1981), the use of ordinary distillation is not recommended. A MSA may be used, if it improves the relative volatility between the key components. (b) When a MSA is used, remove it in the separator immediately following the one into which it is used (Hendry and Hughes, 1972; Rudd et al., 1973, pp. 174–180; Seader and Westerberg, 1977).

2. Heuristic M2 (avoid vacuum distillation and refrigeration). All other things being equal, avoid excursions in temperature and pressure, but aim higher rather than lower (Rudd et al., 1973, pp. 182–183). If vacuum operation of ordinary distillation is required, liquid–liquid extraction with various solvents might be considered. If refrigeration is required (e.g., for separating materials of low boiling points with high relative volatilities as distillate products), cheaper alternatives to distillation such as absorption might be considered (Souders, 1964; Seader and Westerberg, 1977; Nath and Motard, 1981).

3. Heuristic D1 (favor smallest product set). Favor sequences that yield the minimum necessary number of products. Avoid sequences that separate components that should ultimately be in the same product (Thompson and King, 1972; King, 1980, p. 720). In other words, when multicomponent products are specified, favor sequences that produce these products directly or with a minimum of blending, unless relative volatilities are appreciably lower than those for a sequence that requires additional separators and blending (Seader and Westerberg, 1977; Henley and Seader, 1981, p. 541).

4. Heuristic S1 (remove corrosive and hazardous components first). Remove corrosive and hazardous materials first (Rudd et al., 1973, p. 170).

5. Heuristic S2 (perform difficult separations last). All other things being equal, perform the difficult separations last (Harbert, 1957; Rudd et al., 1973, pp. 171–174). In particular, separations where relative volatilities of the key components are close to unity should be performed in the absence of nonkey components. In other words, try to select sequences that do not cause nonkey components to be present in separations where the key components are close together in relative volatility or separation factor (Heaven, 1969; King, 1980, p. 715).

6. Heuristic C1 (remove most plentiful component first). A product composing a large fraction of the feed should be separated first, provided that the separation factor or relative volatility is reasonable for the separation (Nishimura and Hiraizumi, 1971; Rudd et al., 1973, pp. 167–169; King, 1980, p. 715).

7. Heuristic C2 (favor 50/50 split). If component compositions do not vary widely, sequences that give a more nearly 50/50 or equimolal split of the feed between the distillate ($D$) and bottoms ($B$) products should be favored, provided that the separation factor or relative volatility is reasonable for the split (Harbert, 1957; Heaven, 1969; King, 1980, p. 715). If it is difficult to judge which split is closest to 50/50 and with a reasonable separation factor or relative volatility, then

perform the split with the highest value of the coefficient of ease of separation (CES) first.

The coefficient of ease of separation (CES) as proposed in heuristic C2 is defined as

$$\text{CES} = f \times \Delta \tag{6.3}$$

where $f$ = the ratio of the molal flow rates of products (distillate and bottoms) $B/D$ or $D/B$, depending on which of the two ratios $B/D$ and $D/B$ is smaller than or equal to unity; and $\Delta = \Delta T =$ boiling-point difference between the two components to be separated, or $\Delta = (\alpha - 1) \times 100$ with $\alpha$ being the relative volatility or separation factor of the two components to be separated.

In applying the preceding method to separation sequencing, method heuristics M1 and M2 first decide the separation methods to be used. Design heuristic D1 and species heuristics S1 and S2 then give guidelines about the forbidden splits resulting from product specifications, as well as the essential first and last separations. Finally, the actual initial sequences are synthesized by using composition heuristics C1 and C2 with the help of the coefficient of ease of separation (CES).

Since CES involves the relative volatility or separation factor for the two components to be separated, the application of the new heuristic method depends indirectly on the separation temperature and pressure. A good correlation for the optimum overhead pressure ($P_D$) as a function of the normal feed bubble-point ($T_{FB}$) in ordinary distillation has been presented by Tedder and Rudd (1978) as follows:

$$\ln P_D = \frac{1751}{T_{FB} + 273} - 6.777 \tag{6.4}$$

In Equation (6.4), $P_D$ is in MPa and $T_{FB}$ in °C such that $0.007 < P_D \leq 6.89$ and $-72°C < T < 699°C$. Also, a systematic procedure for specifying other operating pressures in ordinary distillation can be found in Henley and Seader (1981, pp. 432–434).

## 2.  Illustrative Examples

### a.  EXAMPLE 1: SEPARATION OF PRODUCTS FROM THERMAL CRACKING OF HYDROCARBONS

Consider the multicomponent separations involved in the large-scale thermal cracking of hydrocarbons to manufacture ethylene and propylene (Rudd et al.,

1973, pp. 183–185; King, 1980, pp. 708–710). The feed mixture is:

| Species | Moles/Hr | Normal Boiling Point, $T°C$ | $\Delta T$ | CES |
|---|---|---|---|---|
| A: Hydrogen | 18 | −253 | | |
| | | | 92 | 23.0* |
| B: Methane | 5 | −161 | | |
| | | | 57 | 19.6 |
| C: Ethylene | 24 | −104 | | |
| | | | 16 | 14.6 |
| D: Ethane | 15 | −88 | | |
| | | | 40 | 18.1 |
| E: Propylene | 14 | −48 | | |
| | | | 6 | 1.1 |
| F: Propane | 6 | −42 | | |
| | | | 41 | 4.0 |
| G: Heavies | 8 | −1 | | |

* The CES value for the split A/BCDEFG is found from

$$CES = \left( \frac{D}{B} \text{ or } \frac{B}{D} \right) \times \Delta T$$

$$= [18/(5 + 24 + 15 + 14 + 6 + 8)] \times 92$$

$$= 23.0$$

It is desired to separate the feed into the following six products: AB, C, D, E, F, and G. The separation sequencing by the heuristic method can be done as follows.

1. **Heuristics M1 and M2:** use ordinary distillation with refrigeration at high pressure.
2. **Heuristic D1:** avoid splitting AB as it is a single product.
3. **Heuristic S1:** not applicable.
4. **Heuristic S2:** perform splits C/D and E/F last, owing to their small $\Delta T$ of 6–16°C.
5. **Heuristic C1:** not applicable.
6. **Heuristic C2:** for separating ABCDEFG, the best split is AB/CDEFG, which has the largest CES of 19.6 and also retains AB as a single product:

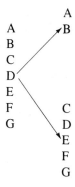

For separating CDEFG, splits C/D and E/F are performed last so that the remaining splits to be chosen are CD/EFG and CDEF/G. Split CD/EFG is done first since it has a larger CES of 28.7:

|      | CD/EFG | CDEF/G |
|------|--------|--------|
| f    | 28/39  | 8/59   |
| ΔT   | 40     | 41     |
| CES  | 28.7   | 5.6    |

The resulting sequence, which performs splits C/D and E/F last, is:

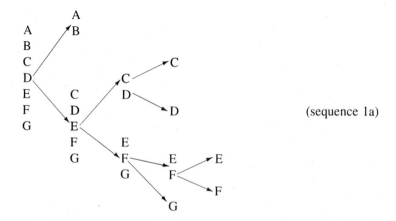

(sequence 1a)

This is exactly the same as the one presently being practiced in the industry (Rudd et al., 1973, p. 185; King, 1980, p. 718).

The second sequence can be obtained by making the split ABCD/EFG first (which has the second largest CES of 18.1, splitting ABCDEFG into two products) and performing the difficult splits C/D and E/F last, as in the sequence described above. This second sequence is:

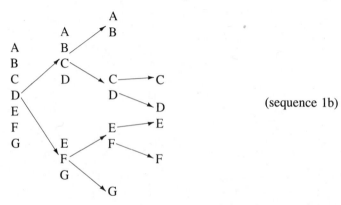

(sequence 1b)

This sequence is used by the industry in the thermal cracking of naphthas (King, 1980, p. 718).

## b. Example 2: n-Butylene Purification by Ordinary and Extractive Distillation

Consider the multicomponent separations in the industrial purification of $n$-Butylene (Hendry and Hughes, 1972). The feed mixture is:

| Species | Mol % | Relative volatility* | | | |
|---------|-------|-------|-------|-------|-------|
| | | $(\alpha)_I$ | $(a)_{II}$ | $(CES)_I$ | $(CES)_{II}$ |
| A: Propane | 1.47 | 2.45 | | 2.163 | |
| B: 1-Butene | 14.75 | 1.18 | 1.17 | 3.485 | 3.29† |
| C: $n$-Butane | 50.29 | 1.03 | 1.70 | 3.485 | 35.25 |
| D: $trans$-Butene-2 | 15.62 | | | | |
| E: $cis$-Butene-2 | 11.96 | 2.50 | | 9.406 | |
| F: $n$-Pentane | 5.90 | | | | |

*$(\alpha)_I$ = Adjacent relative volatility at 65.6°C and 1.03 MPa for separation method I, ordinary distillation; $(\alpha)_{II}$ = adjacent relative volatility at 65.6°C and 1.03 MPa for separation method II, extractive distillation.

†The CES value for the split AB/CDEF by method II (extractive distillation) is found from
$$CES = (D/B \text{ or } B/D) \times (\alpha_{II} - 1) \times 100$$
$$= [(1.47 + 14.75)/(50.29 + 15.62 + 11.96 + 5.90)] \times (1.17 - 1)$$
$$= 3.29$$

The rank lists (RL) of decreasing adjacent relative volatility corresponding to separation methods I and II are given by:

RL(I):ABCDEF        RL(II):ACBDEF

The desired products of the separation are A, C, BDE, and F. The separation sequencing by the new heuristic method can be done as follows.

1. **Heuristic M1**: use extractive distillation for split C/DE and ordinary distillation for all other splits.
2. **Heuristic M2**: use low temperature and ambient-to-moderate pressure.
3. **Heuristic D1**: avoid splitting DE as both D and E are in the same product, and blend together B and DE to obtain a multicomponent product BDE.

4. Heuristic S1: not applicable.

5. Heuristic S2: since split C/DE is difficult and requires extractive distillation, it should be performed last in the absence of A, B, and F.

6. Heuristic C1: although C is a large fraction of the feed, it should not be separated first because of the preceding heuristic S2. Further, it is preferable to carry out the extractive distillation for splitting C/DE at the end of the sequence. This will avoid having the mass separating agent as a possible contaminant in the intermediate separations of the sequence.

7. Heuristic C2: for separating ABCDEF, split ABC/DEF is performed last so that the remaining splits to be chosen are A/BCDEF, AB/CDEF, and ABCDE/F. The last split is chosen since it has the largest $(CES)_I$ of 9.406:

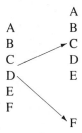

To separate ABCDE, the possible splits are A/BCDE and AB/CDE. Since

|  | A/BCDE | AB/CDE |
|---|---|---|
| $f$ | 1.47/92.63 | 16.22/77.88 |
| $(\alpha - 1) \times 100$ | 145 | 18 |
| CES | 2.301 | 3.749 |

split AB/CDE is preferred over A/BCDE. The resulting sequence, which splits A/B and C/DE last, is as follows:

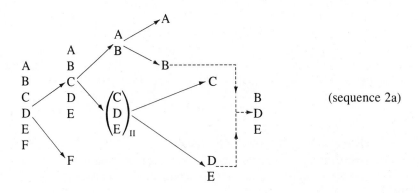

(sequence 2a)

A second sequence is obtained by splitting A/BCDE first, instead of AB/CDE:

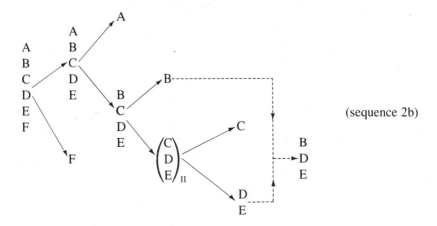

(sequence 2b)

As there are no other sequences with an initial split ABCDE/F that satisfy the constraints imposed by heuristics M1–C1, the third sequence is to be found by examining the alternative initial splits for separating ABCDEF. Thus, if the second best initial split for separating ABCDEF by ordinary distillation, AB/CDEF, with the second largest $(CES)_I$ of 3.485, is done first, the resulting sequence, which splits A/B and C/DE last, is:

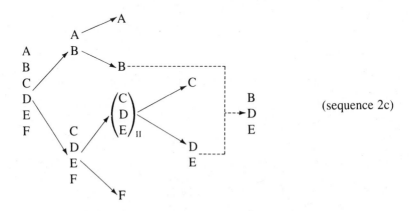

(sequence 2c)

Alternatively, if the third best initial split for separating ABCDEF by ordinary distillation A/BCDEF, with the third largest $(CES)_I$ of 2.163, is performed first, two other sequences, which split C/DE last, can be found as follows:

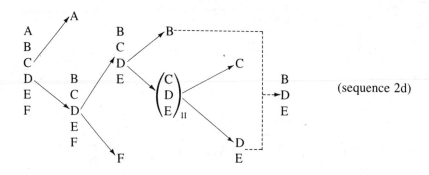

(sequence 2d)

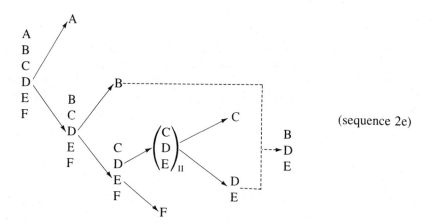

(sequence 2e)

Note that sequence 2d is better than sequence 2e, according to the CES values, in separating BCDEF:

|                        | BCDE/F       | B/CDEF       |
| ---------------------- | ------------ | ------------ |
| $f$                    | 5.90/92.62   | 14.75/83.77  |
| $(\alpha - 1) \times 100$ | 84.06     | 24.85        |
| CES                    | 5.355        | 4.375        |

Table 6.2 compares the sequences synthesized by different methods for the present problem. The additional sequence (2f), which involves replacing the split B/C

in sequence 2e by extractive distillation, is as follows:

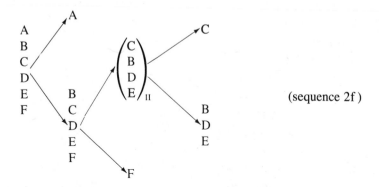

(sequence 2f )

Table 6.2 shows that sequence 2a is cheaper than the initial sequences obtained by the ordered heuristic methods of Seader and Westerberg and Nath and Motard, namely, sequences 2d and 2f, respectively. Based on the costs reported by Hendry (1972), sequence 2a is only 0.8% higher in cost than the best sequence (2c) synthesized by the heuristic–evolutionary methods of Seader and Westerberg and Nath and Motard. Also, both the new heuristic method and the ordered heuristic method of Seader and Westerberg have generated much cheaper initial sequences (2a and 2d) than that (2f) by the ordered heuristic method of Nath and Motard.

## 3. Remarks on Heuristic Methods

It is important to note that although heuristic S1 (remove corrosive and hazardous components first) has not been explicitly applied in the preceding examples, this heuristic is useful important in deciding the essential first splits in many separation-sequencing problems in which the feed mixtures contain corrosive and hazardous components. An example of applying heuristic S1 involves the multicomponent separation sequencing in the manufacture of detergents from petroleum (Rudd et al., 1973, pp. 281–295). In this problem, it is essential to remove the corrosive hydrogen chloride from the mixture of chlorinated products and unreacted species (decane, monochlorodecane, dichlorodecane, chlorine, and hydrogen chloride). For this reason, heuristic S1 has been included in the ordered heuristic method of Nadgir and Liu (1983).

This method has also been tested with many other multicomponent separation-sequencing problems reported in the literature. The specific synthesis problems solved by the method have included essentially all of the problems described by Thompson and King (1972), Westerberg and Stephanopoulos (1975), Stephanopoulos and Westerberg (1976), Rodrigo and Seader (1975), Gomez and Seader (1976), Seader and Westerberg (1977), Nath (1977), and Nath and Motard (1981). Also, many test problems presented in the textbooks of Rudd et al. (1973, Chapter 5) and Henley and Seader (1981, Chapter 14) have been solved. Based on reported

**TABLE 6.2  A Comparison of Reported Sequences for Example 2**

| Sequence | Ordered Heuristic Method Nadgir and Liu (1983) | Algorithmic Method Hendry and Hughes (1972)[a] | Heuristic–Evolutionary Methods Seader and Westerberg (1977) | Nath and Motard (1981) |
|---|---|---|---|---|
| 2a | Initial sequence | $867,400/yr (0.8%) | | |
| 2b | Second sequence | $878,200/yr (1.8%) | | |
| 2c | Third sequence | $860,400/yr (best) | Final sequence $860,400/yr (best) | Final sequence $658,737/yr (best) |
| 2d | Fourth sequence | $878,000/yr (2.0%) | Initial sequence $878,000/yr (2.0%) | |
| 2e | Fifth sequence | $872,400/yr (1.5%) | Second sequence $872,400/yr (1.5%) | Second sequence $669,844/yr (1.7%) |
| 2f | Final sequence | $1,095,600/yr (27.3%) | | Initial sequence $1,171,322/yr (77.8%) |

[a] As reported in Hendry (1972) and quoted in Henley and Seader (1981, p. 547).

separation sequences and costs for the problems solved, the initial sequences synthesized by the method are either identical to industrial sequences or cheaper than those obtained by Seader and Westerberg (1977) and Nath and Motard (1981). These initial sequences are also either identical to or at most a few percent higher in cost than optimum sequences obtained by other algorithmic (e.g., Hendry and Hughes, 1972), heuristic–algorithmic (e.g., Thompson and King, 1972), and heuristic–evolutionary (Seader and Westerberg, 1977; Nath and Motard, 1981) methods. Thus the ordered heuristic method of Nadgir and Liu as described above offers the advantages of simplicity and effectiveness, coupled with good performance.

## C. Evolutionary Synthesis of Multicomponent Separation Sequences

### 1. An Introduction to Evolutionary Synthesis

Consider the six-component separation-sequencing problem in the industrial purification of *n*-butylene by ordinary (method I) and extraction distillation (method II). This problem was described by Hendry and Hughes (1972), and was solved previously by heuristic methods as Example 2 in Section 6.2.B.2. Two of the sequences presented in that example are as follows:

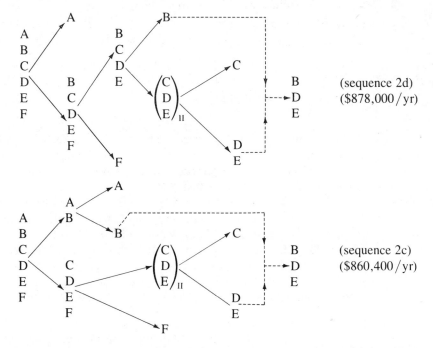

The problem of evolutionary synthesis may be posed as follows: "Starting with an initial flowsheet (e.g., sequence 2d), how does one systematically modify it to obtain an improved flowsheet (e.g., sequence 2c)?"

In the context of separation sequencing, we can evolutionarily synthesize several good sequences by the following steps:

1. Generating an initial sequence by using heuristics (see Section 6.2.B) or other methods.
2. Proposing a set of rules to modify the initial sequence.
3. Developing a strategy to apply the evolutionary rules.
4. Suggesting a means (such as using relative or actual separation cost) to compare initial and modified sequences.

## 2. Combined Heuristic and Evolutionary Methods

### a.   METHOD OF SEADER AND WESTERBERG FOR LOCAL EVOLUTIONS

In the method of Seader and Westerberg (1977), the initial separation sequence is generated by the following ordered heuristics (see also Section 6.2.B.1 and Appendix 6.1):

*1. Heuristic M1 (favor ordinary distillation).
*2. Heuristic M2 (avoid vacuum and refrigeration).
 3. Heuristic S4 (perform easy separations first).
*4. Heuristic C1 (remove most plentiful component first).
 5. Heuristic D3 (favor direct sequence).
 6. Heuristic M3 (remove mass separating agent first).
*7. Heuristic D1 (favor smallest product set).

Note that those heuristics indicated by asterisks have also been included in the ordered heuristic method of Nadgir and Liu (1983) described in Section 6.2.B.1, but in a different ranked order. Heuristics S4 and D3, however, suffer from the serious drawback of leading to a large total separation load and an unbalanced separator (i.e., a separator deviating from a 50/50 split). This drawback is discussed further in Appendix 6.1. Therefore, we recommend that unless an initial sequence synthesized by another method is given, the ordered heuristic method of Nadgir and Liu (1983), described in Section 6.2.B.1, should be used to generate initial sequences for possible improvement by an evolutionary method.

   Seader and Westerberg (1977) suggest the use of the following two evolutionary rules that were originally proposed by Stephanopoulos and Westerberg (1976):

1. Evolutionary rule 1: "Charge the separation method for a given separation task."
2. Evolutionary rule 2: "Interchange the relative positions of two adjacent separators by, for example, moving a separation task to the left (i.e., upstream) in a given separation sequence."

As an illustration, we apply evolutionary rule 2 to interchange the order of the first two splits (A/B and E/F) in sequence 2d of the preceding example, while keeping the same remaining splits:

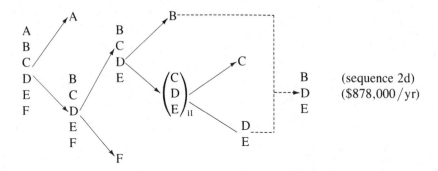

(sequence 2d)
($878,000/yr)

This evolution results in an improved sequence 2a:

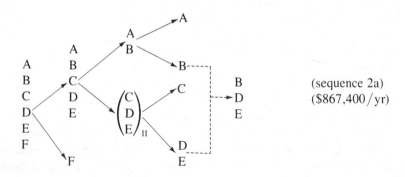

(sequence 2a)
($867,400/yr)

Sequence 2a can be further improved by interchanging the order of the first two splits (E/F and B/C). This leads to:

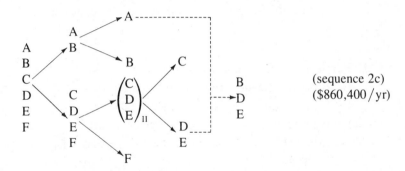

(sequence 2c)
($860,400/yr)

By contrast, interchanging the order of the second and third splits (E/F and B/C) in sequence 2d actually leads to a more expensive sequence, 2e:

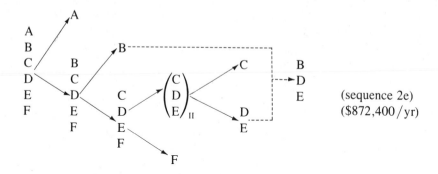

(sequence 2e)
($872,400/yr)

As illustrated by the preceding example, there are two disadvantages of the evolutionary method. First, it is not always possible to uniquely construct a modified sequence when moving a certain separation task to the upstream in a given sequence. This is particularly true when two or more separation methods are used. As a result, we must examine many alternative sequences by interchanging the locations of splits in an initial sequence. Second, when comparing the modified sequences with an initial sequence, it is often necessary to carry out design calculations and equipment costing for different separation tasks involved in a given sequence.

It should also be noted that in the preceding example, evolutionary rules are used to modify two neighboring splits in a given sequence. The evolutions do not alter the downstream separation tasks or splits, only the upstream ones. Such evolutions are commonly called *local evolutions*. In the discussion that follows, *global evolutions* are introduced. These involve replacing all downstream separation tasks when applying an evolutionary rule to modify an initial sequence.

## b.   METHOD OF NATH AND MOTARD FOR GLOBAL EVOLUTIONS

Nath and Motard suggest the use of the following heuristics and design approximations to synthesize an initial separation sequence (see also Section 6.2.B.1 and Appendix 6.1):

*1. Heuristic D1 (favor smallest product set).
*2. Heuristic M1′ (favor ordinary distillation).
 3. Heuristic S4 (perform easy separations first).
*4. Heuristic M3 (remove mass separating agent first).
*5. Heuristic M1″ (a separation method with a separating factor $\alpha_{LK,HK} < \alpha_{\min}$ is not acceptable).

**\*6.** Heuristic M2' (avoid vacuum).

**7.** Set split fractions of key components to prespecified values.

**8.** Set the operating reflux ratio equal to 1.3 times the minimum reflux ratio for each column.

Note that those heuristics indicated by asterisks are also used in the ordered heuristic method of Nadgir and Liu (1983), described in Section 6.2.B.1, but in a different ranked order. Also, a key drawback of the method of Nath and Motard is their use of heuristic S4 (perform easy separations first). It is recommended that the method of Nadgir and Liu be used to synthesize initial sequences for possible evolutionary improvements.

A number of evolutionary rules have been proposed by Nath and Motard:

**1.** Evolutionary rule 1: "Challenge heuristic D1 that favors the smallest product set."

**2.** Evolutionary rule 2: "Examine the neighboring separation sequences if (a) their coefficient of difficulty of separation (CDS) values [defined by Equation (6.22), Appendix 6.1] are within 10% of each other, and (b) refrigeration is required to condense the reflux."

**3.** Evolutionary rule 3: "Challenge heuristic M1'. Examine the possibility of using separations with a mass separating agent (MSA) such as extractive distillation, instead of separations using only an energy separating agent such as ordinary distillation."

**4.** Evolutionary rule 4: "Challenge heuristic M3. Examine the neighboring sequences to see if the removal of a MSA could be delayed."

**5.** Evolutionary rule 5: "Challenge heuristic S4 favoring easy separations first, if (a) $R_{min}$ (minimum reflux ratio) of the immediate, subsequent separator is much greater than that of the separator under consideration, or (b) the cost of the immediate, subsequent separator is much greater than that of the separator under consideration."

When modifying an initial sequence by the evolutionary rules given above, Nath and Motard suggest that all downstream separation tasks or splits are to be eliminated and a new sequence be generated to replace the initial sequence. This type of evolutionary strategy may be called *global evolutions*. The latter generally require more design calculations and equipment costing compared with the local evolutions described in Section 6.2.C.2.a. Among the five evolutionary rules proposed above, rules 1, 3, and 4 could be particularly beneficial in improving the initial sequences of certain separation problems. These rules are illustrated by comparison of sequences A1–A5 in Appendix 6.1.

Table 6.3 lists the advantages and disadvantages of both heuristic and evolutionary methods and compares them with optimization methods.

**TABLE 6.3  A Comparison of Three Common Methods for Synthesis of Multicomponent Separation Sequences**

| Advantages | Disadvantages |
|---|---|
| *Heuristic Methods* | |
| Straightforward to apply by hand. | Heuristics often contradict or overlap one another.[a] |
| No mathematical background or computational skill needed. | Strategy-dependent (which heuristics are used first?).[a] |
| Easy to generate an initial sequence for other methods. | |
| *Evolutionary Methods* | |
| May reveal new sequences through evolutions. | Need an initial sequence generated by other methods. |
| | Strategy-dependent (which evolutionary rules are used first?). |
| | Need quantitative performance criteria (may involve design calculations and equipment costing). |
| | Limited by problem size (many sequences need to be compared). |
| *Optimization Methods* | |
| Can be computerized. | Ignore corrosive, hazardous, and cryogenic properties of feed. |
| Easy to find suboptimal sequences. | Cost-equation dependent. |
| | Limited by problem size. |

[a] These disadvantages are valid in most heuristic methods prior to that by Nadgir and Liu (1983).

## 6.3.  HEAT-EXCHANGER NETWORKS (HENs)

### A.  Problem Statement: Multiobjective Synthesis of HENs

An important process–design problem is the synthesis of networks of exchangers, heaters, and/or coolers, to transfer the excess energy from a set of hot streams to another set of streams that require heating (called cold streams). Figure 6.1 illustrates a typical crude preheat-exchanger network around the topping tower in a petroleum refinery (Huang and Elshout, 1976). In the figure, the exchangers, heaters, and coolers are designated by E, H, and C, respectively; and four hot distillate-product streams, denoted by $S_{h1}$–$S_{h4}$, are used to heat up the cold crude oil $S_{c1}$ before it enters the distillation column. A natural-gas stream and a cooling-water stream are available as a heating-utility stream (denoted by $S_{hu}$) and a cooling-utility stream (denoted by $S_{cu}$), respectively.

The conventional definition of the problem of synthesizing a HEN was first presented by Masso and Rudd (1969), and this has already been described in Sec-

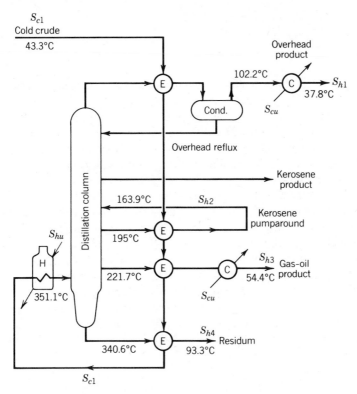

**FIGURE 6.1**  A typical heat-exchanger network for crude preheat recovery (Huang and Elshout, 1976).

tion 5.3.B. Pehler and Liu (1981) introduced the problem of the multiobjective synthesis of HENs, which is briefly summarized below.

There are $N_h$ hot streams $S_{hi}$ ($i = 1,2, \ldots ,N_h$) to be cooled and $N_c$ cold process streams $S_{cj}$ ($j = 1,2, \ldots ,N_c$) to be heated. Associated with each stream are its input temperature $T_i$ (in °C or °F), output temperature $T_i^*$ and heat-capacity flow rate $W_i$ (i.e., the average heat capacity multiplied by the mass flow rate in kW / °C or Btu / hr-°F). There are also available $N_{hu}$ heating-utility streams and $N_{cu}$ cooling-utility streams. The synthesis problem is to create several steady-state optimum and suboptimum networks of units (exchangers, heaters, and coolers) so that the specified stream outlet temperatures are reached. These optimum and suboptimum networks should achieve or nearly achieve at least the following multiple-objective criteria:

- Approaching a practical minimum loss in thermodynamic available energy during heat exchange among hot and cold streams (i.e., achieving the most efficient or nearly reversible exchange of heat among hot and cold streams).
- Minimizing the number of units (i.e., exchangers, heaters, and coolers).

- Minimizing the investment cost of units.
- Minimizing the operating cost of utilities (e.g., heating steam and cooling water).

These criteria were suggested or utilized in part by Hohmann and Lockhart (1976), Nishida et al. (1977), Linnhoff and Flower (1978a,b), and Umeda et al. (1978).

An important feature of the multiobjective synthesis problem is that some of the objective criteria may conflict with others. For example, minimizing the loss of available energy during the heat-exchange process requires maximizing the heat-transfer area, which tends to maximize the network-investment cost (Umeda et al., 1978). Also, minimizing the number of units does not necessarily lead to minimizing the investment cost of units. This follows because the investment cost of units depends not only on the number of units, but also on the total heat-transfer area of units and on how this total area is distributed among the different units. (Nishida et al., 1977; Linnhoff and Flower, 1978b). Consequently, to synthesize several optimum and suboptimum networks, a multistep evolutionary strategy is recommended in this chapter, with each step emphasizing one of the criteria.

In general, the investment costs of the $i$th exchanger, heater, and cooler, denoted by $C_{Ei}$, $C_{Hi}$, and $C_{Ci}$, respectively, are correlated by the empirical expression $C_{Ei} = aA_{Ei}^b$, $C_{Hi} = aA_{Hi}^b$, and $C_{Ci} = aA_{Ci}^b$, where $a$ and $b$ are both constants, with $b$ varying typically from 0.6 to 0.7. The total network-investment and utility-operating cost to be minimized is expressed as:

$$J = \delta \left( \sum_i aA_{Ei}^b + \sum_i aA_{Hi}^b + \sum_i aA_{Ci}^b \right) + \sum_k \sum_l u_k S_{ukl} \qquad (6.5)$$

In the equation, $\delta$ is the annual rate of return on investment cost, ranging typically from 0.1 to 0.15.

The common simplifying assumptions that have been included in the early development of techniques for synthesizing HENs are as follows:

- The use of single-pass countercurrent shell-and-tube exchangers.
- No phase change of process streams.
- Equal values of overall heat-transfer coefficients for exchanges between two process streams, and between process and utility streams.
- Temperature-independent heat-capacity flow rates of process streams.
- Constant minimum-approach temperatures for exchanges between two process streams, and between process and utility streams.

Significant progress has been made over the past decade toward the multiobjective synthesis of HENs, and it is *not necessary* for a network-synthesis problem to incorporate the preceding simplifying assumptions. Many recent synthesis techniques can solve problems with multipass shell-and-tube exchangers, varying overall heat-transfer coefficients and phase changes of process streams (see Section 6.3.I.1).

Appendix 6.2 summarizes the specifications of process and utility streams and design data for the HEN synthesis problems illustrated in this chapter.

## B.  Characteristics of Minimum-Cost and Energy-Optimum HENs: What Can One Learn from Process Analysis?

### 1.  An Illustrative Example

A number of important characteristics of minimum-cost and energy-optimum networks can be learned from comparing several reported solutions, shown in Figure 6.2, to a four-stream synthesis problem, the 4SP2 problem (Ponton and Donaldson, 1974) specified in Appendix 6.2.

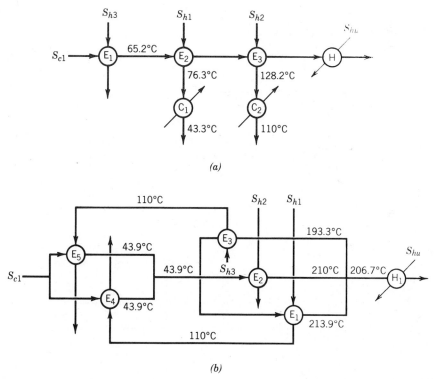

*(a)*

*(b)*

**FIGURE 6.2**  An illustration of different network solutions to the 4SP2 problem specified in Appendix 6.2. (a) An acyclic network: investment = $8,633/yr and utility = $63,787/yr. (b) A cyclic network: investment = $7,617/yr and utility = $12,736/yr. (c) An unsplitting network with an optimum location of heating utility: investment = $9,298/yr and utility = $12,736/yr. (d) A splitting network with only four units: investment = $6,835/yr and utility = $12,736/yr. (e) A cyclic and unsplitting network with a steam heater serving as a preheater: investment = $10,980/yr and utility = $12,736/yr.

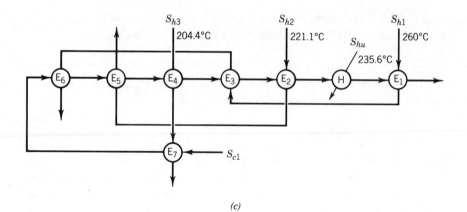

(c)

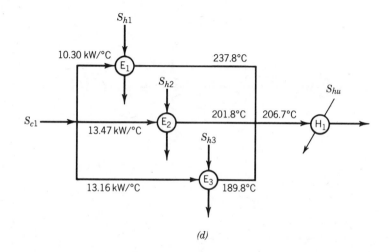

(d)

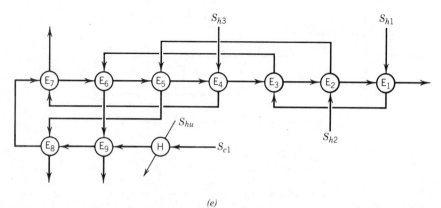

(e)

**FIGURE 6.2** (Continued)

## a.   Cyclic and Acyclic Networks

A cyclic network, in which one stream is exchanged more than once with another stream, generally has a lower utility-operating cost compared with an acyclic network. For example, the utility-operating cost of the acyclic network shown in Figure 6.2*a* is five times more expensive than that of the cyclic network illustrated in Figure 6.2*b*. This follows because in the acyclic network, significant portions of the heat contents of hot and cold process streams are left to be exchanged with heating and/or cooling utilities. By contrast, in the cyclic network, essentially the whole heat contents of hot streams can be transferred to that of cold streams, thus resulting in less use of utilities. Therefore, it is important to consider synthesizing a cyclic network and to exchange as much heat as technically possible among hot and cold process streams before using utilities for heating and/or cooling.

The operating cost of utilities for the cyclic network shown in Figure 6.2*b* corresponds to the minimum cost of utilities for the 4SP2 problem. In Section 6.3.D, we show that for a given HEN synthesis problem, the minimum heating- and cooling-utility requirements can be determined *prior to* the actual synthesis of networks.

## b.   Splitting and Unsplitting Networks

For certain HEN synthesis problems in which the heat-capacity flow rate of any cold (or hot) process stream is markedly greater than those of hot (or cold) process streams, the use of stream splitting generally leads to networks of fewer units and lower investment costs. For instance, in the 4SP2 problem, the heat-capacity flow rate of $S_{c1}$ (36.91 kW/°C) is greater than those of $S_{h1}$, $S_{h2}$, and $S_{h3}$ (10.55–26.37 kW/°C). A comparison of Figures 6.2*b* and *c* shows that while both networks have an identical and minimum utility-operating cost, the splitting network illustrated in Figure 6.2*b* requires two less exchangers (that is, five exchangers) and a lower investment cost, compared with the unsplitting network depicted in Figure 6.2*c*.

The number of units required for a splitting network for the 4SP2 problem, as depicted in Figure 6.2*b*, can be reduced to four (i.e., three exchangers and one heater). Figure 6.2*d* gives an example of a splitting network with only four units. It is shown in Section 6.3.E that for a given HEN synthesis problem, we can determine the most probable minimum (i.e., quasi-minimum) number of units (exchangers, heaters, and/or coolers) required *prior to* the actual synthesis of networks.

## c.   Optimal Locations of Utilities

Figure 6.2*e* shows an energy-optimum and nearly minimum-cost network for the 4SP2 problem reported by Ponton and Donaldson (1974). A comparison of this network with that given in Figure 6.2*c* illustrates how heating and/or cooling utilities are best located in a network. It is shown in Section 6.3.F that in a thermodynamically efficient network, hot process and utility streams are to be exchanged consecutively with cold process and utility streams in a decreasing order of their stream temperatures. For instance, in the network shown in Figure 6.2c, the input temperatures of hot process and utility streams arranged in a decreasing order of magnitude are $T_{h1} = 260°C$, $T_{hu} = 235.6°C$, $T_{h2} = 221.1°C$, and $T_{h3} = 204.4°C$. Thus the preceding thermodynamic principle suggests that the heating-utility (steam) stream should be placed between two exchangers $E_1$ and $E_2$ in which $S_{h1}$ and $S_{h2}$ are matched with $S_{c1}$, respectively. By comparison, the location of the steam heater in the network given in Figure 6.2e does not follow the thermodynamic principle. In fact, this steam heater is placed on the lowest-temperature portion of $S_{c1}$, resulting in a network of higher investment cost.

It is worthwhile to note another observation that is important for specifying the locations of utilities. For the 4SP2 problem, only the heating utility is needed. For other HEN synthesis problems that require *both* heating and cooling utilities, it was indicated in Section 5.3.B and illustrated in Figure 5.6 that a special stream temperature exists, called *the pinch point for heat exchange* (Umeda et al., 1978; Linnhoff and Turner, 1981). In particular, a hot-stream pinch represents the temperature of hot process streams above which no cooling utility should be used; and a cold-stream pinch corresponds to the temperature of cold process streams below which no heating utility should be used. This is further discussed in Section 6.3.D.

## 2. Some Observations from Process Analysis

Based on the preceding example, we suggest several basic requirements in a simple and practical approach to synthesizing minimum-cost and energy-optimum networks.

1. The approach is able to minimize the investment cost of the network by (a) minimizing the number of units (exchangers, heaters, and/or coolers), (b) allowing for the possible use of stream splitting to reduce the number of units, and (c) determining the optimal locations of heating and/or cooling utilities.

2. The approach is able to minimize the operating cost of utilities by (a) exchanging as much heat as technically possible among hot and cold process streams prior to using heating and/or cooling utilities, (b) determining the minimum amounts of heating and/or cooling utilities required, and (c) allowing for the possible generation of cyclic networks.

3. The approach is able to synthesize several energy-optimum and nearly minimum-cost networks for further design considerations (such as the sensitivity of network performance to changes in input temperatures and/or flow rates of process streams) and to overcome the combinatorial difficulties associated with the synthesis of large-scale networks.

## C. Problem Representations

The first step in the multiobjective synthesis of HENs is to represent the given synthesis problem by an appropriate diagram. Three common diagrams for graphically representing the HEN synthesis problem can be illustrated with reference to a four-stream problem, the 4TC2 problem (Linnhoff and Flower, 1978a) specified in Appendix 6.2.

## 1. Heat-Content Diagram

Figure 6.3*a* shows the heat-content diagram for representing the 4TC2 problem. This diagram is essentially a plot of stream temperature versus heat-capacity flow rate (in kW/°C or Btu/hr-°F) (Nishida et al., 1971, 1977; Pehler and Liu, 1981). On the diagram, each stream is represented by a block. The area of a given block (in kW or Btu/hr) corresponds to the amount of heat to be removed from or added to the stream in order for it to reach its desired output temperature.

For convenience, both process and utility streams should be drawn on the diagram such that heating utilities and hot streams are located in a decreasing order of their input temperatures above the horizontal axis; while cooling utilities and cold streams are located in a decreasing order of their output temperatures below the horizontal axis. In most cases, the heating utility is a saturated steam at a constant temperature; and both the output temperature and heat-capacity flow rate of cooling utility are unknown before its exchange with hot streams are specified. Thus, in representing a given synthesis problem, a heating or cooling utility is initially represented by a point on the diagram, with the ordinate value indicating its known input temperature.

Figure 6.3*b* illustrates a heat-content diagram corresponding to a network solution to the 4TC2 problem depicted in Figure 6.3*c*. The procedure to obtain this solution is discussed in Section 6.3.F. Three important features can be seen from Figure 6.3*b*. First, the cooling of the low-temperature portion of a hot stream, such as $S_{h2}$ by a water cooler, as well as the heating of the high-temperature portion of a cold stream, such as $S_{c1}$ by a steam heater, are both represented by hatched utility blocks. Second, the multiple heat exchange of $S_{h1}$ with $S_{c1}$ and then with $S_{c2}$ is represented by horizontally dividing hot block $S_{h1}$. Third, the splitting of $S_{h2}$ into two portions to exchange separately with $S_{c1}$ and $S_{c2}$ is represented by vertically dividing hot block $S_{h2}$.

## 2. Temperature–Enthalpy Diagram (Heat-Availability Diagram)

Figure 6.3*d* shows the temperature–enthalpy ($T$–$Q$ or $T$–$H$) diagram for representing the 4TC2 problem. This diagram is a plot of stream temperature versus stream enthalpy (Whistler, 1948). The enthalpy value (in kW or Btu/hr) in the abscissa represents the product of stream heat-capacity flow rate (kW/°C or Btu/hr-°F) and hot-stream temperature drop or cold-stream temperature rise. Since the enthalpy value is defined with respect to a thermodynamic reference state, its value is relative. This implies that we are free to horizontally shift a temperature–enthalpy curve on a $T$–$Q$ diagram. Also, a temperature–enthalpy curve representing

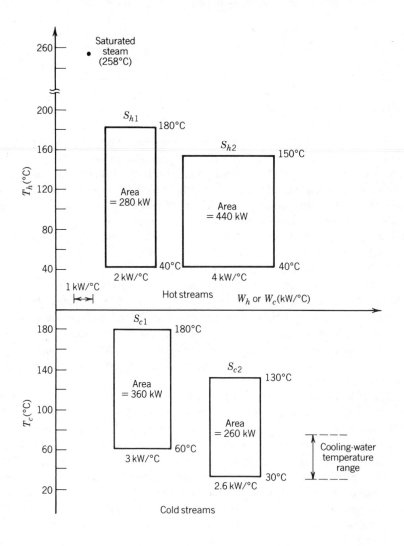

*(a)*

**FIGURE 6.3** Representation of the 4TC2 problem specified in Appendix 6.2 and examples of network solutions. (*a*) A heat-content diagram representing the problem. (*b*) A heat-content diagram corresponding to the network solution shown in Figure 6.3*c*. (*c*) A network solution represented by the heat-content diagram in Figure 6.3*b*. (*d*) A temperature–enthalpy diagram representing the problem. (*e*) A temperature–enthalpy diagram corresponding to the network solution shown in Figure 6.3*c*. Between 150 and 80°C, $S_{h2}$ is split into two streams ($S_{h2,a}$ and $S_{h2,b}$) which are later merged together as $S_{h2,c}$ between 80 and 40°C. Between 30 and 150°C, $S_{c2}$ is split into two streams ($S_{c2,a}$ and $S_{c2,b}$). (*f*) An enlarged diagram of a portion of Figure 6.3*e*, illustrating the representation of stream splitting. Between 150 and 80°C, the original unsplit portion of $S_{b2}$ is indicated by ($S_{h2,a}$ + $S_{h2,b}$). After splitting, it is represented by two streams $S_{h2,a}$ and $S_{h2,b}$, which are later merged together as $S_{h2,c}$ between 80 and 40°C. (*g*) A temperature-interval diagram representing the problem. (*h*) A temperature-interval (or grid) diagram corresponding to the network shown in Figure 6.3*c*.

**176**

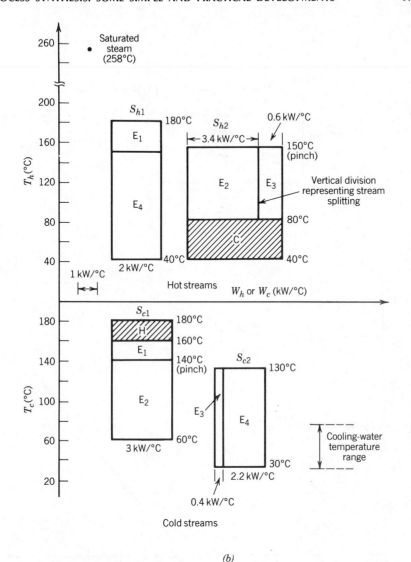

*(b)*

**FIGURE 6.3** *(Continued)*

a stream can be broken into parts to indicate different combinations of heat exchanges with process and/or utility streams through exchangers, heaters, and/or coolers. For example, Figure 6.3*e* illustrates the *T–Q* diagram corresponding to the network solution to the 4TC2 problem shown in Figure 6.3*c*. To indicate how the stream splitting is represented on a *T–Q* diagram, we illustrate portions of Figure 6.3*e* in an enlarged diagram, Figure 6.3*f*. By comparing Figures 6.3*e* and *f* with Figure 6.3*b*, we see that it is easier to visualize the stream splitting on a heat-content diagram than on a *T–Q* diagram.

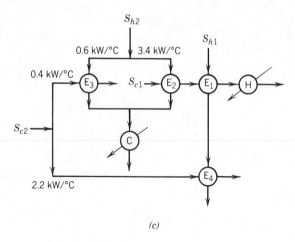

(c)

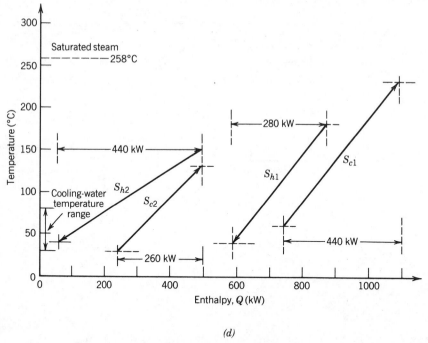

(d)

**FIGURE 6.3** (*Continued*)

### 3. Temperature–Interval Diagram (Grid Diagram)

Figure 6.3g shows the use of temperature–interval (*TI*) diagram (or grid diagram) to represent the 4TC2 problem. This method of problem representation was first proposed by Hohmann (1971), who called it the temperature-range diagram. It

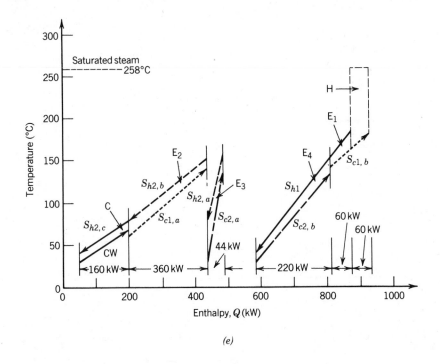

*(e)*

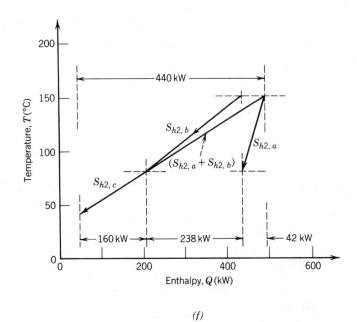

*(f)*

**FIGURE 6.3** *(Continued)*

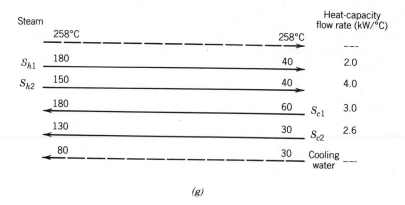

(g)

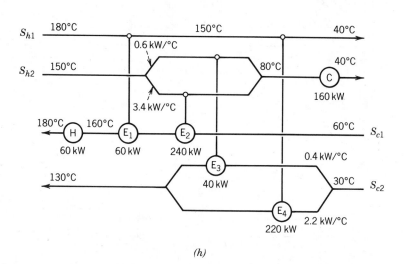

(h)

**FIGURE 6.3** (Continued)

was later systematized by Linnhoff and Flower (1978a). Figure 6.3h depicts the
TI diagram representing the network solution shown in Figure 6.3c. In this representation, a stream match, or the heat exchange, between two process streams,
is designated by placing a pair of circles on each of the streams and connecting
them with a straight line. Numerical values below the circles refer to the heat loads
of exchangers, heaters, or coolers. Also, for simplicity, both heating steam and
cooling water originally included in Figure 6.3g have been eliminated in Figure
6.3h.

## 4. Comparison of Three Methods

The key advantages and disadvantages of the three common methods of problem representation are summarized as follows.

## a. HEAT-CONTENT DIAGRAM

### (1) Advantages

- Stream splitting and cyclic networks are easily represented.
- Temperature-dependent heat-capacity flow rates are easily represented.
- Stream-dependent minimum-approach temperatures can be explicitly designated on the diagram.
- Phase changes of process stream are easily represented, with the latent heat of condensation or vaporization being treated as a pseudo-heating or pseudo-cooling utility, respectively (Pehler and Liu, 1981; see Section 6.3.I).

### (2) Disadvantages

- Indirect representation of the physical network structure.
- The maximum amount of heat exchange among process streams as well as the minimum heating- and cooling-utility requirements may be graphically determined on the diagram, but sometimes with a great deal of effort.

## b. TEMPERATURE–ENTHALPY DIAGRAM

### (1) Advantages

- Process streams displayed on the diagram need not have constant heat capacities. Temperature–enthalpy curves representing the temperature-dependence of stream heat-capacities can be drawn and manipulated as easily as straight lines.
- Temperature–enthalpy $(T-Q)$ curves for all hot and cold streams can be merged to form composite $T-Q$ curves that allow for determination of the maximum heat-recovery levels and minimum utility-consumption figures (Umeda et al., 1978; Colbert, 1981; see Section 6.3.I).

### (2) Disadvantages

- Indirect representation of the physical network structure.
- Stream splitting is not easily represented.
- Process streams with small heat-capacity flow rates are represented by the $T-Q$ curves with steep slopes. Consequently, displaying branches of a $T-Q$ curve sometimes requires replotting or enlarging the horizontal scale of each branch.

c.  Temperature–Interval Diagram

*(1)  Advantages*

- Direct representation of the physical network structure.
- The use of the problem table related to the temperature-internal diagram facilitates the determination of the maximum heat-recovery levels and minimum utility-consumption figures (see Section 6.3.D).

*(2)  Disadvantages*

- Temperature-dependent heat-capacity flow rates are not easily represented.
- Phase changes of process streams are not easily represented.

In view of the different simplifying assumptions included in a given network synthesis problem, it is suggested that certain benefits can be gained through the combined use of the methods listed above. This will allow the designer to take advantage of the attractive features of each method. This point is discussed further below.

## D.  Determination of the Minimum Heating- and Cooling-Utility Requirements and the Pinch Point

The second step in the multiobjective synthesis of HENs is to determine the minimum heating- and cooling-utility requirements and the pinch point prior to the actual synthesis of the networks. The minimum utility requirements are determined based on (1) the heat-capacity flow rates and input/output temperatures of hot and cold process streams, and (2) the minimum-approach temperatures of exchanges between two process streams and between process and utility streams.

The simplest method to determine the minimum utility requirements is by using the problem table originally proposed by Hohmann (1971), who called it the feasibility table, and later systematized by Linnhoff and Flower (1978a). Table 6.4 presents a problem table for the 4TC2 problem illustrated in Figure 6.3 and specified in Appendix 6.2.

To set up a problem table, we first identify the subnetwork boundary temperatures by adding $\Delta T_{min}$ from given input/output temperatures of cold streams, and subtracting $\Delta T_{min}$ from given input/output temperatures of hot streams. For the 4TC2 problem, the input/output temperatures as illustrated in Figure 6.3 and specified in Appendix 6.2 are:

$$T_c \text{ (cold stream): 30, 60, 130, and 180°C}$$

$$T_h \text{ (hot streams): 40, 40, 150, and 180°C}$$

TABLE 6.4  A Problem Table for Determining the Minimum Utility Requirements for the 4TC2 Problem

| Temperature Interval (TI) or Subnetwork (SN) | Hot Streams | | $T(°C)$ | Cold Streams | | Subnetwork Minimum Heating Requirement (kW) | Accumulated Utility Requirement (kW) | | Network Minimum Utility Requirement (kw) | |
|---|---|---|---|---|---|---|---|---|---|---|
| | $S_{h1}$ | $S_{h2}$ | | $S_{c1}$ | $S_{c2}$ | | Heating[d] | Cooling[c] | Heating | Cooling |
| SN(1) | | | 180 (190) 180 (170) | | | $3.0 \times 10$ $= 30$ | 30 | −100 | 60[d] | −30 |
| SN(2) | | | 150 (140) | 3.0 | | $(3.0\text{-}2.0) \times 30$ $= 30$ | 60[d] | −130 | 30 | 0[g] |
| SN(3) | 2.0[a] | | 130 (140) | | | $(3.0\text{-}2.0\text{-}4.0) \times 10$ $= -30$ | 30 | −160[e] | 0[g] | −30 |
| SN(3) | | 4.0 | 60 (70) | | 2.6 | $(3.0 + 2.6\text{-}2.0\text{-}4.0) \times 70$ $= -28$ | 2 | −130 | 30 | −58 |
| SN(4) | | | 40[f] 30[f] | | | $(2.6\text{-}2.0\text{-}4.0) \times 30$ $= -102$ | −100 | −102 | 58 | −160[e] |

[a] Heat-capacity flow rate, kW.

[b] Accumulated heating requirement obtained by adding together the subnetwork minimum heating requirements for different subnetworks from the highest to lowest temperature.

[c] Accumulated cooling requirement obtained by adding together the subnetwork minimum cooling utility requirements for different subnetworks from the lowest to highest temperature.

[d] Minimum heating-utility requirement = 60 kW (by the time subnetwork 2 is reached from top down). Why not introduce the heating utility to the highest temperature level, subnetwork 1 (steam temperature = 256 > 180°C)?

[e] Minimum cooling utility requirement = −160 kW (by the time subnetwork 3 is reached from bottom up). Why not introduce the cooling utility to the lowest-temperature level, subnetwork 5 (cooling-water = 30–80°C)?

[f] Here $T_h$ (hot stream) = 40°C, $T_c$ (cold stream) = 30°C, and $T_h = T_c + \Delta T_{min} = T_c + 10°C$.

[g] The hot-stream pinch corresponds to $T_h$ = 150°C, and the cold-stream pinch is $T_c$ = 140°C.

With $\Delta T_{min} = 10°C$, the resulting boundary temperatures are:

$$T_b \text{ (cold streams): 40, 70, 140, and 190°C}$$

$$T_b \text{ (hot streams): 30, 30, 140, and 170°C}$$

Arranging these temperatures in a decreasing order gives the following set of boundary temperatures which are listed in Table 6.4 to define five subnetworks (or temperature intervals):

$$T_b \text{ (all streams): 190, 170, 140, 70, 40, and 30°C}$$

As found in Table 6.4, the minimum heating-utility requirement $Q_{h,min}$ is 60 kW, and the minimum cooling-utility requirement $Q_{c,min}$ is $-160$ kW. The last two columns of Table 6.4 show the heat flows or exchanges between different subnetworks. These heat flows can be visualized more clearly in Figures 6.4a–c. In particular, these figures show that there is no heat flow from subnetwork 2 to subnetwork 3, that is, across a hot-stream temperature $T_h$ of 150°C and a cold-stream temperature $T_c$ of 140°C with their difference of 10°C being equal to $\Delta T_{min}$.

The temperatures $T_h = 150°C$ and $T_c = 140°C$ are very significant in the synthesis of HENs. These temperatures are called *the pinch-point temperatures* for heat exchange (Umeda et al., 1978; Linnhoff and Turner, 1981; see also Figure 5.6 in Section 5.3.B). The point represents the temperature of hot process streams above which no cooling utility should be used. In other words, it is more economical to cool down hot process streams above this temperature by exchanging with cold process streams. The pinch point also represents the temperature of cold process streams below which no heating utility should be used. Instead, hot process streams should be used to heat up cold process streams below the pinch point.

It should be stressed that *pinch-point temperatures exist only in HEN synthesis problems that require both heating and cooling utilities.* They do not exist in those problems that need only heating or cooling utility.

For the 4TC2 problem illustrated in Table 6.4, a constant $\Delta T_{min}$ of 10°C has been specified for all stream matches. In practice, there are situations where $\Delta T_{min}$ is "match-dependent." For example, we might expect the $\Delta T_{min}$ value for heat exchange between a gas and a liquid stream to be different from that for heat exchange between two liquid streams. To take into account such match-dependent $\Delta T_{min}$ values in the determination of minimum utility requirements, we introduce the concept of $\Delta T_{min}$ "contribution" for an individual stream (Townsend and Linnhoff, 1983) and slightly modify the problem-table approach as illustrated in Table 6.4.

Suppose that a constant $\Delta T_{min}$ value ($\Delta T_{min,1}$) is used to set up a problem table, as in the case of Table 6.4, and this value applies to all stream matches with only one exception—a larger $\Delta T_{min}$ value ($\Delta T_{min,2}$) to be maintained when a hot stream $S_{hi}$ (or a cold stream $S_{cj}$) exchanges with a cold (or hot) process/utility stream. The difference between these two $\Delta T_{min}$ values may be designated as the $\Delta T_{min}$ contribution for $S_{hi}$ (or $S_{cj}$), that is:

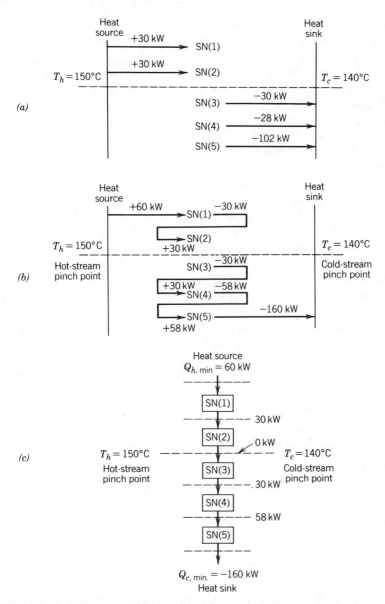

**FIGURE 6.4** (a–c)  Illustrations of subnetwork heat flows and the determination of minimum heating- and cooling-utility requirements and the pinch point corresponding to the 4TC2 problem shown in Table 6.4.

$$\Delta T_{\min,2} - \Delta T_{\min,1} = \Delta T_{\min} \text{ contribution for } S_{hi} \text{ (or } S_{cj}) \qquad (6.6)$$

To specify this hot stream $S_{hi}$ (or cold stream $S_{cj}$) in a problem table with a constant $\Delta T_{\min}$ value of $\Delta T_{\min,1}$, we first identify the "modified" input and output temperatures of $S_{hi}$ (or $S_{cj}$). This is done by subtracting the calculated value of $\Delta T_{min}$ contribution from the original input and output temperatures of $S_{hi}$ (or by adding the calculated value of $\Delta T_{min}$ contribution to the original input and output temperatures of $S_{cj}$). We then represent $S_{hi}$ (or $S_{cj}$) with its modified input and output temperatures in the problem table and proceed to determine the minimum utility requirements.

Consider, for example, the 4TC2 problem again. Suppose that $S_{h1}$ maintains a $\Delta T_{min}$ of 20°C when it exchanges with a cold process/utility stream, and that a constant $\Delta T_{min}$ of 10°C applies to all other stream matches. Equation (6.6) suggests that the $\Delta T_{min}$ contribution for $S_{h1}$ is 10°C. As illustrated in Table 6.4, $S_{h1}$ is to be cooled down from 180 to 40°C. By subtracting the $\Delta T_{min}$ contribution of 10°C to its specified input and output temperatures, we find that $S_{h1}$ has a modified input temperature of 170°C and output temperature of 30°C. Table 6.5 shows the problem table for this situation. It gives a $Q_{h,min}$ of 80 kW and a $Q_{c,min}$ of −180 kW. The table also indicates a cold-stream pinch temperature of 140°C and a hot-stream pinch temperature of 150°C. However, the latter applies only to $S_{h2}$. For $S_{h1}$, the correct pinch temperature should be 160°C, because we must add the $\Delta T_{min}$ contribution of 10°C for $S_{h1}$ to the observed hot-stream pinch temperature of 150°C.

The key conclusion of the preceding development is as follows. To synthesize an energy-optimum or a minimum-utility HEN, we should only use the minimum heating- and/or cooling-utility requirement(s) as found from the problem table. Furthermore, for synthesis problems requiring both heating and cooling utilities, no cooling utility should be used above the hot-stream pinch temperature and no heating utility should be used below the cold-stream pinch temperature.

## E. Determination of the Most Probable Minimum (Quasi-Minimum) Number of Units

The third step in the multiobjective synthesis of HENs is to determine the most probable minimum (quasi-minimum) number of units (exchangers, heaters, and coolers), denoted by $N_{min}$, prior to the actual synthesis of the networks. According to Hohmann (1971), $N_{min}$ in a given synthesis problem can be found from

$$N_{min} = N_h + N_c + N_{hu} + N_{cu} - 1 \qquad (6.7)$$

In the equation, $N_h$ and $N_{hu}$ are the numbers of hot process and utility streams, respectively; $N_c$ and $N_{cu}$ are the numbers of cold process and utility streams, respectively.

The numbers of hot and cold utility streams $N_{hu}$ and $N_{cu}$ to be substituted into Equation (6.7) must be related to the question of whether heating and cooling utilities are needed, as found from the problem table. As an example, for the 4SP2 problem illustrated in Figure 6.2 and specified in Appendix 6.2, for which no cooling utility is needed and $N_h = 3$, $N_c = 1$, $N_{hu} = 1$, and $N_{cu} = 0$, the most

**TABLE 6.5  A Problem Table for Determining the Minimum Utility Requirements for the 4TC2 Problem with Two $\Delta T_{min}$ Values[a]**

| Temperature Interval (TI) or Subnetwork (SN) | Hot Streams | | T(°C) | | Cold Streams | | Subnetwork Minimum Heating Requirement (kW) | Accumulated Utility Requirement (kW) | | Network Minimum Utility Requirement (kW) | |
|---|---|---|---|---|---|---|---|---|---|---|---|
| | $S_{h1}$ | $S_{h2}$ | (190) | 180 | $S_{c1}$ | $S_{c2}$ | | Heating | Cooling | Heating | Cooling |
| SN(1) | | | 170 | (160) | 3.0 | | 60 | 60 | −100 | 80 | 20 |
| SN(2) | | | 150 | (140) | | | 20 | 80 | −160 | 20 | 0[b] |
| SN(3) | | | (140) | 130 | | | −30 | 50 | −180 | 0[b] | −30 |
| SN(4) | 2.0 | 4.0 | (70) | 60 | | 2.6 | −28 | 22 | −150 | 30 | −58 |
| SN(5) | | | 40 | 30 | | | −102 | −80 | −122 | 58 | −160 |
| SN(6) | | | 30 | (20) | | | −20 | −100 | −20 | 160 | −180 |

[a] Heat-capacity flow rate in kW. $S_{h1}$ maintains a $\Delta T_{min}$ of 20°C when it exchanges with a cold process/utility stream, and a constant $\Delta T_{min}$ of 10°C applies to all other stream matches.

[b] The hot-stream pinch corresponds to $T_{h1}$ = 160°C and $T_{h2}$ = 150°C and the cold-stream pinch corresponds to $T_{c1}$ = 140°C.

probable minimum number of units (exchangers and heaters, and no coolers) is

$$N_{min} = N_h + N_c + N_{hu} + N_{cu} - 1$$
$$= 3 + 1 + 1 + 0 - 1$$
$$= 4$$

This result provides one with an advanced knowledge of the "target" number of units in the synthesis of an initial network and in the improvement of a given network.

Equation (6.7) for $N_{min}$ assumes that any stream match (i.e., heat exchange) can be made between a hot and a cold stream. It ignores both the $\Delta T_{min}$ constraint and the possible existence of pinch-point temperatures (which implies that no stream match is to be made across the pinch). It has been suggested that Equation (6.7) should be applied separately above and below the pinch (Linnhoff et al., 1982). Note that the pinch exists only if both heating and cooling utilities are needed in a HEN synthesis problem.

Consider again, for example, the 4TC2 problem analyzed in Table 6.4. Figure 6.5 summarizes the key results from Table 6.4 with a $\Delta T_{min}$ of 10°C for all stream matches. Based on the figure and Equation (6.7), we can write the following:

1. Considering the pinch: allowing no heat exchange across the pinch and utilizing only minimum amounts of heating and cooling utilities

$$N_{min} \text{ (above the pinch)} = N_h + N_c + N_{hu} + N_{cu} - 1$$
$$= 1 + 1 + 1 + 0 - 1$$
$$= 2 \text{ (1 exchanger and 1 heater)} \qquad (6.8)$$

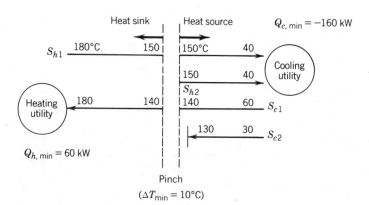

**FIGURE 6.5** An illustration of key results of problem-table analysis of the 4TC2 problem given in Table 6.4.

$$N_{\min} \text{ (below the pinch)} = N_h + N_c + N_{hu} + N_{cu} - 1$$

$$= 2 + 2 + 0 + 1 - 1$$

$$= 4 \text{ (e.g., 3 exchangers and 1 cooler)} \tag{6.9}$$

$$N_{\min} \text{ (total)} = 2 + 4 = 6 \text{ (e.g., 4 exchangers, 1 heater, and 1 cooler)}$$

**2.** Ignoring the pinch: permitting heat exchange across the pinch or for all temperature ranges, and not utilizing minimum amounts of heating and cooling utilities

$$N_{\min} \text{ (across the pinch)} = N_h + N_c + N_{hu} + N_{cu} - 1$$

$$= 2 + 2 + 1 + 1 - 1$$

$$= 5 \text{ (e.g., 3 exchangers, 1 heater, and 1 cooler)} \tag{6.10}$$

The preceding results suggest a trade-off between minimizing the number of units (i.e., synthesizing a minimum-cost or nearly minimum-cost network) and minimizing the utility consumption (i.e., synthesizing a minimum-utility or an energy-optimum network). The importance of this trade-off in the synthesis of HENs is illustrated further in Sections 6.3.F and G.

## F. Systematic Synthesis of Initial HENs

### 1. Thermodynamic Matching Rule

The fourth step in the multiobjective synthesis of HENs is to create an initial network that maximizes the thermodynamic efficiency of heat exchange among hot and cold process/utility streams and minimizes the investment cost of units (exchangers, heaters, and coolers). This step depends to a great extent on how hot and cold streams are to be exchanged and where heaters and coolers should be placed in a network. To approach the most efficient or nearly reversible exchange of heat among hot and cold streams, we recommend the use of the following thermodynamic matching rule: "The hot process and utility streams, and the cold process and utility streams are to be matched consecutively in a decreasing order of their average stream temperatures." The application of this matching rule to determine the order of heat exchanges among hot and cold streams is illustrated in this section via two examples.

The thermodynamic matching rule was first presented as corollary 3 of Nishida et al. (1977, p. 81) and its thermodynamic basis can be learned from a thermodynamic-availability analysis of the network synthesis problem described in Pehler and Liu (1981). A similar matching rule to guide the exchanges among hot and cold streams was presented earlier by Nishida et al. (1971) and later by Ponton and Donaldson (1974). However, the matching rule, as described in these publications, was not intended to guide the exchange among hot process and cold utility (or cold process and hot utility) streams. Instead, it was assumed that if a process stream could not reach its desired output temperature, it would reach its output temperature by an exchange with a utility stream in an auxiliary heater or cooler. In other words, the exchange or matching of a process stream with a heating- or cooling-utility stream was restricted to its last matching step. The latter approach, however, would exclude many cheaper networks. This observation was discussed previously in Section 6.3.B.1 in conjunction with the comparison between Figure 6.2c and e.

For HEN synthesis problems where multiple utility sources (e.g., high-, medium-, and low-pressure steam streams) are available for heating and/or cooling, as is commonly found in gas and oil processing, it is important that both process and utility streams are matched according to the thermodynamic matching rule. In other words, utility-stream selections and heater/cooler placement cannot be considered separately from process stream matches. If this guideline is not followed, it is most likely that the resulting network will be thermodynamically inefficient and highly expensive.

Based on the work of Pehler and Liu (1981), we recommend that the thermodynamic matching rule should not only be applied in the initial generation of an energy-optimum and nearly minimum-cost network, but also in the evolutionary synthesis of an energy-optimum and minimum-cost network. As shown in Section 6.3.G, the matching rule can be readily applied to a modified network, obtained after deleting or merging certain units, which may no longer be very energy efficient or approach a practical minimum loss in available energy. This application provides explicit thermodynamic guidance on how to systematically shift and/or unsplit the remaining units in the modified network so as to reach a practical minimum loss in available energy.

## 2. Example 1: The Pinch-Design Method

Consider a four-stream HEN synthesis problem, as represented in the heat-content diagram shown in Figure 6.6a and in the problem table given in Table 6.6, with a $\Delta T_{min}$ of 20°F (11.1°C). A saturated heating steam at 456°F (235.6°C) and a cooling-water stream with a temperature range of 100–180°F (37.8–82.2°C) are available as utilities.

Table 6.6 shows that the hot-stream pinch temperature is 340°F and the cold-stream pinch temperature is 320°F. The basic idea of the pinch-design method (Linnhoff and Hindmarsh, 1983) is to decompose the HEN synthesis problem into two subproblems, one above and the other below the pinch. No heat exchange between hot and cold streams across the pinch is allowed, and this minimizes the heating- and cooling-utility consumptions.

### a. ABOVE THE PINCH POINT

Before synthesizing an initial network above the pinch point, we know the following.

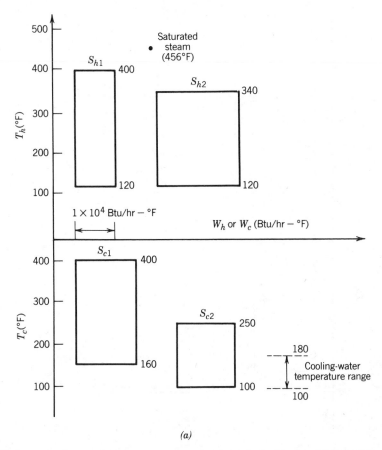

(a)

**FIGURE 6.6** (a) A heat-content diagram representing Example 1 of Section 6.3.F. (b) A heat-content diagram representing a network solution to Example 1. (c) A temperature-interval diagram representing the network solution to Example 1. (d) A comparison of two options for matching $S_{c1}$ with $S_{h1}$ and with $S_{h2}$ below the pinch point in Example 1.

- Minimum heating-utility requirement (Table 6.6) $= 60 \times 10^4$ Btu/hr.
- Minimum number of units:

$$N_{\min} \text{ (above the pinch)} = N_h + N_c + N_{hu} + N_{cu} - 1$$

$$= 1 + 1 + 1 + 0 - 1$$

$$= 2$$

- No cooling utility should be utilized above the pinch point.
- There are only two process streams:

$$S_{h1}: W_{h1} = 1 \times 10^4 \text{ Btu/hr (from 400 to 340°F)}$$
$$S_{c1}: W_{c1} = 1.5 \times 10^4 \text{ Btu/hr (from 320 to 400°F)}$$

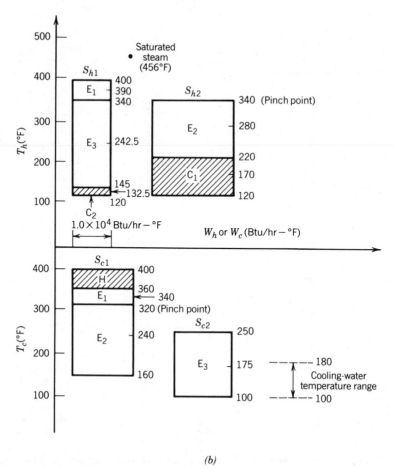

*(b)*

**FIGURE 6.6** *(Continued)*

- Saturated steam (456°F) is the highest-temperature hot stream.

As shown in the heat-content diagram of Figure 6.6*b* and in the temperature–interval diagram of Figure 6.6*c*, we place a steam heater on the highest-temperature portion of $S_{c1}$ and then match $S_{h1}$ and $S_{c1}$ through exchanger $E_1$. The order of this stream matching follows the thermodynamic matching rule.

b.  BELOW THE PINCH POINT

The following information is known below the pinch point.

- Minimum cooling-utility requirement (Table 6.6) = $-225 \times 10^4$ Btu/hr.
- Minimum number of units:

$$N_{\min} \text{ (below the pinch)} = N_h + N_c + N_{hu} + N_{cu} - 1$$
$$= 2 + 2 + 0 + 1 - 1$$
$$= 4$$

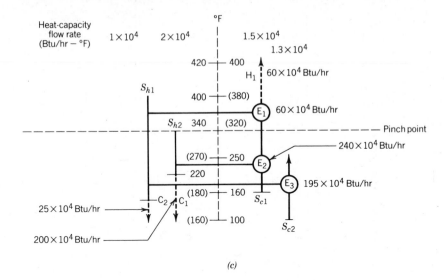

(c)

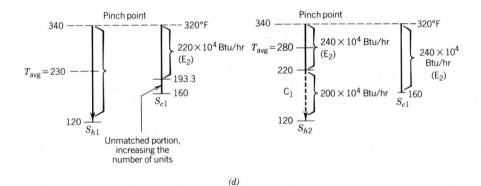

(d)

**FIGURE 6.6**  (Continued)

- No heating utility should be used below the pinch point.
- There are four process streams:

$$S_{h1}: W_{h1} = 1 \times 10^4 \text{ Btu/hr-}°F \text{ (from 340 to 120°F)}$$
$$S_{h2}: W_{h2} = 2 \times 10^4 \text{ Btu/hr-}°F \text{ (from 340 to 120°F)}$$
$$S_{c1}: W_{c1} = 1.5 \times 10^4 \text{ Btu/hr-}°F \text{ (from 160 to 320°F)}$$
$$S_{c2}: W_{c2} = 1.3 \times 10^4 \text{ Btu/hr-}°F \text{ (from 100 to 250°F)}$$

- Cooling water is the lowest-temperature cold stream.

Since $S_{c1}$ has the highest temperature between two cold streams ($S_{c1}$ and $S_{c2}$), the thermodynamic matching rule suggests that $S_{c1}$ (instead of $S_{c2}$) should be matched first with either $S_{h1}$ or $S_{h2}$. These two options are compared in Figure 6.6d. We

**TABLE 6.6  A Problem Table for Determining the Minimum Utility Requirements for Example 1 of Section 6.3.F**

| Temperature Interval (TI) or Subnetwork (SN) | Hot Streams $S_{h1}$ | Hot Streams $S_{h2}$ | T(°F) 400 | T(°F) 420 | Cold Streams $S_{c1}$ | Cold Streams $S_{c2}$ | Subnetwork Minimum Heating Requirement[c] | Accumulated Utility Requirement[c] Heating | Accumulated Utility Requirement[c] Cooling | Network Minimum Utility Requirement[c] Heating | Network Minimum Utility Requirement[c] Cooling |
|---|---|---|---|---|---|---|---|---|---|---|---|
| SN(1) |  |  | 400 (380) | 400 | 1.5 ↑ |  | 30 | 30 | −165 | 60 | −30 |
| SN(2) |  |  | 340 (320) |  |  |  | 30 | 60 | −195 | 30 | $0^b$ |
| SN(3) | $1.0^a$ | 2.0 | 250 | (270) |  | 1.3 ↑ | −105 | −45 | −225 | $0^b$ | −105 |
| SN(4) |  |  | 160 | (180) |  |  | −18 | −63 | −120 | 105 | −123 |
| SN(5) |  |  | 120 | 100 |  |  | −102 | −165 | −102 | 123 | −225 |

[a] Heat-capacity flow rate ($\times\,10^{-4}$) Btu/hr-°F. $1 \times 10^4$ Btu/hr-°F = 5.275 kW/°C.

[b] Hot-stream pinch = 340°F (171.1°C); cold-stream pinch = 320°F (160°C); $\Delta T_{\min}$ = 20°F (11.1°C).

[c] Utility requirement ($\times\,10^{-4}$) Btu/hr.

see that (1) the thermodynamic matching rule favors exchanging heat between $S_{c1}$ and $S_{h2}$ (which has a higher average stream temperature $T_{avg} = 280°F$) over that between $S_{c1}$ and $S_{h1}$ (which has a lower average stream temperature $T_{avg} = 230°F$) and (2) the exchange between $S_{c1}$ and $S_{h1}$ may lead to the use of an extra unit (i.e., an additional exchanger) for heating $S_{c1}$ from 160 to 193.3°F. We thus conclude that $S_{c1}$ should be matched with $S_{h2}$. We next match $S_{c2}$ with $S_{h1}$, leaving the unmatched and lowest-temperature portion of $S_{h1}$ to be cooled from 220 to 120°F by cooling water. Figures 6.6b and c illustrate the resulting network solution on the heat-content and temperature-interval diagrams, respectively.

To see whether the network solution satisfies or closely follows the thermodynamic matching rule, we check the average stream temperatures for hot- and cold-stream matches. From Figure 6.6b, we find that the ranked order of average stream temperatures of hot-stream matches is as follows:

|  | $E_1$ | $E_2$ | $E_3$ | $C_1$ | $C_2$ |
|---|---|---|---|---|---|
| $T_{avg}$: | 370°F | 280°F | 242.5°F | 170°F | 132.5°F |

By comparison, the ranked order of average temperature levels of cold-stream matches is as follows:

|  | $H_1$ | $E_1$ | $E_2$ | $E_3$ |
|---|---|---|---|---|
| $T_{avg}$: | Highest | 340°F | 240°F | 175°F |

This comparison indicates that heat exchanges among hot and cold streams in the network solution appear to be thermodynamically efficient.

It is worthwhile to use this example problem to illustrate two generally applicable heuristics for heat exchanges that are immediately adjacent to the pinch point (i.e., immediately above and below the pinch). As seen from Figure 6.6c, the match immediately above the pinch shows

$$W_{h1} = 1.0 \times 10^4 < W_{c1} = 1.5 \times 10^4 \text{ Btu/hr-°F}$$

Also, from Figure 6.6d, the match immediately below the pinch indicates

$$W_{h2} = 2 \times 10^4 > W_{c1} = 1.5 \times 10^4 \text{ Btu/hr-°F} \quad \text{(adopted match)}$$

$$W_{h1} = 1 \times 10^4 < W_{c1} = 1.5 \times 10^4 \text{ Btu/hr-°F} \quad \text{(rejected match)}$$

These results can be generalized to give the following heuristics for deciding the proper matches immediately adjacent to the pinch point (Linnhoff et al., 1982):

$$W_h \leq W_c \quad \text{(above the pinch)}$$
$$W_h \geq W_c \quad \text{(below the pinch)} \tag{6.8}$$

In summary, by applying the step-by-step procedure described in Sections 6.3.D–6.3.F, we have obtained an initial network solution to the example problem that satisfies the following multiple-objective criteria:

- Using minimum amounts of heating and cooling utilities.
- Using the minimum number of units both above and below the pinch.
- Approaching an efficient heat exchange among hot and cold streams (i.e., following the thermodynamic matching rule).

## 3. Example 2: The Use of Stream Splitting

This example introduces the concept of using stream splitting to overcome the difficulty caused by $\Delta T_{min}$ violations in the synthesis of an initial network with a minimum number of units. It also illustrates the effect of permitting heat flow across the pinch point on increasing both heating- and cooling-utility consumptions. The latter could possibly lead to a network with less units than that obtained by considering the pinch.

Consider again the 4TC2 problem specified in Appendix 6.2 and illustrated in Figures 6.3–6.5. A step-by-step application of the synthesis procedure described in Sections 6.3.C–6.3.F gives the following information.

### a. PROBLEM REPRESENTATION

See the heat-content diagram shown in Figure 6.3a and the temperature–interval diagram given in Figure 6.3g.

### b. DETERMINATION OF THE MINIMUM HEATING- AND COOLING-UTILITY REQUIREMENTS AND THE PINCH POINT

See Table 6.4 with a $\Delta T_{min}$ of 10°C and Figures 6.4 and 6.5. We find that $Q_{h, min}$ = 60 kW, $Q_{c, min}$ = −160 kW, the hot-stream pinch corresponds to $T_h$ = 150°C, and the cold-stream pinch cold-stream pinch $T_c$ = 140°C.

### c. DETERMINATION OF THE MOST PROBABLE MINIMUM NUMBER OF UNITS

See the numbers given by Equations (6.8)–(6.10) in Section 6.3.E.

### d. SYNTHESIS OF AN INITIAL HEN

The synthesis of a network featuring both minimum utility consumptions and minimum number of units is illustrated in Figure 6.7. Above the pinch temperatures of $T_h$ = 150°C and $T_c$ = 140°C, we know the following information.

- Minimum heating-utility requirement (Table 6.3) = 60 kW.
- Minimum number of units (Equation 6.8) = 2.
- No cooling utility should be utilized above the pinch point.

- There are only two process streams:

$$S_{h1}: W_{h1} = 2.0 \text{ kW}/°\text{C (from 180 to 150°C)}$$
$$S_{c1}: W_{c1} = 3.0 \text{ kW}/°\text{C (from 140 to 180°C)}$$

- Saturated steam (258°C) is the highest-temperature hot stream.

Following a situation similar to Example 1 of this section, we place a steam heater H and an exchanger $E_1$ above the pinch. This is illustrated in Figure 6.7$a$.

Below the pinch temperatures of $T_h = 150°\text{C}$ and $T_c = 140°\text{C}$, the following information is known.

- Minimum cooling-utility requirement (Table 6.3) = $-160$ kW.
- Minimum number of units (Equation 6.9) = 4.
- No heating utility should be utilized below the pinch point.
- There are four process streams:

$$S_{h1}: W_{h1} = 2.0 \text{ kW}/°\text{C (from 150 to 40°C)}$$
$$S_{h2}: W_{h2} = 4.0 \text{ kW}/°\text{C (from 150 to 40°C)}$$
$$S_{c1}: W_{c1} = 3.0 \text{ kW}/°\text{C (from 60 to 140°C)}$$
$$S_{c2}: W_{c2} = 2.6 \text{ kW}/°\text{C (from 30 to 130°C)}$$

- Cooling water is the lowest-temperature cold stream.

As illustrated in Figure 6.7$a$, we follow the thermodynamic matching rule and place a water cooler with $Q = Q_{c,\min} = -160$ kW at the lowest-temperature portion of $S_{h2}$. The latter hot stream is chosen since $S_{h2}$ has a lower average temperature level than $S_{h1}$. To obtain a network solution with only 4 units below the pinch, we can have at most three exchangers among four process streams. Figure 6.7$a$ shows an example of a network with 3 exchangers and 1 cooler below the pinch, utilizing only a minimum amount of cooling utility. This network almost satisfies the goal of having a minimum of 4 units below the pinch, but it contains a $\Delta T_{\min}$ violation on $E_4$.

Figures 6.7$b$ and $c$ illustrate the step-by-step application of *an important tool (i.e., stream splitting)* to overcome the difficulty caused by $\Delta T_{\min}$ violations in the synthesis of a network solution with only 3 exchangers and 1 cooler below the pinch. The resulting network, shown in Figure 6.7$c$, is identical to that illustrated in the heat-content diagram of Figure 6.3$b$ , in the physical network structure of Figure 6.3$c$, and in the temperature–interval diagram of Figure 6.3$h$. This network features both minimum utility consumptions and minimum number of units by considering the pinch constraint (i.e., permitting no heat flow across the pinch).

Figure 6.8$a$ shows an unsplitting network solution to the 4TC2 problem obtained by merging exchangers $E_3$ (40 kW) and $E_4$ (220 kW) in Figure 6.7$a$ as a

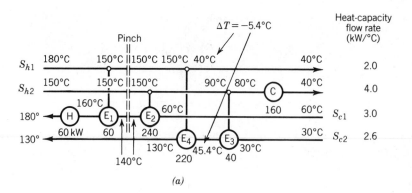

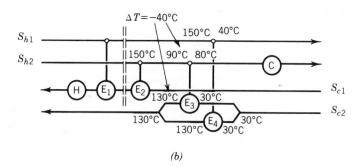

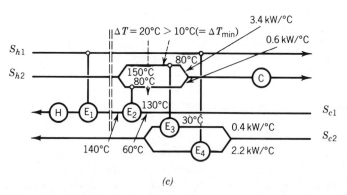

**FIGURE 6.7** The development of a network solution for the 4TC2 problem, Example 2 of Section 6.3.F, featuring both minimum utility consumption and minimum number of units. (a) An unsplitting network with a $\Delta T_{min}$ violation on $E_4$ below the pinch. (b) The use of stream splitting on $S_{c2}$ to reduce its inlet temperature to $E_4$ from 45.4 to 30°C, thus removing the $\Delta T_{min}$ violation on $E_4$. This stream splitting, however, also raises the outlet temperature of $S_{c2}$ from 45.4 to 130°C, resulting in a new $\Delta T_{min}$ violation on $E_3$. (c) The use of stream splitting on $S_{h2}$ to raise its inlet temperature to $E_3$ from 90 to 150°C, thus removing the $\Delta T_{min}$ violation on $E_3$. This network is identical to that shown in Figure 6.3h.

**198**

single exchanger $E_3'$ (260 kW). A simple procedure to identify the changes in heat loads on different units from Figure 6.7a–6.8a is described later in Section 6.3.G. The network of Figure 6.8a uses *more* heating and cooling utilities than minimum values; but it contains only 5 units, as suggested by Equation (6.10), that is, *one less* than that obtained by considering the pinch. Figure 6.8b illustrates a very important observation when we ignore the pinch or permit heat flow across the pinch. The latter *always* increases *both* heating- and cooling-utility consumptions. The significance of Figure 6.8b can be better understood by comparing it with Figure 6.5.

The preceding example clearly demonstrates a trade-off between minimizing the number of units and minimizing the utility consumption in the synthesis of

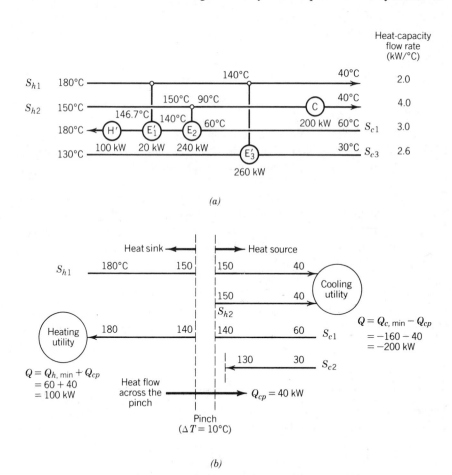

(a)

(b)

**FIGURE 6.8** (a) A network solution to the 4TC2 problem, Example 2 of Section 6.3F, obtained by permitting heat flow across the pinch point that features more utility consumptions than minimum values and a less number of units than that obtained by considering the pinch point. (b) An illustration of the effect of permitting heat flow across the pinch point on increasing *both* heating- and cooling-utility consumptions for the 4TC2 problem (see also Figure 6.5).

HENs. This trade-off is illustrated further in the evolutionary improvement of HENs in the following section.

## G. Evolutionary Synthesis of Improved HENs

### 1. An Introduction to Evolutionary Synthesis

The fifth step in the multiobjective synthesis of HENs is to evolutionarily improve an initial network. Consider, for example, a 7-stream problem, the 7SP1 problem (Masso and Rudd, 1969) specified in Appendix 6.2. Figures 6.9a–c show an initial network for the 7SP1 problem proposed by Masso and Rudd and its representations by the heat-content and temperature–interval diagrams. A number of questions may be raised about this given initial network.

- Does the 7SP1 problem require both steam heater and water cooler as illustrated in Figures 6.9a–c?
- Does the given initial network include a minimum number of units (exchangers, heaters, and/or coolers)?
- If the answer to these two questions is negative, how do we systematically modify the given initial network to obtain an improved network with both minimum utility consumption and investment cost?

To provide definitive answers to the questions posed above, we introduce a simple and practical approach to the evolutionary synthesis of HENs. This approach combines the best features of several reported studies on the subject, including those by Nishida et al. (1977), Linnhoff and Flower (1978a), Pehler and Liu (1981, 1984), and Su and Motard (1984).

The basic strategies of our approach are (1) to achieve a *thermodynamically* efficient heat exchange among hot and cold streams, and (2) to obtain an *economically* attractive network with a minimum number of units (exchangers, heaters, and/or coolers). Because of these considerations, our method may be called *the thermoeconomic approach*.

### 2. The Thermoeconomic Approach

The thermoeconomic approach to the evolutionary synthesis of energy-optimum and minimum-cost HENs consists of the following three key steps.

a. DETERMINATION OF THE MINIMUM HEATING- AND COOLING-UTILITY REQUIREMENTS, OF THE PINCH POINT, AND OF THE MOST PROBABLE MINIMUM NUMBER OF UNITS, $N_{min}$

As an exercise, the reader can use the design data given in Appendix 6.2 to set up a problem table for the 7SP1 problem with a $\Delta T_{min}$ of 11.1°C (20°F) and verify

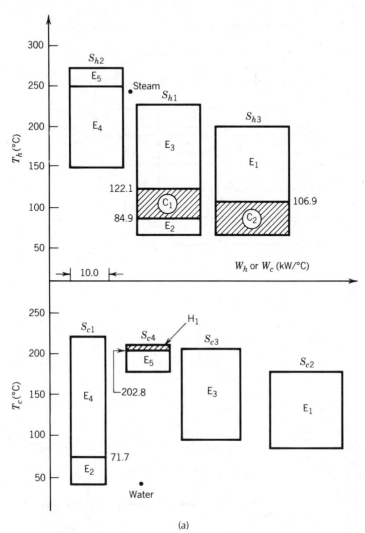

(a)

**FIGURE 6.9** (a) A heat-content diagram representing an initial network for the 7SP1 problem shown in Figure 6.9c. (b) A temperature-interval diagram representing an initial network for the 7SP1 problem. (c) An initial network for the 7SP1 problem (Masso and Rudd, 1969). (d) An illustration of the use of a path between a steam heater $H_1$ and a water cooler $C_1$ to apply evolutionary rule 1 to delete $H_1$ and to systematically modify the heat loads of different units on the path. Resulting heat loads are shown within parentheses. (e) An improved network for the 7SP1 problem.

the following results (see Sections 6.3D and 6.3F):

- Minimum heating-utility requirement = 0 kW (no steam heating should be used!).
- Minimum cooling-utility requirement = 1219.8 kW.
- The pinch-point temperature does *not* exist.

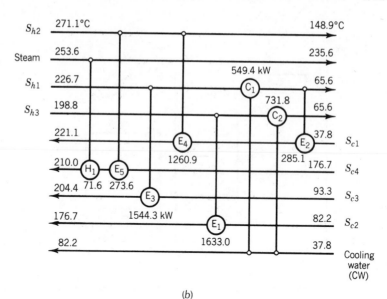

(b)

(c)

**FIGURE 6.9** (Continued)

Also, for the 7SP1 problem, Equation (6.7) becomes

$$N_{min} = N_h + N_c + N_{hu} + N_{cu} - 1$$

$$= 3 + 4 + 0 + 1 - 1$$

$$= 7 \text{ (exchangers plus coolers; } no \text{ } heaters\text{)} \qquad (6.11)$$

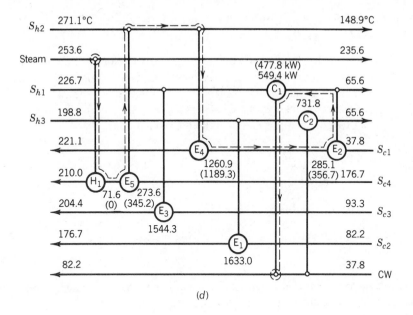

(d)

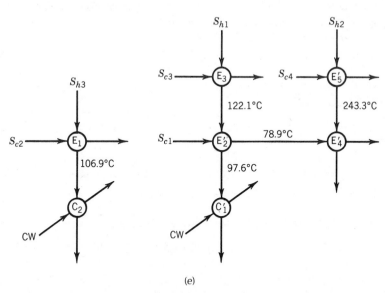

(e)

**FIGURE 6.9**  (Continued)

From Figures 6.9$a$–$c$, the given initial network for the 7SP1 problem has 5 exchangers, 2 coolers, *and 1 heater*!

It is clear that to improve or "retrofit" the given initial network for the 7SP1 problem through better heat integration requires the deletion of the unnecessary heater.

b.  Minimization of the Number of Units to Reduce the Network-
Investment Cost Through Systematic Merging of Units

The basic idea of this step is actually very simple and can be illustrated clearly from the relationship expressing the investment cost as a function of heat-transfer areas of exchangers, heaters, and coolers. For example, since the cost factor $b$ in the investment-cost expression of exchangers $C_{Hi} = aA_{Hi}^b$ is normally taken as 0.6, the following two inequalities can be written (Nishida et al., 1977):

$$a(A_{H1}^b + A_{H2}^b + \ldots + A_{Hm}^b)$$
$$\geq a(A_{H1} + A_{H2} + \ldots + A_{Hm})^b \quad 0 \leq b \leq 1 \quad (6.12)$$
$$a(A_{H1}^b + A_{H2}^b + \ldots + A_{Hm}^b)$$
$$\geq a(A_{H1} + A_{H2})^b + a(A_{H3} + \ldots + A_{Hi})^b$$
$$+ a(A_{H,i+1} + \ldots + A_{Hm})^b \quad 0 \leq b \leq 1 \quad (6.13)$$

These inequalities imply that without increasing the total heat-transfer area of exchangers, the investment cost of exchangers can be reduced if (1) several exchangers can be combined together as a single one, or (2) a smaller number of exchangers is to be used. Here, Equation (6.12) corresponds to the case where $m$ exchangers are merged as a single exchanger; Equation (6.13) represents the case where $m$ exchangers are reduced to three exchangers. Obviously, the same idea is also applicable to reducing the investment cost of heaters or coolers.

c.  Application of the Thermodynamic Matching Rule to the
Modified Network with a Fewer Number of Units*

To effectively implement the second and third steps of the thermoeconomic approach, we introduce below a number of basic evolutionary rules (Pehler and Liu, 1981, 1984) as well as the concepts of path and loop breaking (Su and Motard, 1984). These rules are to be applied sequentially in their numerical order; that is, rule 1 should be applied before rule 2, and so on.

## 3.  Evolutionary Rules and Path/Loop Breaking

*Rule 1.* "Delete, shift or merge units so as to reduce the number of units in a selected local subnetwork (i.e., a selected number of units in an initial network) and to minimize the required changes to adjacent subnetworks due to changes in heating or cooling load. When selecting candidates for unit modifications, choose initially a redundant heater or cooler exceeding the minimum heating or cooling requirement and then a redundant exchanger with a small heat load exceeding the quasi-minimum number of units. Avoid modifying any single-exchanger match between two hot and cold streams. If a selected unit modification in a given sub-

---

*This entails systematic shifting, and/or unsplitting the remaining units in order to achieve a thermo-dynamic efficient (nearly reversible) heat exchange between streams.

network results in extensive structural modifications in adjacent or other subnetworks, then another candidate for unit modifications should be evaluated."

This evolutionary rule is illustrated with the following rule in an example of improving or "retrofitting" the initial network for the 7SP1 problem shown in Figures 6.9*a–c*.

*Rule 2.* "Shift heaters or coolers to approach a thermodynamically efficient, heat exchange between streams. (a) When shifting a heater (cooler) between matches on a given cold (hot) process stream, always shift the heater (cooler) from the low-temperature (high-temperature) portion to the high-temperature (low-temperature) portion of the cold-stream (hot-stream) match. (b) When merging two heaters (coolers) matched with two different cold (hot) process streams, always shift the heater (cooler) from one stream to the other so that the resulting merged heater (cooler) will have a higher (lower) arithmetic average of its input and output temperatures."

### a.  EXAMPLE 1: APPLICATION OF EVOLUTIONARY RULES 1 AND 2A TO THE 7SP1 PROBLEM

Based on the discussion in Section 6.3.G.2, we know that the initial network represented by Figures 6.9*a–c* uses more cooling utility for coolers $C_1$ and $C_2$ (1281.2 kW) than the minimum amount required (1204.4 kW). It also includes one more unit, that is, an unnecessary heater ($H_1$, 71.6 kW) than the most probable minimum number of 7 units found from Equation (6.11). Further, the thermodynamic matching rule suggests that $C_1$ leads to a thermodynamically inefficient use of the lowest-temperature cooling utility, as it is being placed on the intermediate-temperature portion of $S_{h1}$. Thus the following "retrofitting" or evolutionary changes can be made with reference to the heat-content diagram of Figure 6.9*a* or the temperature–interval diagram of Figure 6.9*b*.

- Deleting the unnecessary heater $H_1$ (evolutionary rule 1) by enlarging exchanger $E_5$ on $S_{c4}$ to compensate for the elimination of the heating utility.
- $E_4$ on both $S_{h2}$ and $S_{c1}$ is reduced to accommodate for the increase in the heat load of $E_5$ on $S_{h2}$.
- $E_2$ on both $S_{h1}$ and $S_{c1}$ is enlarged, which leads to a decrease in the heat load of $C_1$ on $S_{h1}$.
- The reduced $C_1$ on $S_{h1}$ is shifted to the lowest-temperature portion of $S_{h1}$ (evolutionary rule 2a), resulting in a thermodynamically efficient use of the lowest-temperature cooling utility.

The preceding network evolutions can be visualized more easily with reference to the temperature-interval (or grid) shown in Figure 6.9*d*, in which the concept of path is illustrated. Here, a *path* is defined as a system of connections of units in a HEN that form a continuous pathway between two units (exchanges, heaters and/or coders) (Linnhoff et al., 1982). The quantitative effect of deleting heater $H_1$ on the heat loads of different connected units in the initial network can be determined fairly easily by "breaking" the path from heater $H_1$ to cooler $C_1$ shown in Figure 6.9$d$:

Path: $(H_1, E_5, E_4, E_2, C_1)$
Heat Load: $(71.6, 273.6, 1260.9, 285.1, 549.4)$ kW

$\downarrow$ Evolutionary rule 1
$\;$ (Path breaking)

Broken path: $(0, E_5' = H_1 + E_5, E_4' = E_4 - H_1, E_2' = E_2 + H_1, C_1'$
$= C_1 - H_1)$

Heat load: $(0, 345.2, 1189.3, 356.7, 477.8)$ kW

The heat loads on different modified units on the "broken" path are shown in parenthesis in Figure 6.9$d$.

In making the preceding evolutions of the initial network, it is important to check if any $\Delta T_{min}$ violations arise from the evolutions. In other words, the feasibility of any evolutionary change *must* be considered.

Next, we apply evolutionary rule 2a to move cooler $C_1'$ (477.8 kW) to the lowest-temperature portion of $S_{h1}$. This is equivalent to exchanging the positions of $C_1'$ (477.8 kW) and $E_2'$ (356.7 kW) on the low-temperature portion of $S_{h1}$, as seen in Figure 6.9$d$. The resulting improved network is depicted in Figure 6.9$e$. This network features several desirable characteristics:

- Using a minimum amount of utilities (in fact, cooling utility only), 1204.4 kW.
- Using a minimum number of 7 units.
- Exchanging heat between hot and cold streams in a thermodynamically efficient manner.

## b. EXAMPLE 2: APPLICATION OF EVOLUTIONARY RULES 1 AND 2b TO THE 5SP1 PROBLEM

We illustrate the use of evolutionary rules 1 and 2b with reference to an initial network reported by Masso and Rudd (1969) for a 5-stream problem that is specified as the 5SP1 problem in Appendix 6.2. Figures 6.10$a$–$b$ show the heat-content and temperature-interval (or grid) diagrams for representing this initial network.

By using the design data specified in Appendix 6.2, the reader can carry out a problem-table analysis of the 5SP1 problem with a $\Delta T_{min}$ of 11.1°C (20°F) and obtain the following results:

- The minimum heating utility required = 881.5 kW
- The minimum cooling utility needed = 0 kW
- The most probable minimum number of units = 5 (exchangers and heaters)
- The pinch-point temperature does *not* exist.

The initial network represented in Figure 6.10$a$–$b$ utilizes a minimum amount of heating utility ($H_1 + H_2 = 881.5$ kW), but it contains one more unit (an exchanger or a heater) than the most probable minimum number of 5 units. Also, in the initial network, exchanger $E_3$ represents a "single-exchanger match" between $S_{c2}$ and $S_{h2}$. Thus we should try to avoid making changes involving exchanger $E_3$ (and exchanger $E_4$ on $S_{h2}$).

The following evolutionary changes are proposed with reference to Figures 6.10a and b:

- Shifting $H_1$ from $S_{c1}$ to $S_{c3}$ and merging it with $H_2$ to form a composite heater $H = H_1 + H_2$ (rules 1 and 2b).

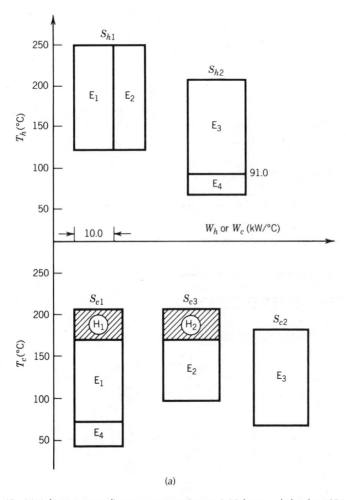

(a)

**FIGURE 6.10** (a) A heat-content diagram representing an initial network for the 5SP1 problem (Masso and Rudd, 1969). (b) A temperature-interval diagram representing an initial network for the 5SP1 problem. (c) An improved network for the 5SP1 problem. (d) An illustration of a loop ($H_1$, $E_1$, $E_2$, $H_2$) in the initial network for the 5SP1 problem. Heat loads resulting from shifting $H_1$ from $S_{c1}$ to $S_{c3}$ and merging it with $H_2$ to form an enlarged heater $H = H_1 + H_2$ are shown within parentheses. (e) An illustration of a broken loop (H, $E_1'$, $E_2'$) in the improved network for the 5SP1 problem shown in Figure 6.10c.

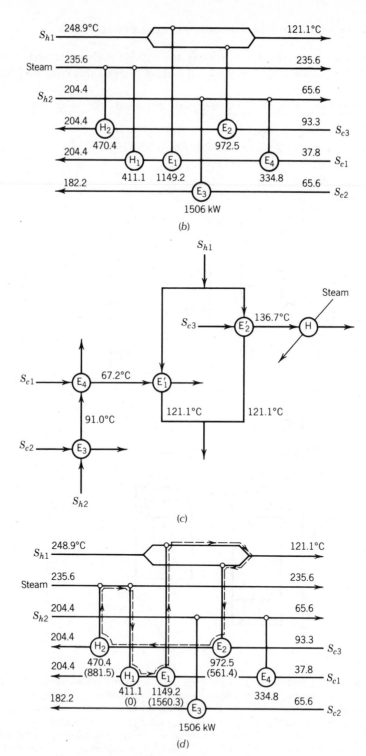

FIGURE 6.10 (*Continued*)

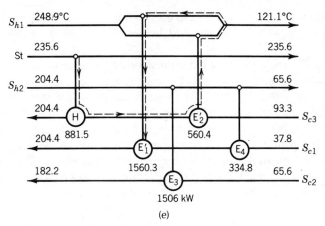

**FIGURE 6.10** (Continued)

- Reducing the load of $E_2$ on $S_{c3}$ to compensate for the increased heater load $(H = H_1 + H_2)$.
- Enlarging $E_1$ on $S_{h1}$ by changing the splitting ratio of $S_{h1}$.

The resulting improved network is shown in Figures 6.10c and e.

It is instructive to consider further the evolutions involved in the 5SP1 problem on a temperature-interval (or grid) diagram and introduce the concept of "loop" (Su and Motard, 1984). Figure 6.10d illustrates a loop in the initial network. Here, a *loop* is defined as a system of connections of different units in a HEN that form a closed pathway. Shifting $H_1$ from $S_{c1}$ to $S_{c3}$ and merging it with $H_2$ to form an enlarged heater $H = H_1 + H_2$ are essentially equivalent to "breaking" the loop in Figure 6.10d, with the resulting temperature–interval (or grid) diagram shown in Figure 6.10e:

Loop:      $(H_1, E_1, E_2, H_2)$ (Figure 6.10d )
Heat load:  (411.1, 1149.2, 972.5, 470.4) kW

│ Evolutionary rules 1 and 2b
↓ (Loop breaking)

Broken loop:  $(0, E_1' = E_1 + H_1, E_2' = E_2 - H_1, H = H_2 + H_1)$
(Figure 6.10e )
Heat load:  (0, 1560.3, 560.4, 881.5) kW

As in the preceding example, we should always check whether any $\Delta T_{\min}$ violations arise from the evolutions or loop breaking. Finally, the improved network shown in Figures 6.10c and e represents an energy-optimum and minimum-cost network solution to the 5SP1 problem that is identical to the best solution reported in the literature (Nishida et al., 1977).

*Rule 3.* "Reduce the number of units by deleting repeated matches between two hot and cold process streams in a given network. In particular, if a given network contains a local subnetwork in which a hot (cold) stream matches the same cold (hot) stream that it has matched before, delete either of these repeated matches.''

Figures 6.11a and b illustrate the application of evolutionary rule 3 to a simple HEN to reduce the number of units by deleting repeated matches. In particular, these figures show that $S_{h1}$ and $S_{c1}$ are heat-exchanged twice through $E_1$ and $E_3$, and $S_{h2}$ and $S_{c2}$ are both involved in "single-exchanger matches" through $E_2$ and $E_4$. Thus, as seen in Figure 6.11a, $E_3$ on both $S_{h1}$ and $S_{c1}$ is shifted upward through $E_4$ and merged with $E_1$. On $S_{h1}$, $E_4$ is moved downward to the low-temperature portion. On the temperature–interval (or grid) diagram of Figure 6.11b, evolutionary rule 3 is used to "break" a loop ($E_1$, $E_3$) through merging together $E_1$ and $E_3$. This loop breaking results in an improved network.

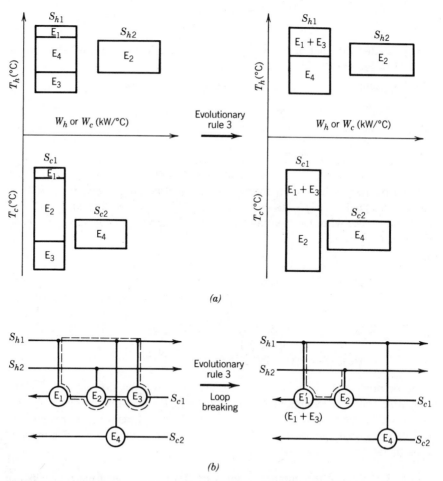

FIGURE 6.11 An illustration of the use of evolutionary rule 3 (a) on a heat-content diagram and (b) on a temperature-interval (or grid) diagram via loop breaking.

*Rule 4.* "Unsplit a given splitting network to minimize the number of units and reduce the loss in available energy during the heat-exchange process. When unsplitting a given splitting network, always match the hot and cold process streams in the resulting network in a decreasing order of their arithmetic averages of input and output temperatures."

Figures 6.12*a* and *b* illustrate the application of evolutionary rule 4 to "unsplit" a given splitting network so as to minimize the number of units and to approach a thermodynamically efficient heat exchange. In Figure 6.12*a*, the following evolutionary changes are shown:

- On both $S_{h1}$ and $S_{c1}$, combine $E_1$ with $E_2$.
- On $S_{h1}$ convert ($E_1 + E_2$) and $E_3$ from a splitting to an unsplitting arrangement by matching $S_{h1}$ with $S_{c1}$ through ($E_1 + E_2$), and then exchanging $S_{h1}$ with $S_{c2}$ through $E_3$. Note that the order of matching of $S_{h1}$ with $S_{c1}$ and then with $S_{c2}$ follows the relative magnitude of average temperature levels of ($E_1 + E_2$) and $E_3$.

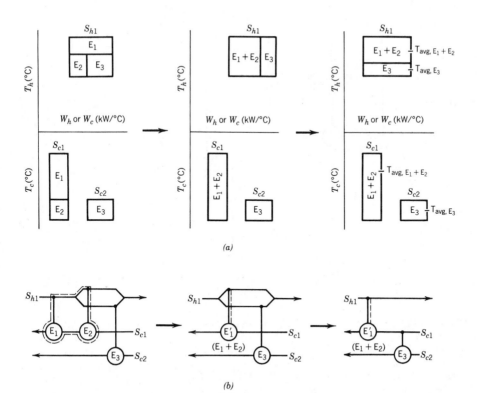

**FIGURE 6.12** An illustration of the use of evolutionary rule 3 (a) on a heat-content diagram and (b) on a temperature-interval (or grid) diagram via loop breaking.

Figure 6.12*b* shows the evolutions described above on a temperature–interval (or grid) diagram and illustrates the use of loop breaking to obtain an improved network.

## 4. An Illustrative Example: The 10SP1 Problem

To further illustrate the applications of evolutionary rules and path/loop breaking, we consider a 10-stream problem, 10SP1, proposed by Pho and Lapidus (1973). Detailed stream and design data for the problem are given in Appendix 6.2.

Figures 6.13*a* and *b* show an initial network for the 10SP1 problem reported

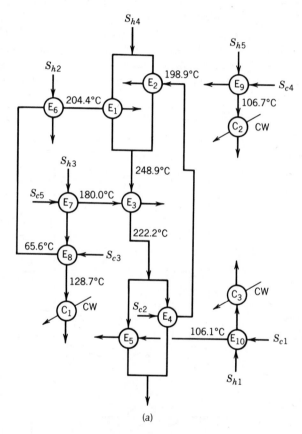

(a)

**FIGURE 6.13** (a) An initial network for the 10SP1 problem (Nishida et al., 1977). (b) A temperature-interval (or grid) diagram representing the initial network shown in Figure 6.13a. (c) A heat-content diagram representing the initial network in Figure 6.13a. (d) A heat-content diagram representing an improved network shown in Figure 6.13f. (e) A comparison of two options for shifting the enlarged cooler $C_3''$ ($= C_3 + E_1$) shown in Figure 6.13c from $S_{h1}$ to $S_{h3}$ and to $S_{h5}$. (f) An improved network for the 10SP1 problem. (g) An illustration of the application of path/loop breaking to improve an initial network for the 10SP1 problem given in Figures 6.13 a and b. This involves: (i) deleting $E_2$ from loop A, ($E_2$, $E_4$); (ii) deleting $E_1$ from loop B, ($E_1$, $E_5$, $E_{10}$, $C_3$, $C_1$, $E_8$, $E_6$); and (iii) deleting $C_3$ from path C, ($C_1$, $C_3$). (h) A temperature-interval (or grid) diagram representing an improved network with a minimum of 10 units obtained after path/loop breaking shown in Figure 6.13g. See also Figure 6.13f.

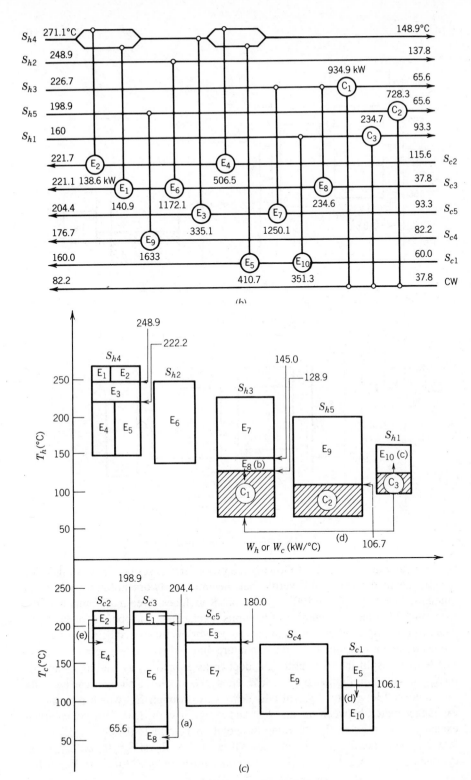

**FIGURE 6.13** (Continued)

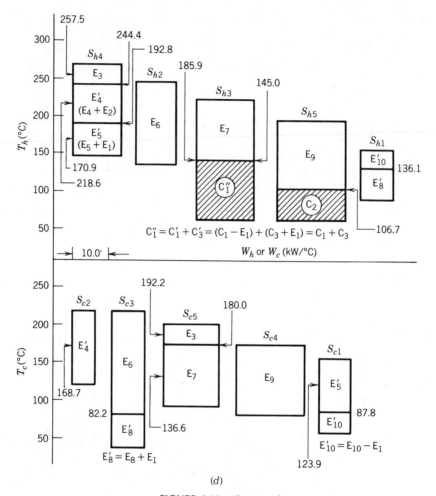

**FIGURE 6.13** (Continued)

by Nishida et al. (1977). By following the procedures described in Sections 6.3D and E, the reader can easily verify that the initial network utilizes a minimum amount of cooling utility (1903.4 kW), but it includes three more units than the most probable minimum number of ten units (exchangers and coolers; no heaters). The evolutionary synthesis of an improved network is illustrated below using both the heat-content diagram and the temperature–interval (or grid) diagram.

We first unsplit the two splitting subnetworks $E_1$ and $E_2$ as well as $E_4$ and $E_5$ on $S_{h4}$, as shown in Figures 6.13c and d. Next, the following changes can be made on Figure 6.13c: (a) on $S_{c3}$, shift $E_1$ downward through $E_6$ (which is a single-exchanger match between $S_{h2}$ and $S_{c3}$) and merge it with $E_8$ to form an enlarged exchanger $E_8' = E_8 + E_1$ according to evolutionary rule 1; (b) on $S_{h3}$, $E_8$ is enlarged to take into account the increased heat load of $E_8$ on $S_{c3}$; (c) on $S_{h1}$, $C_3$ is enlarged to $C_3' = C_3 + E_1$ and $E_{10}$ is simultaneously reduced to $E_{10}' = E_{10} - E_1$;

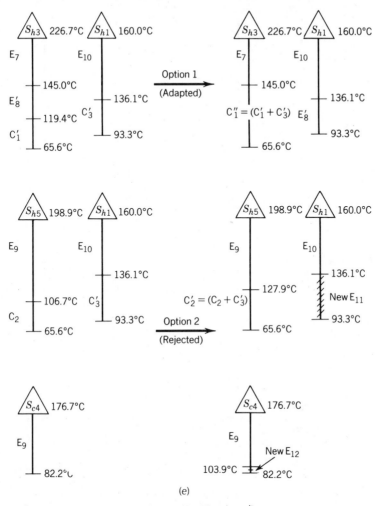

(e)

**FIGURE 6.13** (Continued)

(d) on $S_{c1}$, $E_{10}$ is also reduced to $E'_{10} = E_{10} - E_1$; and (e) on $S_{c2}$, $E_2$ is merged with $E_4$ to form an enlarged exchanger $E'_4 = E_4 + E_2$. These changes (a)–(e), shown in Figures 6.13c, have effectively eliminated $E_1$ and $E_2$ on cold streams $S_{c3}$ and $S_{c2}$, respectively, and also have not led to extensive structural changes to adjacent or other subnetworks (evolutionary rule 1). The corresponding modifications for eliminating $E_1$ and $E_2$ from hot streams can be seen clearly by comparing $S_{h4}$ in Figures 6.13c and d. In particular, these modifications are as follows: (a) shift $E_2$ downward through $E_3$ to merge it with $E_4$ to form an enlarged exchanger $E'_4 = E_4 + E_2$; (b) shift $E_1$ downward through $E_2$ to $E_4$ and merge it with $E_5$ to form an enlarged exchanger $E'_5 = E_5 + E_1$; and (c) shift $E_3$ upward to the highest-temperature portion of $S_{h4}$. Note that the directions (upward and downward) of these shifts and the corresponding arrangements of $E_3$, $E'_4$ ($= E_4 + E_2$), and $E'_5$ ($= E_5 + E_1$)

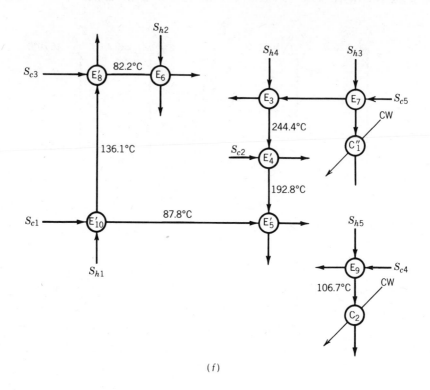

(f)

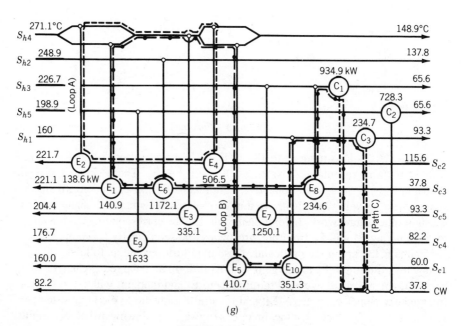

(g)

**FIGURE 6.13** (Continued)

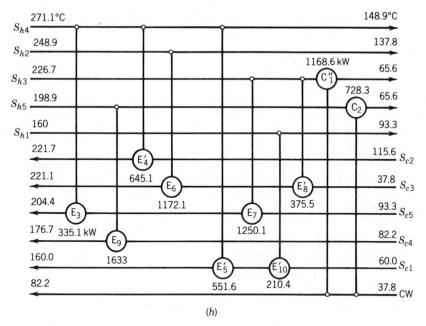

(h)

**FIGURE 6.13**  (Continued)

on $S_{h4}$ are uniquely determined by comparing the arithmetic averages of input and output temperatures of $E_3$, $E_4'$, and $E_5'$ on $S_{c5}$, $S_{c2}$, and $S_{c1}$, respectively, as shown in Figure 6.13d, according to the thermodynamic matching rule. The last change to be made is shown as (d) in Figure 6.13c. This involves shifting the enlarged cooler $C_3' = C_3 + E_1$ from $S_{h1}$ to $S_{h3}$ and simultaneously shifting the enlarged exchanger $E_8' = E_8 + E_1$ from $S_{h3}$ to $S_{h1}$. The resulting composite cooler on $S_{h3}$, $C_1''$, has the same heat load as $C_1' + C_3'$, as shown in Figure 6.13d. Figure 6.13f shows the improved network represented by the heat-content diagram of Figure 6.13d. This network features a maximum thermodynamic efficiency during heat exchange, a minimum amount of cooling utility, and a minimum number of units. It is also identical to the minimum-cost network reported in Nishida et al. (1977).

Figure 6.13e illustrates two options available to shifting the enlarged cooler $C_3' = C_3 + E_1$ from $S_{h1}$ to either $S_{h3}$ or $S_{h5}$, previously shown in Figure 6.13c. Evolutionary rule 2b suggests that $C_3'$ should be shifted from $S_{h1}$ to $S_{h5}$ (option 2) instead of $S_{h3}$ (option 1), so as to result with a composite cooler $C_2' = C_2 + C_3'$ with a lower arithmetic average of its input and output temperatures. This recommendation, however, is rejected because of evolutionary rule 1. In particular, Figure 6.13e shows that the structural changes required after merging $C_3'$ with $C_1'$ to form an enlarged cooler $C_1'' = C_1' + C_3'$ on $S_{h3}$ and shifting $E_8'$ from $S_{h3}$ to $S_{h1}$ (option 1) can be localized in a single subnetwork. Further, no cold-stream matches are affected by these changes. Therefore, option 1 closely follows the guidelines provided by evolutionary rule 1. By contrast, as option 2 shows, when $C_3$ is merged with $C_2$ to form an enlarged cooler $C_2' = C_2 + C_3'$ on $S_{h5}$, two new exchangers

$E_{11}$ and $E_{12}$ must be introduced to take into account the unmatched heat load. This addition of new exchangers represents a violation of evolutionary rule 1. Consequently, even though option 2 represents a thermodynamically more efficient use of the cooling utility compared with option 1, it is not adopted because of the increased network investment associated with adding two new exchangers $E_{11}$ and $E_{12}$. This comparison clearly demonstrates the importance of sequentially applying different evolutionary rules, presented in Section 6.3.G.3, so as to take into account both thermodynamic and economical aspects of the multiobjective HEN synthesis problem.

The preceding evolutionary changes can also be made on a temperature-interval (or grid) diagram representing the initial network, as shown in Figure 6.13$g$. By referring to the latter figure and applying the evolutionary rules and concepts of path and loop breaking, we make the following changes.

Loop A: ($E_2$, $E_4$) (Figure 6.13$g$)
Heat load: (138.6, 506.5) kW
  | Evolutionary rule 1
  ↓ (Loop breaking)
Broken loop: (0, $E_4' = E_4 + E_2$) (Figure 6.13$h$)
Heat load: (0, 645.1) kW

Path C: ($C_1$, $C_3$) (Figure 6.13$g$)
Heat load: (934.9, 234.7) kW
  | Evolutionary rule 2b
  ↓ (Path breaking)
Broken path: ($C_1''$, 0) (Figure 6.13$h$ and $d$)
Heat load: (1168.6, 0) kW

Loop B: ($E_1$, $E_5$, $E_{10}$, $\underline{C_3}$, $\underline{C_1}$, $E_8$, $\underline{E_6}$) (Figure 6.13$g$)
Heat load: (140.9, 410.7, 351.3, 934.9, 234.7, 234.6, 1172.1)
  | Evolutionary rules 3 and 4
  ↓ (Loop breaking)
Broken loop: (0, $E_5'$, $E_{10}'$, $\underline{C_1''}$, $E_8'$, $\underline{E_6}$) (Figure 6.13$h$)
Heat load: (0, 551.6, 210.4, 1168.6, 375.5, 1172.1)

Note that in breaking up loop B, heat loads of coolers $C_3$ and $C_1$ are not to be changed. These coolers have already been merged together as a single cooler $C_1''$ (= $C_3 + C_1$) (see Figure 6.13$d$) in breaking up path C. Also, since $E_6$ is involved in a single-exchanger match, we should not change its heat load during evolutions. The other heat loads resulting from breaking up loop B are found as follows:

$$E_5' = E_5 + E_1 = 410.7 + 140.9 = 551.6 \text{ kW}$$

$$E_{10}' = E_{10} - E_1 = 351.3 - 140.9 = 210.4 \text{ kW}$$

$$E_8' = E_8 + E_1 = 234.6 + 140.9 = 375.5 \text{ kW}$$

Figure 6.13$h$ gives a temperature-interval (or grid) diagram representing the improved network obtained after the path/loop breaking shown in Figure 6.13$g$.

## 5.  Reducing the Number of Units by Stream Splitting and Bypassing

Thus far we have introduced two useful applications of stream splitting in the synthesis of HENs. The first application is concerned with using stream splitting to overcome the difficulty caused by $\Delta T_{min}$ violations in the synthesis of an initial HEN, as discussed in Section 6.3.F.3 and illustrated in Figures 6.7$b$ and $c$. The other application of stream splitting is aimed at reducing the number of units. We see in Section 6.3.B.1 and Figures 6.2$b$ and $c$ that when the heat-capacity flow rate of a cold (or hot) process stream is markedly greater than those of hot (or cold) process streams, the use of stream splitting leads to networks of fewer units and lower investment costs. To help the reader better understand how the latter conclusion is reached, we present in Figure 6.14$a$ an initial network for a 4-stream problem that requires the use of heating utility only.

As seen in Figure 6.14$a$, the initial network has a total of 1 heater and 6 exchangers. The most probable number of units for the network is, however, only 4 (1 heater and 3 exchangers). To reduce the number of units, we identify the following three loops in the initial network of Figure 6.14$a$:

Loop 1: ($E_1$, $E_5$)     Heat load: (621, 1665) kW<br>
Loop 2: ($E_2$, $E_4$)     Heat load: (1858, 1073) kW<br>
Loop 3: ($E_3$, $E_6$)     Heat load: (1495, 1055) kW

By merging together the two exchangers in each of these loops (evolutionary rule 1) and using stream splitting, we obtain an improved network with a minimum number of 4 units, as shown in Figure 6.14$b$. This example demonstrates that the combined use of loop breaking and stream splitting is an effective technique to reduce the number of units in the synthesis of HENs.

Wood et al. (1984) show that stream bypassing is another useful tool to reduce the number of units in the synthesis of HENs. Figure 6.15$a$ shows a 3-unit, splitting network for a 3-stream HEN synthesis problem. In the figure, a loop consisting of $E_1$ and $E_3$ is identified. To reduce the number of units from 3 to the most probable minimum number of 2, Wood et al. suggest the use of stream bypassing together with loop breaking and stream splitting. This results in an improved network with only 2 units, as shown in Figure 6.15$b$. It is apparent from both Figures 6.14 and 6.15 that *the combined use of loop breaking and stream splitting/bypassing* is an important technique in the evolutionary synthesis of improved HENs.

## H.  Systematic Synthesis of Multipass Shell-and-Tube HENs

## 1.  Thermal Design of Minimum-Cost Multipass Units

### a.  THERMAL DESIGN OF MULTIPASS UNITS

Multipass shell-and-tube exchangers, heaters, and coolers are often used in the process industries. Such equipment allows for a great deal of flexibility to a given

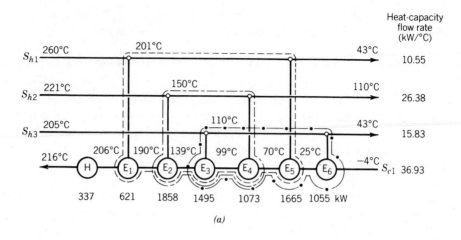

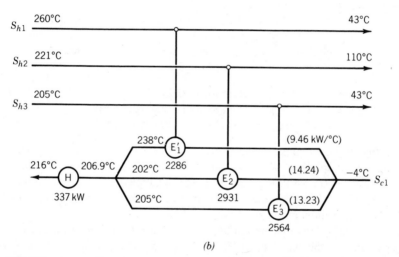

**FIGURE 6.14** An illustration of reducing the number of units through loop breaking and stream splitting in a four-stream problem with a $\Delta T_{min}$ of 11°C. (a) The initial network. (b) The improved network.

heat-transfer process design. For example, multiple shell passes may be used (1) to improve the mean temperature difference between hot and cold streams for a given exchanger in which the stream flows are not parallel or countercurrent, and (2) to decrease the amount of floor space required for a given exchanger. Multiple tube passes may be used (1) to increase the fluid velocity in the tubes, thereby increasing the overall heat transfer coefficient, and (2) to increase or decrease the available heat transfer area without increasing or decreasing the shell length. The numbers of shell and tube passes are limited by the maximum allowable pressure drop and by space considerations. Further, the number of tube passes is also lim-

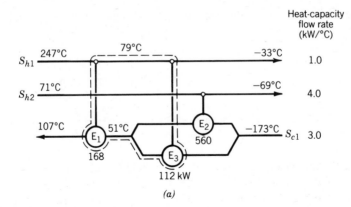

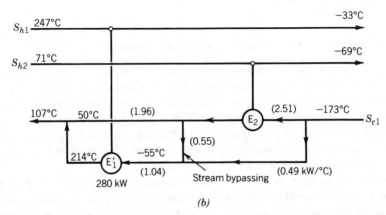

**FIGURE 6.15** An illustration of reducing the number of units through loop breaking and stream splitting/bypassing in a three-stream problem. Heat-capacity flow rates of splitting fractions are given in parentheses (Wood et al., 1984). (a) The initial network. (b) The improved network.

ited by the amount of fluid passing through the tubes and by the maximum permissible shell diameter.

The basic equation for the thermal design (i.e., finding the heat-transfer area) of multipass exchangers, heaters, or coolers is (Bell, 1984)

$$A = \frac{Q}{U(\text{MTD})} = \frac{Q}{UF_N(\text{LMTD})} \tag{6.14}$$

Here $F_N$ is a correction factor which is so determined that when it is multiplied by the logarithmic-mean temperature difference (LMTD) for a single-pass countercurrent exchanger, the product $F_N(\text{LMTD})$ represents the true mean temperature difference (MTD) of an equivalent multipass exchanger. It is well-known that $F_N$ is commonly correlated graphically as a function of the number of shells $N$, the number of tube passes, the capacity ratio $R$, and the thermal efficiency $P$. The latter are defined by

$$R = \text{capacity ratio} = \frac{W_c}{W_h} = \frac{T_h - T_h^*}{T_c^* - T_c}$$

$$P = \text{thermal efficiency} = \frac{T_c^* - T_c}{T_h^* - T_c} \tag{6.15}$$

For a thermodynamically feasible and efficient multipass exchanger (heater or cooler), the number of shells $N$ to be used in series should be chosen such that $F_N$ is at least 0.8 at the given values of $R$ and $P$. Also, the use of up to six shells in series is quite common in the chemical process industries, particularly in crude-unit preheat-recovery trains.

## b.  MINIMUM-COST MULTIPASS EXCHANGERS, HEATERS, AND COOLERS

For the purpose of estimating the heat-transfer area and the corresponding investment cost of a multipass exchanger (heater or cooler) with $N$ shells in series, it is not important to know the exact number of tube passes as long as the latter is at least twice the number of shells. This is evident by observing the available graphical correlations of $F_N$ as functions of $R$ and $P$, such as those found in Bell (1984), which show only a negligible difference among the values of $F_N$ at different tube passes for the same values of $N$, $R$, and $P$.

To find the number of shells that corresponds to a minimum-cost multipass exchanger, it is important to recognize two points. First when replacing a single-pass countercurrent exchanger of a heat-transfer area $A$ ($m^2$) and an investment cost $aA^b$ (\$) with a multipass exchanger of $N$ shells in series and a MTD correction factor $F_N$, the heat-transfer area and the corresponding investment cost of each of the shells in series are commonly assumed to be identical, being equal to $(A/F_N N)$ $m^2$ and $a(A/F_N N)^b$ \$, respectively. The total investment cost for the $N$ shells in series is then $N \cdot a(AF_N N)^b$ \$. Second, for thermodynamically feasible designs of multipass exchangers with $0.8 \le F_N \le 1.0$ and $N = 2$–6, it can be shown that

$$aA^b < N \cdot a\left(\frac{A}{F_N N}\right)^b < (N+1) \cdot a\left[\frac{A}{F_{N+1}(N+1)}\right]^b \tag{6.16}$$

for $0 < b < 1$ ($b$ is typically 0.6). This inequality suggests that the investment cost of multipass exchange of $(N+1)$ shells is always higher than that of $N$ shells, and the latter is always higher than that of a single-pass countercurrent exchanger designed at the same values of $R$ and $P$. Consequently, a key step in the thermal design of minimum-cost multipass exchangers is to find the required number of shells to be used in series.

c. ESTIMATING THE NUMBER OF SHELLS ($N$)

Bell (1978) has proposed a graphical procedure, which is similar to the construction of operating lines in stagewise-process design, to estimate the value of $N$. This procedure utilizes the inlet and outlet temperatures of both hot and cold streams, as illustrated in Figure 6.16 in which $N$ is about 3. Liu et al. (1985) have found that for $N > 3$, the operating-line method of Bell frequently cannot be used to predict feasible designs of multipass exchangers. Specifically, by following the procedure used in the development of the well-known Kremser equation in stagewise process design (McCabe and Smith, 1976), it can be shown that the construction of operating lines in Figure 6.14 can be represented analytically by

$$\frac{RP}{1 - RP} = \frac{T_h - T_h^*}{T_h^* - T_c} = R + R^2 + \cdots + R^N = \frac{R(1 - R^N)}{1 - R} \quad (6.17)$$

Simplifying Equation (6.17) gives an explicit expression for finding the number of shells at given values of $R$ and $P$ by the operating-line method of Bell,

$$N = \frac{\ln\left(\dfrac{1 - R}{1 - RP}\right)}{\ln R} \quad (6.18)$$

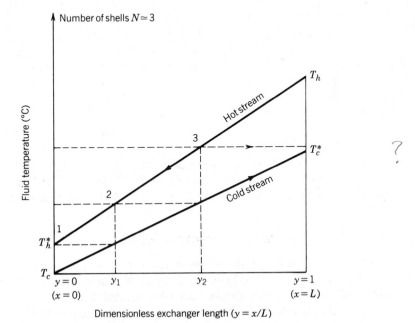

FIGURE 6.16 An illustration of the operating-line method of Bell (1978) for finding the number of shells in a multipass exchanger (heater or cooler).

and the relationship

$$P = \frac{1 - R^N}{1 - R^{N+1}} \tag{6.19}$$

To evaluate the applicability of Bell's method, $P$ is calculated from Equation (6.19) for given values of $R$ and $N$, and the values of $F_N$ corresponding to a wide range of values of $P$, $R$ and $N$ are then read from available graphical correlations (Bell, 1984). The results show that for $N = 4$ to $6$, $R \geq 2.0$ for $P \leq 0.50$, Bell's method results in $F_N < 0.8$ and does not predict feasible designs of multipass exchangers.

Based on the preceding discussion and the inequality given by Equation (6.16), the following procedure is recommended for estimating the number of shells to be used in series.

1. Assume a value of $N$ ($\leq 6$).

2. Determine the value of $F_N$ from given values of $R$ and $P$ by using available graphical correlations (Bell, 1984).

3. If the resulting $F_N < 0.8$, we increase the value of $N$ by increments of 1 until $F_N$ is at least 0.8. For the value of $N$ that corresponds to a transition from $F_N < 0.8$ to $F_N \geq 0.8$, this value of $N$ is the number of shells to be used in a minimum-cost multipass exchanger. In applying this procedure, it is important to check the graphical correlations of $F_N = f(R, P, N)$ to ensure that $F_N$ is not very sensitive to small changes in values of $R$ and $P$ near the given values of $R$ and $P$ at the selected value of $N$. In other words, $F_N$ should always be checked to see that it does not fall into an asymptotic $- F_N$ region in the graphical correlation of $F_N = f(R, P, N)$ at given values of $R$ and $P$ (Taborek, 1979); and we should consider the lower limit of $F_N = 0.8$ as an *apparent, but not absolute*, minimum value of $F_N$. If $F_N$ is very sensitive to small changes in values of $R$ and $P$, the chosen value of $N$ may not predict a feasible design of multipass exchanger even when $F_N > 0.8$. In the latter case, we should further increase the value of $N$ by increments of 1 and check again if $F_N$ is still very sensitive to small changes in $R$ and $P$ at the selected value of $N$.

4. When $F_N$ is very sensitive to small changes in values of $R$ and $P$ for $N = 2$ to $6$, or when $F_N < 0.8$ at given values of $R$ and $P$ for $N = 2$ to $6$, the use of a single-pass countercurrent exchanger (heater or cooler) is recommended.

## 2.  Systematic Synthesis of Multipass HENs

A simple approach to the systematic synthesis of multipass HENs can now be presented (Liu et al., 1985). First, the given HEN synthesis problem is solved by following the step-by-step procedure described in Sections 6.3.C–6.3.G, assuming initially that only single-pass countercurrent units (exchangers, heaters, and coolers) are to be used. The necessary heat-transfer areas of different single-pass units in the resulting energy-optimum and minimum-cost network are determined. Second, with the resulting heat-transfer areas of single-pass units, the minimum-cost multipass units that achieve the same objectives (with the same values of $R$ and $P$) are determined by finding the required number of shells $N$ and the corresponding value of $F_N$. The investment costs of different multipass units are then determined from $N \cdot a(A/F_N N)^b$.

## 3. Illustrative Examples

Figure 6.17 illustrates the minimum-cost (final) and two nearly minimum-cost single-pass networks for a four-stream problem 4SP1 (Lee et al., 1970), synthesized by the procedure presented in Sections 6.3.C–6.3.G. (See Appendix 6.2 for specifications of stream and design data for the 4SP1 problem.) These networks are energy-optimum and have an identical, minimum utility-operating cost of $10,426

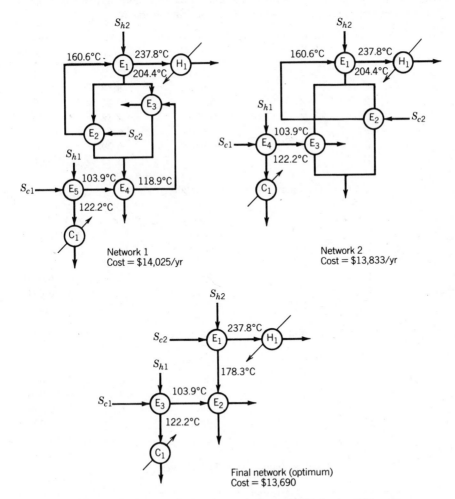

**FIGURE 6.17** Optimum and suboptimum single-pass HENs for the 4SP1 problem.

per year. By using the proposed approach, the results of synthesizing the corresponding multipass networks with an apparent minimum $F_N$ of 0.8 are summarized in Table 6.7. The utility-operating costs of all multipass networks are identical and equal to the minimum utility cost of the single-pass networks. The total annual costs of these energy-optimum multipass networks differ only by a few percent; this small cost difference is similar to that observed for the single-pass networks shown in Figure 6.15.

Based on the comparison described above and the results obtained for other reported synthesis problems with 5–10 process streams, including problems 5SP1, 6SP1, 7SP1 (Masso and Rudd, 1969), and 10SP1 (Pho and Lapidus, 1973), which

**TABLE 6.7  Design and Cost: Multipass Versions of the Optimum and Suboptimum Single-Pass Networks Shown in Figure 6.15 for the 4SP1 Problem[a]**

| Network | Unit | Number of Shells, $N$ | LMTD Correction Factor, $F_N$ | Area (m$^2$) | Investment Cost ($) |
|---|---|---|---|---|---|
| No. 1 | $E_1$ | 3 | 0.874 | 23.06 | 16,103 |
|  | $E_2$ | 2 | 0.924 | 8.38 | 7,215 |
|  | $E_3$ | 2 | 0.925 | 10.06 | 8,045 |
|  | $E_4$ | 2 | 0.993 | 4.32 | 4,643 |
|  | $E_5$ | 2 | 0.980 | 6.64 | 6,057 |
|  | $C_1$ | 2 | 0.976 | 6.29 | 6,071 |
|  | $H_1$ | 1 | — | 3.72 | 3,203 |
|  | (Total network cost = $15,560/yr) | | | | $51,337 |
| No. 2 | $E_1$ | 3 | 0.874 | 23.06 | 16,103 |
|  | $E_2$ | 2 | 0.860 | 10.07 | 8,410 |
|  | $E_3$ | 2 | 0.887 | 12.92 | 9,587 |
|  | $E_4$ | 2 | 0.980 | 6.64 | 6,057 |
|  | $C_1$ | 2 | 0.976 | 6.29 | 6,071 |
|  | $H_1$ | 1 | — | 3.72 | 3,203 |
|  | (Total network cost = $15,370/yr) | | | | $49,431 |
| Final | $E_1$ | 4 | 0.888 | 29.28 | 20,652 |
|  | $E_2$ | 2 | 0.828 | 19.88 | 12,939 |
|  | $E_3$ | 2 | 0.980 | 6.64 | 6,056 |
|  | $C_1$ | 2 | 0.976 | 6.29 | 6,071 |
|  | $H_1$ | 1 | — | 3.72 | 3,203 |
|  | (Total network cost = $15,317/yr) | | | | $48,921 |

[a]From Liu et al. (1985).

are specified in Appendix 6.2, a number of general conclusions and observations can be summarized as follows:

1. For all problems solved, $F_N$ for each exchanger, heater, or cooler was not sensitive to small changes in values of $R$ and $P$ for $N = 2$ to 6 under the given problem specifications and the use of an apparent minimum $F_N$ of 0.8 appeared to be feasible.

2. The relative cost ranking of different energy-optimum single-pass networks synthesized by the thermoeconomic approach presented in Section 6.3.G.2 is generally preserved in the corresponding multipass networks obtained by the proposed method for multipass HEN synthesis.

## I. Other Practical Aspects of HEN Synthesis

### 1. Deviations from Common Simplifying Assumptions

The common simplifying assumptions that have been included in the early development of techniques for synthesizing HENs were summarized immediately following Equation (6.5) in Section 6.3.A. In this section, we briefly describe techniques applicable to solving HEN synthesis problems that deviate from those simplifying assumptions.

a. Use of Multipass Shell-and-Tube Units (Exchangers, Heaters, and Coolers)

See Section 6.3.H.

b. Phase Changes of Process Streams and Temperature-Dependent Heat-Capacity Flow Rate

Figure 6.18a shows a heat-content diagram representing a HEN synthesis problem with a condensation of a hot stream $S_{h1}$ and a temperature-dependent heat-capacity flow rate that exists in the vapor-phase of $S_{h1}$. As seen in the figure, $S_{h1}$ is initially in a vapor phase with an input temperature above the average condensation temperature of the stream. The vapor stream is cooled to the average condensation temperature by exchanging its sensible heat with a cold process stream $S_{c1}$ through exchanger $E_1$. At the average condensation temperature, the latent heat of phase change (condensation) for $S_{h1}$, denoted by $\lambda_{h1}$ in Figure 6.18a, is available as *a pseudo-heating utility* to heat up cold process stream $S_{c1}$ to a higher temperature.

Figure 6.18b illustrates the case where a cold process stream $S_{c1}$ is vaporized during heat exchange. In this case, the latent heat of vaporization of $S_{c1}$, denoted by $\lambda_{c1}$, serves as *a pseudo-cooling utility* to cool down a hot process stream $S_{h1}$.

In Tables 6.4–6.6, we illustrate the use of a problem table to determine the minimum heating- and cooling-utility requirements and the pinch point for HEN synthesis problems without phase changes of process streams. In those tables, a

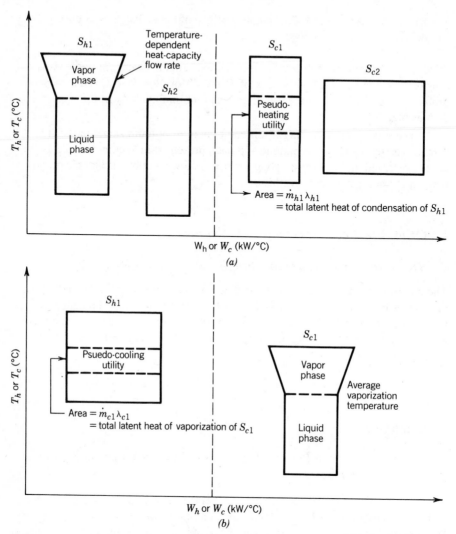

**FIGURE 6.18** (a) A heat-content diagram representing the condensation of $S_{h1}$ and the concept of pseudo-heating utility. $\dot{m}_{h1}$ and $\lambda_{h1}$ are the mass flow rate and latent heat of condensation of $S_{h1}$, respectively. (b) A heat-content diagram representing the vaporization of $S_{c1}$ and the concept of pseudo-cooling utility. $\dot{m}_{c1}$ and $\lambda_{c1}$ are the mass flow rate and latent heat of vaporization of $S_{c1}$, respectively.

*single*, vertical "stream line" directed with an arrow is used to represent the input- and output-temperature levels as well as the heat-capacity flow rate of a process stream. When a process stream encounters a phase change, the stream line will have discontinuous, *multiple* vertical segments in a problem table. Each segment represents a phase condition with its own initial and final temperature levels as well as its own heat-capacity flow rate. For example, three segments are needed to represent the following cold process stream: (1) liquid phase from 37.8 to 105°C with a heat-capacity flow rate $W_c$ of 5 kW/°C; (2) two-phase condition from 105 to 108°C with $W_c = 10$ kW/°C; and (3) vapor phase from 108 to 230°C with $W_c = 4$ kW/°C. The same procedure as illustrated in Tables 6.4–6.6 can be used to carry out a problem-table analysis of such HEN synthesis problems with phase changes and temperature-dependent heat-capacity flow rates.

### C. VARIABLE OVERALL HEAT-TRANSFER COEFFICIENTS FOR EXCHANGES BETWEEN TWO PROCESS STREAMS AND BETWEEN PROCESS AND UTILITY STREAMS

In Appendix 6.2, the design data specified for illustrative examples presented in this chapter include three *constant*, overall heat-transfer coefficients ($U$) for exchangers, heaters, and coolers. In practice, the value of $U$ may depend on the nature of stream matches involved. It is obvious that the value of $U$ for a match between a gas and a liquid process stream will be different from that between two liquid process streams. This difference then raises an important question: "Will the difference in values of $U$'s affect the applicability of the synthesis procedure discussed thus far? In particular, will the thermodynamic matching rule presented in Section 6.3.F.1 remain effective in the synthesis of HENs with variable $U$'s?" We answer this question below by briefly summarizing the results and recommendations from an evaluation of stream-matching rules for HEN synthesis with variable $U$'s (Liu, 1982a).

Umeda et al. (1978) proposed the following stream-matching rule: "Heat exchange is made between the two streams with the nearest values of heat-transfer coefficient." These authors have further observed that "It is very often found that simpler heat exchanger networks are found by this rule; and there are no practical difficulties in applying this method to obtain better networks." It is not obvious, however, as to what is meant by "the nearest values" of the overall heat-transfer coefficient $U$'s. For example, when a wide range of $U$ values is available for the possible process/process and process/utility stream matches in an exchanger network, it is not clear whether the design engineer should match process and/or utility streams in an order of decreasing or increasing $U$'s. In applying the matching rule of Umeda et al. (1978), it is assumed that the exchange or matching of a process stream with a heating or cooling utility is restricted to its last matching step.

The other approach to HEN synthesis with variable $U$'s was proposed by Hohmann (1971), using the concept of a stream contention. As defined by Hohmann, stream contention occurs in a stream system if two hot streams have the same potential heat flux [i.e., $U(T_h - T_c)$] with respect to the same portion of a given cold stream. Throughout the range of the concentration, the two hot streams will be able to supply heating at the same cost in terms of area. Contention can also exist for two cold streams and a single hot stream, or for several hot and cold streams. According to Hohmann, a stream-contention or heat-flux matching rule is to match process streams of equal heat fluxes over specified temperature intervals.

Table 6.8 shows the results of applying the heat-flux and thermodynamic matching rules to a simple 3-stream problem 3SP3 (Hohmann, 1971), specified in Appendix 6.2. The same design data given for "other problems" in the appendix are used, except that the values of $U$ are specified in Table 6.8. As can be seen from the table, the heat-flux matching rule generates a more expensive network with two more exchangers than that synthesized by the thermodynamic matching rule. Note that a practical difficulty associated with applying the heat-flux matching rule is to find the input and output temperatures of process streams for intermediate exchangers that lead to stream matches with equal heat fluxes (see, for example, the output temperature of $S_{c1}$ from exchanger II, 142.8°C and the input temperature of $S_{c2}$ to exchanger III, 141.1°C, in Table 6.8). This step requires trial-and-error calculations, which would become prohibitive for synthesis problems with more than 6 process streams.

Table 6.9 shows the results obtained for a 4-stream problem, the 4SP1 problem (Lee et al., 1970) specified in Appendix 6.2. The same design data specified in the appendix are used, except that the values of $U$ are given in Table 6.10 for three cases evaluated. The latter values are chosen in such a manner that any possible prejudices caused by the numerical ordering (ranking) of $U$'s can be avoided.

An examination of the results of comparative evaluations shown in Tables 6.8 and 6.9 suggests that *regardless of the values of* U *associated with process and/ or utility stream matches*, the use of the thermodynamic matching rule almost always leads to minimum-cost or nearly minimum-cost HENs. Because of its simplicity and effectiveness demonstrated, the thermodynamic matching rule is thus recommended as a *general* stream-matching heuristic for the systematic synthesis of energy-optimum and minimum-cost HENs.

d.  MATCH-DEPENDENT MINIMUM-APPROACH TEMPERATURES FOR EXCHANGES BETWEEN TWO PROCESS STREAMS AND BETWEEN PROCESS AND UTILITY STREAMS

See Section 6.3.D, particularly the discussion related to Table 6.5.

## 2.  *The Double-Temperature-of-Approach (DTA) Concept*

An objective of this chapter is to present an organized discussion and illustration of the technical background and practical insight needed to better understand and apply the current and developing computer-aided design (CAD) packages for HENs. In the reference section, we give the literature sources and mailing addresses of developers of a number of CAD packages. The latter include:

- HENS (Elshout and Hohmann, 1979)
- HEXTRAN (Simulation Sciences, Inc., 1984)
- ADVENT (Gautnam and Smith, 1985)
- MAGNETS (Floudas et al., 1986)
- RESHEX (Saboo et al., 1986)

**TABLE 6.8  Results of Applying Thermodynamic and Stream-Contention (Heat-Flux) Matching Rules to the 3SP3 Problem**[a]

| Matching Rule | Exchange Number | Hot Streams Temperature (°C) | Name | $Q$ (kW) | Cold Streams Temperature (°C) | Name | $Q$ (kW) | $U$ (kW/m² – °C) | Area (m²) | Investment Cost ($) |
|---|---|---|---|---|---|---|---|---|---|---|
| Thermodynamic matching rule | I | 204.4–160.0 | $S_{h1}$ | −206.52 | 126.1–160.0 | $S_{c1}$ | +206.52 | 0.56781 | 9.34 | 556 |
|  | II | 204.4–160.0 | $S_{h1}$ | −258.3 | 126.1–160.0 | $S_{c2}$ | +258.3 | 0.85171 | 7.98 | 499 |
|  |  |  |  |  |  |  |  |  | 17.32 | 1055 |
| Heat-flux (stream-contention) matching rule (Hohmann, 1971) | I | 204.4–194.4 | $S_{h1}$ | −104.34 | 142.8–160.0 | $S_{c1}$ | +104.34 | 0.56781 | 3.83 | 326 |
|  | II | 194.4–170.6 | $S_{h1}$ | −247.08 | 126.1–142.8 | $S_{c1}$ | +102.39 | 0.56781 | 3.75 | 322 |
|  | III | 194.4–170.6 | $S_{h1}$ |  | 141.1–160.0 | $S_{c2}$ | +144.79 | 0.85171 | 5.33 | 397 |
|  | IV | 170.6–160.0 | $S_{h1}$ | −113.43 | 126.1–141.1 | $S_{c2}$ | +113.43 | 0.85171 | 4.18 | 344 |
|  |  |  |  |  |  |  |  |  | 17.09 | 1419 |

[a]From Liu (1982a).

231

TABLE 6.9   Evaluation of Stream Matching Rules for Variable Overall

| | | Case 1 | | |
|---|---|---|---|---|
| Matching rule | Investment Cost ($) | Utility Operating Cost ($/yr) | Total Cost ($/yr) | Units Used |
| Rule 1 (thermody-namic matching rule) | 39,400 | 10,414 | 14,354 | 3E, 1H, 1C[b] |
| Rule 2 (match in or-der of decreasing $U$'s) | 38,810 | 10,414 | 14,295 | 3E, 1H, 1C |
| Rule 3 (match in or-der of increasing $U$'s) | 27,537 | 42,836 | 45,590 | 3E, 1H, 2C |

[a]From Liu (1982a).

[b]3E, 1H, and 1C mean that 3 exchangers, 1 heater, and 1 cooler are used.


TABLE 6.10   Values of Overall Heat-Transfer Coefficients ($U$) Used in the Evaluation of Stream Matching Rules for the 4SP1 Problem[a]

| $U$ (kW/m$^2\cdot$°C) | Case 1 | Case 2 | Case 3 |
|---|---|---|---|
| $U_1 \equiv U_{S_{h2}/S_{c1}}$ | 0.42586 | 0.56781 | 0.85171 |
| $U_2 \equiv U_{S_{h2}/S_{c2}}$ | 0.56781 | 0.85171 | 0.42581 |
| $U_3 \equiv U_{S_{h1}/S_{c1}}$ | 0.85171 | 0.42586 | 0.56781 |

[a]From Liu (1982a). $U_4 \equiv U_{S_{h1}/S_{c2}} = 0.857$ kW/m$^2\cdot$°C; $U_{heater} = 1.13561$ kW/m$^2\cdot$C°; $U_{cooler} = 0.85171$ kW/m$^2\cdot$°C for all cases.


An important idea incorporated in some CAD packages that we have not yet discussed is *the concept of double-temperature-of-approach (DTA)*, which essentially involves the use of two different approach temperatures in solving a HEN synthesis problem. The first approach temperature is the exchanger minimum-approach temperature (EMAT) that we have used throughout our discussion thus far. The second approach temperature is a new term called the heat-recovery approach temperature (HRAT) (Colbert, 1982; Simulation Sciences, Inc., 1982, 1984; O'Reilly, 1985).

a.   HEAT-RECOVERY APPROACH TEMPERATURE (HRAT)

We introduce below the HRAT by considering a HEN synthesis problem with the following 4 process streams:

$$S_{h1}:\ W_{h1} = 2 \text{ kW}/°\text{C (from 150 to 60°C)}$$
$$S_{h2}:\ W_{h2} = 8 \text{ kW}/°\text{C (from 90 to 60°C)}$$

Heat-Transfer Coefficients (U) for the 4SP1 Problem[a]

| | Case 2 | | | | Case 3 | | |
|---|---|---|---|---|---|---|---|
| Investment Cost ($) | Utility Operating Cost ($/yr) | Total Cost ($/yr) | Units Used | Investment Cost ($) | Utility Operating Cost ($/yr) | Total Cost ($/yr) | Units Used |
| 36,600 | 10,414 | 14,074 | 3E, 1H, 1C | 38,800 | 10,414 | 14,294 | 3E, 1H, 1C |
| 40,454 | 15,604 | 19,649 | 3E, 2H, 1C | 24,990 | 30,870 | 33,370 | 3E, 1H, 1C |
| 35,840 | 16,641 | 20,225 | 3E, 1H, 1C | 39,325 | 11,618 | 15,551 | 3E, 1H, 1C |

$$S_{c1}: \quad W_{c1} = 2.5 \text{ kW}/°C \text{ (from 20 to 125°C)}$$
$$S_{c2}: \quad W_{c2} = 3.0 \text{ kW}/°C \text{ (from 25 to 100°C)}$$

A saturated heating steam at 250°C and a cooling-water stream with a temperature range of 20 to 80°C are available as utilities.

By following the procedure discussed in Section 6.3.C.2, we plot a temperature–enthalpy ($T$–$Q$) diagram for individual process streams, as shown in Figure 6.19a. We next construct a "composite" $T$–$Q$ diagram for all process streams. Figures 6.19b and c illustrate the development of composite $T$–$Q$ "curves" for hot and cold streams. For example, the composite $T$–$Q$ "curve" for cold streams includes three temperature intervals with the following enthalpy values shown in Figure 6.19c:

20–25°C $\quad \Sigma Q = 2.5 \text{ kW}/°C \cdot (25 - 20)°C = 12.5 \text{ kW}$
25–100°C $\quad \Sigma Q = [2.5 + 3.0] \text{ kW}/°C \cdot (100 - 25)°C = 412.5 \text{ kW}$
100–125°C $\quad \Sigma Q = 2.5 \text{ kW}/°C \cdot (125 - 100)°C = 62.5 \text{ kW}$

Figure 6.19d depicts a composite $T$–$Q$ diagram for both hot and cold streams. In the diagram, the composite $T$–$Q$ "curves" for hot and cold streams have been moved together in a horizontal direction with the composite "curve" for hot streams being placed above that for cold streams. In making this horizontal movement to prepare a composite diagram, we need to specify *the closest vertical distance of approach or the minimum temperature of approach* between the two composite $T$–$Q$ "curves" for hot and cold streams. This vertical distance or temperature difference is called *the heat-recovery approach temperature (HRAT)* (Colbert, 1982; Simulation Sciences, Inc., 1982, 1984), and it is very important in deter-

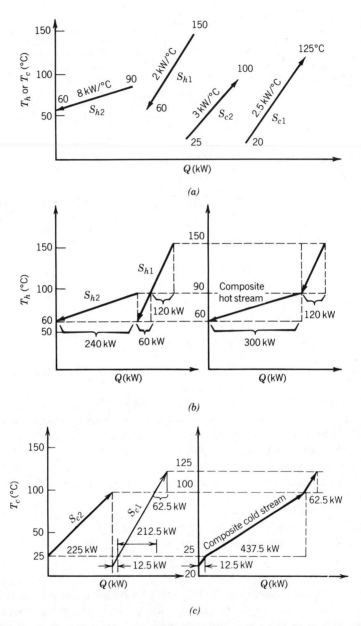

**FIGURE 6.19** An illustration of the composite temperature–enthalpy ($T$–$Q$) diagram and the concept of heat-recovery approach temperature (HRAT) for a four-stream HEN synthesis problem. (a) The $T$–$Q$ diagram for individual process streams. (b) The development of composite $T$–$Q$ "curve" for hot streams. (c) The development of composite $T$–$Q$ "curve" for cold streams. (d) An illustration of the minimum ("threshold") value of heat-recovery approach temperature (HRAT) at which the amount of either heating or cooling utility is zero. (e) An illustration of a heat-recovery approach temperature (HRAT) above its threshold value at which none of the heating- or cooling-utility requirement is zero.

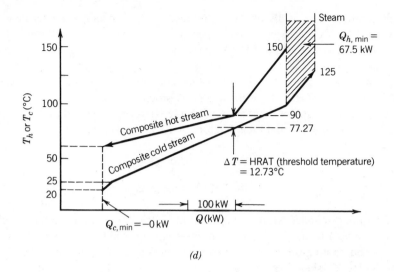

(d)

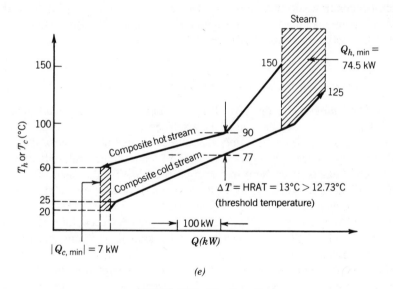

(e)

**FIGURE 6.19** (Continued)

mining the minimum heating- and cooling-utility requirements in a HEN synthesis problem.

For example, in Figure 6.19d, the HRAT is 12.73°C and the corresponding heating- and cooling-utility requirements are:

$$Q_h = 67.5 \text{ kW} \equiv Q_{h,\min}$$

$$Q_c = -0 \text{ kW} \equiv Q_{c,\min}$$

In this example, one of the utility requirements, $Q_c$, happens to be zero and the HRAT of 12.73°C has been given a special name. Specifically, for certain HEN synthesis problems, a value of HRAT may exist such that the required amount of either heating or cooling utility is reduced to zero. The value of HRAT for which this situation takes place is called the "threshold temperature," and it represents a minimum value of HRAT for the synthesis problem.

Figure 6.19e illustrates a situation when the HRAT is increased to 13°C which is slightly above its threshold temperature of 12.73°C. We see that

$$Q_h = 74.5 \text{ kW} > Q_{h,\min}$$

$$|Q_c| = 7.0 \text{ kW} > |Q_{c,\min}|$$

Based on this example, we recognize that the HRAT sets the maximum level of heat recovery to be achieved among process streams, and it also determines the heating- and cooling-utility requirements in a given HEN synthesis problem. As such, the HRAT should be considered as *a heat-duty specification* rather than a stream-temperature specification. We also learn that for certain HEN synthesis problems, there may exist a threshold (minimum) value of HRAT, resulting in the need of no heating or cooling utility.

## b. EXCHANGER MINIMUM-APPROACH TEMPERATURE (EMAT)

As discussed previously, the EMAT is the minimum-approach temperature, or $\Delta T_{\min}$, that we have used thus far in the synthesis of HENs. The assignment of an EMAT (or a $\Delta T_{\min}$) does not imply that all exchangers in a network synthesized have approach temperatures equal to the EMAT value. In general, only one or two exchangers (heaters or coolers) may exhibit the EMAT value, and those are the limiting units in a network. Also, for HEN synthesis problems with phase changes of process streams, the EMAT value may occur with an exchanger.

In such CAD packages as HEXTRAN (Simulation Sciences, Inc., 1982, 1984), both HRAT and EMAT values must be specified for making a computer simulation of a given HEN and for synthesizing a new HEN, and the value of EMAT should always be less than or equal to the HRAT value. In other words, the HRAT (particularly its minimum or threshold value) represents the practical maximum value of the EMAT. Further discussion on the use of both HRAT and EMAT in the synthesis of HENs can be found in Colbert (1982) and O'Reilly (1985).

## 3. The Concept of Downstream Paths

In this section, we introduce a simple new concept that is useful to the analysis and synthesis of HENs subjected to possible variations in stream inlet conditions such as input temperatures and heat-capacity flow rates. The concept, called the downstream path, is described in Linnhoff and Kotjabasakis (1986).

## a.  NETWORK DISTURBANCES AND CONTROLLED PARAMETERS

Consider a four-stream HEN under nominal inlet conditions (i.e., the base-case design), as shown in Figure 6.20$a$. In the figure, input temperatures of process streams are denoted by $T_{h1}$, $T_{h2}$, $T_{c1}$, and $T_{c2}$; and heat-capacity flow rates of process streams are designated $W_{h1}$, $W_{h2}$, $W_{c1}$, and $W_{c2}$. The desired output (target) temperatures of process streams are represented by $T_{h1}^*$, $T_{h2}^*$, $T_{c1}^*$, and $T_{c2}^*$.

We express the deviations from nominal input temperatures and heat-capacity flow rates as network disturbances, denoted by **D**. For example, Figure 6.20$b$

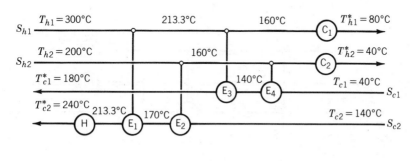

$(a)$

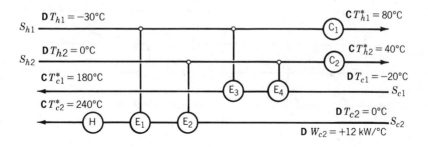

$(b)$

**FIGURE 6.20**  Illustrations of the concept and applications of downstream paths. (a) A four-stream HEN under nominal stream inlet conditions; $W_{h1} = 30$, $W_{h2} = 45$, $W_{c1} = 40$, $W_{c2} = 60$ kW/C°. (b) A four-stream HEN subjected to variations in stream inlet conditions with **D** representing the network disturbances and **C** denoting the controlled parameters; $W_{h1} = 30$, $W_{h2} = 45$, $W_{c1} = 40$, $W_{c2} = 72$ kW/°C. (c) A completely downstream path and a partly upstream path between disturbance **D** $T_{c1}$ and controlled parameter **C** $T_{h1}^*$; (- - - -) partly upstream; (— · — · —) completely downstream. (d) Four downstream paths between network disturbance **D** $W_{c2}$ and controlled parameters **C** $T_{h1}^*$, **C** $T_{h2}^*$, **C** $T_{c1}^*$ and **C** $T_{c2}^*$. (e) A downstream path between network disturbance **D** $W_{c2}$ and controlled parameter **C** $T_{c1}^*$. (f) Breaking up downstream path between **D** $W_{c2}$ and **C** $T_{c1}^*$ shown in Figure 6.20e by merging together exchangers $E_1$ and $E_3$, assuming that temperature and other constraints permit such a network change. Also shown is another downstream path between **D** $W_{c2}$ and **C** $T_{c1}^*$ to be broken up in Figure 6.20g. (g) Breaking up the downstream path between **D** $W_{c2}$ and **C** $T_{c1}^*$ shown in Figure 6.20e by shifting $E_2'$ from $S_{c2}$ to $S_{c1}$ and shifting $E_4$ from $S_{c1}$ to $S_{c2}$. This network change introduces a partly upstream path on $S_{h2}$ from **D** $W_{c2}$ to **C** $T_{c1}^*$.

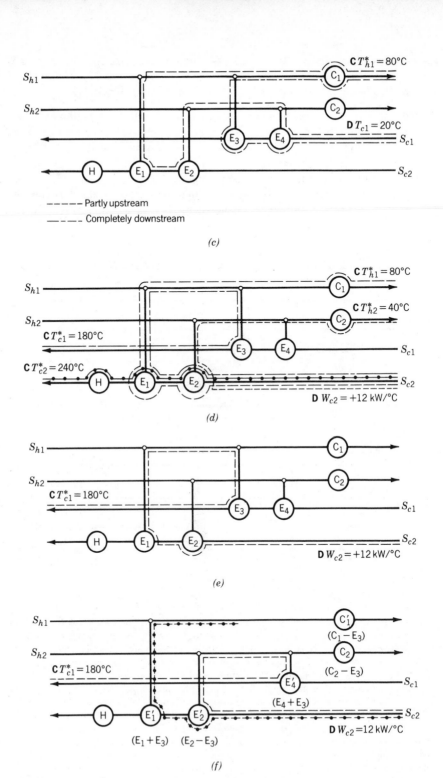

**FIGURE 6.20** (*Continued*)

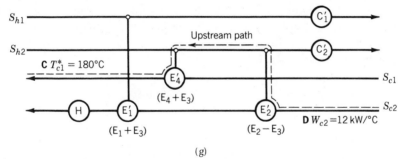

(g)

**FIGURE 6.20**   (*Continued*)

shows the following disturbances:

**D** $T_{h1}$ = $-30°C$ ($T_{h1}$ is reduced from 300 to 270°C)

**D** $T_{c1}$ = $-20°C$ ($T_{c1}$ is reduced from 40 to 20°C)

**D** $W_{c2}$ = $+12$ kW/°C ($W_{c2}$ is increased from 60 to 72 kW/°C)

In the same figure, $T_{h2}$, $T_{c2}$, $W_{h1}$, $W_{h2}$, and $W_{c1}$ remain at their nominal values, those specified in Figure 6.20a.

We wish to investigate whether the network given in Figure 6.20b represents "a good design" even in the presence of the preceding network disturbances. Here a good design implies that the desired output temperatures of process streams, $T_{h1}^*$, $T_{h2}^*$, $T_{c1}^*$, and $T_{c2}^*$ can still be reached. We call these output temperatures controlled parameters, denoted by **C**, and write the following:

$$\mathbf{C}\ T_{h1}^* = 80°C \qquad \mathbf{C}\ T_{h2}^* = 40°C$$
$$\mathbf{C}\ T_{c1}^* = 180°C \qquad \mathbf{C}\ T_{c2}^* = 240°C$$

By using the concepts of network disturbances **D** and controlled parameters **C**, we now pose two questions:

- Will any of the network disturbances **D** affect the controlled parameters **C**?
- If yes, what is the magnitude of the effect and how can we best eliminate that effect?

These questions can be conveniently answered by considering the concept of downstream paths.

## b.   DOWNSTREAM PATHS

As discussed in Section 6.3.G.3, a path is a connection between any two points in the temperature-interval (or grid) diagram representing a HEN. For example, Figure 6.20c shows two paths between the inlet of $S_{c1}$ and the outlet of $S_{h1}$, or between network disturbance $\mathbf{D}\ T_{c1}$ and controlled parameter $\mathbf{C}\ T_{h1}^*$. A path is called downstream if it follows the direction of going from low to high temperatures on a cold stream and of going from high to low temperatures on a hot stream. In Figure 6.20c, we see from $\mathbf{D}\ T_{c1}$ to $\mathbf{C}\ T_{h1}^*$

- Path ($E_4$, $E_3$, $C_1$): a completely downstream path.
- Path ($E_4$, $E_2$, $E_1$, $C_1$): a partly upstream path.

Figure 6.20d illustrates four completely downstream paths from network disturbance $\mathbf{D}\ W_{c2}$ to controlled parameters $\mathbf{C}\ T_{h1}^*$, $\mathbf{C}\ T_{h2}^*$, $\mathbf{C}\ T_{c1}^*$, and $\mathbf{C}\ T_{c2}^*$:

- Path ($E_2$, $E_1$, $C_1$): from $\mathbf{D}\ W_{c2}$ to $\mathbf{C}\ T_{h1}^*$.
- Path ($E_2$, $C_2$): from $\mathbf{D}\ W_{c2}$ to $\mathbf{C}\ T_{h2}^*$.
- Path ($E_2$, $E_1$, $E_3$): from $\mathbf{D}\ W_{c2}$ to $\mathbf{C}\ T_{c1}^*$.
- Path ($E_2$, $E_1$, $H_1$): from $\mathbf{D}\ W_{c2}$ to $\mathbf{C}\ T_{c2}^*$

The primary significance of a downstream path may be stated as follows: A network disturbance $\mathbf{D}$ can affect a controlled parameter $\mathbf{C}$ only if there is a completely downstream path between $\mathbf{D}$ and $\mathbf{C}$. As an illustration, Figure 6.20e indicates the presence of a completely downstream path between $\mathbf{D}\ W_{c2}$ and $\mathbf{C}\ T_{c1}$, that is, path ($E_2$, $E_1$, $E_3$). Therefore, any change in the heat-capacity flow rate of cold stream $S_{c2}$ (denoted by $W_{c2}$) affects the desired output temperature $T_{c1}^*$ of cold stream $S_{c1}$.

The concept of downstream paths may be used to suggest design changes that eliminate the undesirable variations in controlled parameters $\mathbf{C}$ caused by network disturbances $\mathbf{D}$. For example, we may break up the downstream path ($E_2$, $E_1$, $E_3$) shown in Figure 6.20e if exchangers $E_1$ and $E_3$ can be merged together as a single unit $E_1'$ ($= E_1 + E_3$) without violating any temperature and design constraints. The resulting network is illustrated in Figure 6.20f. In the latter figure, there is another downstream path, ($E_2'$, $E_4'$) between $\mathbf{D}\ W_{c2}$ and $\mathbf{C}\ T_{c1}^*$ that is yet to be broken up. Figure 6.20g shows one way to do this by shifting $E_2'$ from $S_{c2}$ to $S_{c1}$ and shifting $E_4'$ from $S_{c1}$ to $S_{c2}$. This network evolution introduces a partly upstream path on $S_{h2}$, thus eliminating the effect of $\mathbf{D}\ W_{c2}$ on $\mathbf{C}\ T_{c1}^*$.

The preceding examples clearly demonstrate that the simple concept of downstream paths is practically significant in the analysis and synthesis of HENs subjected to variations in stream inlet conditions. Further discussion of this concept can be found in Kotjabasakis and Linnhoff (1986).

In the following section, we present two additional examples of HENs subjected to variations in stream inlet conditions and introduce the topic of "resilient" HENs.

## 4.  An Introduction to Resilient HENs

An industrially significant extension of the multiobjective HEN synthesis problem under nominal (fixed or steady-state) stream inlet conditions, discussed from Section 6.3.A to I.2, is one that recognizes the variable nature of stream inlet conditions. Specifically, input temperatures and heat-capacity flow rates of process streams are often subjected to significant transient changes relative to their nominal (steady-state) values. In the literature, a HEN is called *resilient* if it is able to effectively accommodate the changes in stream inlet conditions and also maintain the desired output temperatures of process streams (Marselle et al., 1982; Liu, 1982b).

Traditional industrial practice for introducing resilience in a HEN is to use empirical overdesign that may accommodate "extreme" stream inlet conditions. For instance, these extremes may correspond to the minimum and maximum heat-capacity flow rates of a process stream. When these extremes are chosen, the resulting HEN often performs satisfactorily for a wide range of expected stream inlet conditions. In what follows, we present two examples that illustrate the complex nature of synthesizing a resilient HEN and explain why the traditional approach based on empirical overdesign may often fail to introduce resilience in a HEN (Colberg and Morari, 1987).

Consider a four-stream HEN with a $\Delta T_{min}$ of 10°C, shown in Figure 6.21a, in which the heat-capacity flow rate $W_{h1}$ for hot stream $S_{h1}$ may vary. Further, the extreme values of $W_{h1}$ are

$$1.0 \text{ kW}/°C \leq W_{h1} \leq 1.85 \text{ kW}/°C$$

By following the synthesis procedure presented in Sections 6.3.A–6.3.G, the reader can easily show that the network of Figure 6.21a represents an excellent design featuring both minimum utility consumptions and minimum number of units for

- $W_{h1} = 1.0 \text{ kW}/°C$ with $Q_1 = 20 \text{ kW}$
- $W_{h1} = 1.85 \text{ kW}/°C$ with $Q_1 = 241 \text{ kW}$

Since these heat-capacity flow rates represent the "extreme" inlet conditions for $S_{h1}$, it seems logical for a design engineer to expect that the network of Figure 6.21a can effectively accommodate all variations in $W_{h1}$ between 1.0 and 1.85 kW/°C. This expectation, in fact, results from the traditional approach to introducing "resilience" in a HEN. Unfortunately, this expectation is not true, as shown below (see also Section 8.3.A of Chapter 8).

By referring to the heat loads of exchangers $E_1$ and $E_3$ in Figure 6.21a, we write the following simple energy balances:

$$Q_1 \text{ kW} = (W_{h1} \text{ kW}/°C)(310 - T_{h1,E_1})°C$$

$$= (2 \text{ kW}/°C)(290 - T_{c2,E_1})°C$$

$$240 \text{ kW} = (W_{h1} \text{ kW}/°C)(T_{h1,E_1} - 50)°C$$

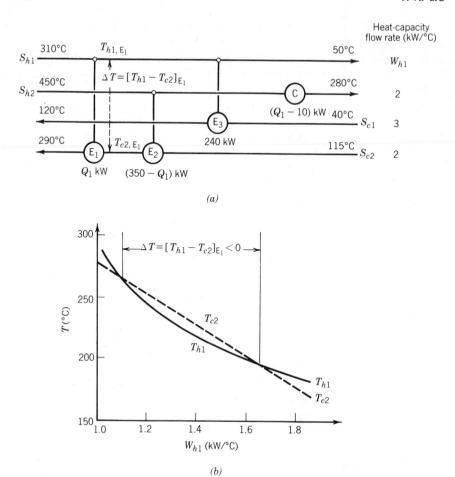

FIGURE 6.21 (a) A four-stream HEN with a $\Delta T_{min}$ of 10°C and a variable heat-capacity flow rate $W_{h1}$ of hot stream $S_{h1}$. (b) Effect of $W_{h1}$ on the approach temperature $\Delta T$ between $S_{h1}$ and cold stream $S_{c2}$ on exchanger $E_1$.

Simplifying these equations gives

$$T_{c2, E_1} = 290 - 0.5W_{h1}(310 - T_{h1, E_1}) \tag{6.20}$$

$$T_{h1, E_1} = \frac{240}{W_{h1}} + 50 \tag{6.21}$$

Equation (6.20) indicates that $T_{c2, E_1}$ is a *linear* function of $W_{h1}$, while Equation (6.21) reveals that $T_{h1, E_1}$ is a *nonlinear* function of $W_{h1}$. Figure 6.21b gives a plot of both $T_{h1, E_1}$ and $T_{c2, E_1}$ versus $W_{h1}$. It shows that the approach temperature $\Delta T$

between $S_{h1}$ and $S_{c2}$ on exchanger $E_1$ becomes *negative*; that is,

$$\Delta T = T_{h1,E_1} - T_{c2,E_1} < 0$$

over a finite range of *intermediate* values of $W_{h1}$ between 1.0 and 1.85 kW/°C. This result clearly suggests that the network of Figure 6.21a designed based on extreme values of $W_{h1}$ represents a technically infeasible network and fails to accommodate all changes in $W_{h1}$ between its extreme values.

Actually, we can make the network of Figure 6.21a feasible for all variations in $W_{h1}$ between 1.0 and 1.85 kW/°C by simply shifting cooler C from the low-temperature portion of $S_{h2}$ to that of $S_{h1}$. The resulting network satisfies the $\Delta T_{min}$ constraint on all exchangers. This shifting of cooler C from $S_{h2}$ to $S_{h1}$ is also consistent with the guidance of the thermodynamic matching rule given in Section 6.3.F.1, since $S_{h1}$ has a lower average of its input and output temperature ($T_{avg,h1} = 180°C$) compared with $S_{h2}$ ($T_{avg,h2} = 365°C$).

Figure 6.22 shows a four-stream HEN with a $\Delta T_{min}$ of 10°C and a variable input temperature $T_{h2}$ of hot stream $S_{h2}$. In the network, the approach temperature between $S_{h2}$ and $S_{c2}$ on exchanger $E_2$ is denoted by $\Delta T$. The heat loads of heater H, exchanger $E_2$, and cooler C, represented by $Q_H$, $Q_E$ and $Q_C$, are to be determined from $T_{h2}$. By following the problem-table analysis described in Section 6.3.D, the reader can easily show that:

- $T_{h2} = 380°C$: $Q_H = 0$, $Q_E = 260$ kW, $Q_C = 120$ kW
  $\Delta T = 20°C > \Delta T_{min}$ (feasible)
- $T_{h2} = 360°C$: $Q_H = 20$ kW, $Q_E = 240$ kW, $Q_C = 0$
  $\Delta T = 5°C < \Delta T_{min}$ (infeasible)
- $T_{h2} = 340°C$: $Q_H = 160$ kW, $Q_E = 100$ kW, $Q_C = 0$
  $\Delta T = 20°C < \Delta T_{min}$ (feasible)

These results illustrate that even though the network structure of Figure 6.22 works well for $T_{h2} = 380$ and 340°C, it fails to satisfy the $\Delta T_{min}$ constraint with $T_{h2} = 360°C$. Also, reducing $T_{h2}$ from 380 to 340°C changes the network from requiring cooling to heating utilities. For some other HEN synthesis problems, it is likely that this shifting of utility requirements can lead to changes in pinch points (Colberg and Morari, 1987).

Based on Figures 6.21 and 6.22, we learn that the traditional approach to introducing resilience in a HEN through empirical overdesign to accommodate "extreme" stream inlet conditions often fails to give technically feasible HENs for "intermediate" stream inlet conditions. Thus resilience of a HEN may not be obtained with additional exchangers or excessive oversizing, but often by a simple redesign of the network structure.

Our discussion on resilience in this section has only touched the surface of a fairly complex topic in process design. In fact, resilience is an increasingly important topic of research in process design and many significant results are being

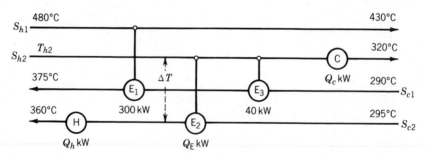

**FIGURE 6.22** A four-stream HEN for illustrating the effect of the input temperature $T_{h2}$ of hot stream $S_{h2}$ on the approach temperature $\Delta T$ between $S_{h2}$ and cold stream $S_{c2}$ on exchanger $E_2$. $Q_H$ and $Q_E$ are, respectively, heat loads (kW) of heater H and exchanger $E_2$ that are to be determined from $T_{h2}$.

published in the literature. For further discussion on resilient HENs, we refer the reader to the simple articles by Liu (1982b) and Marselle et al. (1982), and to the comprehensive review by Colberg and Morari (1987). Other publications on the subject include Parkinson et al. (1982), Floudas and Grossmann (1986), and Saboo et al. (1986). The last paper describes a computer-aided package for the analysis and synthesis of resilient HENs.

## 6.4.  CONCLUDING REMARKS

In this chapter, we have described in some detail the simple and practical developments on two important topics in process synthesis, namely, the synthesis of multicomponent separation sequences and heat-exchanger networks. The heuristic and evolutionary synthesis techniques presented are easy to apply by hand calculations and can be readily taught to undergraduate students. We have also covered the necessary technical background and practical insight needed by the reader to better understand and apply the current and developing computer-aided packages for HENs.

Because of the space limitation, we have not discussed the simple and practical developments on several other topics in process synthesis. These include the synthesis of initial process flowsheets, multieffect evaporator trains, heat and power systems, and combined heat-exchange and refrigeration system. Interested readers may refer to some recent references cited in Chapters 4 and 5.

In closing, we share the suggestion of many chemical engineers that process-design teaching should be more oriented toward synthesis, rather than the usual emphasis upon analysis alone. It is hoped that the simple techniques and illustrative examples presented in this chapter as well as a description of a course on process synthesis published elsewhere (Liu, 1980) will assist the interested faculty in developing courses in processes synthesis and in bringing a better balance between analysis and synthesis in process-design teaching.

## APPENDIX 6.1.   A SURVEY OF PUBLISHED HEURISTICS FOR SYNTHESIS OF MULTICOMPONENT SEPARATION SEQUENCES

The following summarizes the heuristics for synthesis of multicomponent separation sequences published since 1947 according to the categories of method (M), design (D), species (S), and composition (C) heuristics.


### A.  Method Heuristics

1. Heuristic M1 (favor ordinary distillation). All other things being equal, favor separation methods using only energy-separating agents (e.g., ordinary distillation), and avoid using separation methods (e.g., extractive distillation) that require the use of species not normally present in the processing, that is, the mass separating agent (MSA) (Rudd et al., 1973, pp. 174–181). However, if the separation factor or relative volatility of the key components $\alpha_{LK,HK} < 1.10$, the use of ordinary distillation is not recommended (Seader and Westerberg, 1977; Nath and Motard, 1981). A MSA may be used, provided that it improves the relative volatility between the key components.

Nath and Motard (1981), however, have presented an interesting example that challenges heuristic M1. They consider the separation of hexane (A), benzene (B), and cyclohexane (C) using ordinary distillation (method I) and extractive distillation with phenol (D) as solvent (method II). The ranked lists (RL) of decreasing adjacent relative volatilities corresponding to separation methods I and II are given by

$$\text{RL(I): ABC} \qquad \text{RL(II): ACBD}$$

Nath and Motard evaluate the following two sequences:

Sequence A1:  Favoring ordinary distillation
Total annual cost = \$274,803/yr

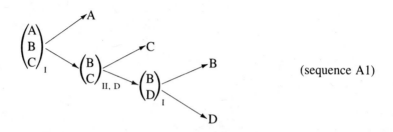

(sequence A1)

Sequence A2:  Favoring extractive distillation
Total annual cost = \$214,675/yr

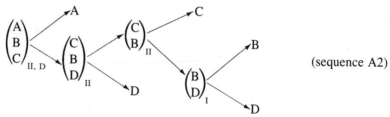

(sequence A2)

By comparing the costs of both sequences, Nath and Motard conclude that it is sometimes beneficial to consider using separations with a MSA, such as extractive distillation.

2. Heuristic M2 (avoid vacuum and refrigeration). All other things being equal, avoid excursions in temperature and pressure, but aim higher rather than lower (Rudd et al., 1973, pp. 182–183). If vacuum operation of ordinary distillation is required, liquid–liquid extraction with various solvents might be considered. If refrigeration is required (e.g., for separating materials of low boiling point with high relative volatilities as distillate products), cheaper alternatives to distillation such as absorption might be considered (Souders, 1964; Seader and Westerberg, 1977; Nath and Motard, 1981).

3. Heuristic M3 (remove mass separating agent first). When a mass separating agent (MSA) is used, remove it in the separator immediately following the one into which it is introduced. In other words, remove the MSA soon after it is used (Hendry and Hughes, 1972; Rudd et al., 1973, pp. 174–180; Seader and Westerberg, 1977).

Heuristic M3 has not been explicitly included in the ordered heuristic method of Nadgir and Liu (1983), as described in Section 6.2.B.1. This follows because it is intuitively obvious that an early removal of MSA eliminates the presence of a possible contaminant in subsequent separations.

Nath and Motard (1981) illustrate the possible benefit of delaying the removal of MSA in certain separation sequences. Specifically, they suggest that the MSA (that is, solvent D) in sequence A2 in the preceding example could be removed last in the sequence. This leads to an improved sequence as follows:

Sequence A3: Delaying the removal of MSA
Total annual cost = $158,699/yr

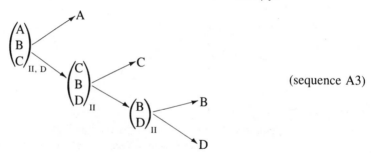

(sequence A3)

## B.  Design Heuristics

1.  Heuristic D1 (favor smallest product set). Favor sequences that yield the minimum necessary number of products. Equivalently, avoid sequences that separate components that should ultimately be in the same product (Thompson and King, 1972; King, 1980, p. 720). In other words, when multicomponent products are specified, favor sequences that produce these products directly or with a minimum of blending, unless relative volatilities are appreciably lower than for a sequence that requires additional separators and blending (Seader and Westerberg, 1977; Henley and Seader, 1981, p. 541).

To follow heuristic D1, it is sometimes necessary to use a separator with a mass separating agent (MSA). An additional separator is then needed to isolate the MSA (heuristic M3). For such situations, there is a possibility that could lead to a superior flowsheet. This possibility involves breaking a multicomponent product that has made the use of a MSA necessary in the first place. The two products defined by breaking the multicomponent product may both be isolated by ordinary separation methods without a MSA. The resulting separation sequence will have the same number of separators as that obtained by following heuristic D1, but it may be superior.

Nath and Motard (1981) have illustrated the preceding observation by considering the separation of a three-component mixture, (A,B,C), into two product streams, (A,B) and (C). Both ordinary distillation (method I) and extractive distillation with solvent D (method II) are to be used according to the following ranked lists (RL) of decreasing adjacent relative volatilities:

RL(I):  ACB        RL(II):  ABCD

Nath and Motard compare the following two sequences:

Sequence A4:  Favoring the smallest product set (heuristic D1)

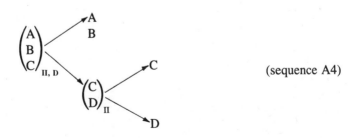

(sequence A4)

Sequence A5:  Breaking the multicomponent product
(violating heuristic D3)

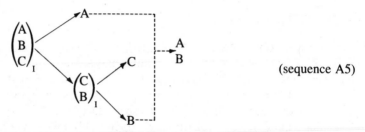

(sequence A5)

They suggest that sequence A5 is generally better than sequence A4, since the former involves only an ordinary distillation without the need for a solvent.

2. Heuristic D2 (cheapest-first heuristic). The next separator to be incorporated into the separator sequence at any point is the one that is cheapest (Thompson and King, 1972).

3. Heuristic D3 (favor direct sequence). During distillation, when neither the relative volatility nor the molar percentage in the feed varies widely, remove the components one by one as distillate products. The resulting sequence is commonly known as the direct sequence, in which the operating pressure tends to be highest in the first separator and is reduced in each subsequent separator (Lockart, 1947; Harbert, 1957; Heaven, 1969; Rudd et al., 1973, pp. 183–184; King, 1980, p. 715).

The use of the direct-sequence rule tends to maximize the total separation load, thus requiring large separators with high investment costs. Further, this heuristic fails to consider the important concept of a balanced separator (column) that features an equimolar (50/50) split of the feed between overhead and bottom products as well as a low-energy requirement (Harbert, 1957; King, 1980, pp. 715–716). Specific examples that demonstrate the advantage of heuristic C2 (favor 50/50 split) over heuristic D3 (favor direct sequence) in minimizing the total separation load and separation cost can be found in Rudd et al. (1973, pp. 167–173).

4. Heuristic D4 (remove valuable or desired product last). When using distillation or similar separation schemes, choose a sequence that will finally remove the most valuable species or desired product as an overhead product, all other things being equal (Harbert, 1957; Rudd et al., 1973, pp. 180–182).

## C.  Species Heuristics

1. Heuristic S1 (remove corrosive and hazardous components first). Remove corrosive and hazardous materials first (Rudd et al., 1973, p. 170).

2. Heuristic S2 (perform high-recovery separations last). Separations involving species with very high specified recovery fractions should be reserved last in a sequence (Heaven, 1969; King, 1980, p. 716).

3. Heuristic S2 (perform difficult separations last). All other things being equal, perform the difficult separations last (Harbert, 1957; Rudd, et al., 1973, pp. 171–

174). In particular, separations where the relative volatility of the key components is close to unity should be performed in the absence of nonkey components. In other words, try to select sequences that do not cause nonkey components to be present in separations where the key components are close together in separation factor or relative volatility (Heaven, 1969; King, 1980, p. 715).

4. Heuristic S4 (perform easy separations first). Favor easy separations first. Specifically, arrange the components to be separated according to their relative volatilities or separation factors in an ordered list. When the adjacent relative volatilities of the ordered components in the feed vary widely, sequence the splits in the order of decreasing adjacent relative volatility (Seader and Westerberg, 1977). Alternatively, arrange the separations to split the feed into the distillate and bottom products in an increasing order of the coefficient of difficulty of separation (CDS) defined by

$$
CDS = \frac{\log\left\{\dfrac{sp_{LK}}{1 - sp_{LK}} \cdot \dfrac{sp_{HK}}{1 - sp_{HK}}\right\}}{\log \alpha_{LK\,HK}} \cdot \frac{D}{D + B} \cdot \left\{1 + \left|\frac{D - B}{D + B}\right|\right\} \quad (6.22)
$$

where $sp_{LK}$ and $sp_{HK}$ are, respectively, the split fractions of the light and heavy key components in the distillate and bottom products; $D$ and $B$ are, respectively, the molal flow rates of the distillate and bottom products; and $\alpha_{LK,HK}$ is the relative volatility of the key components. Specifically, (1) favor large $\alpha_{LK,HK}$ (i.e., heuristic S4, perform easy separations first); (2) favor a balanced column where $D = B$ (i.e., heuristic C2, favor 50/50 split); (3) favor sloppy splits of low recoveries of the keys such that the split ratios $sp_{LK}/(1 - sp_{LK})$ and $sp_{HK}/(1 - sp_{HK})$ are small; and (4) favor less distillate products (i.e., heuristic D3, favor direct sequence) (Nath and Motard, 1981).

As described above by Seader and Westerberg, heuristic S4 favors performing easy separations first based on the ranking of separation factors. This heuristic fails to consider the effect of feed composition and the importance of a balanced column, and it can make subsequent separations difficult or costly. Consider, for example, the separation of a mixture of light olefins and paraffins by ordinary distillation. The feed mixture is (Thompson and King, 1972):

| Species | Mole Fraction | Relative Volatility* $\alpha$ | CES |
|---|---|---|---|
| A: Ethane | 0.20 | | |
| B: Propylene | 0.15 | 3.50 | 62.5 |
| C: Propane | 0.20 | 1.20 | 10.7 |
| D: 1-Butane | 0.15 | 2.70 | 139.1 |
| E: n-Butane | 0.15 | 1.21 | 9.0 |
| F: n-Pentane | 0.15 | 3.00 | 35.3 |

*At 37.8°C and 0.1 MPa.

It is desired to find the sequences for separating the feed into pure components, namely, A–F. The following two sequences have been reported in the literature:

Sequence A6:  Arrange the splits in the order of decreasing
adjacent relative volatility according to heuristic S4
Total annual cost = $1,234,000/yr (Seader and Westerberg, 1977)

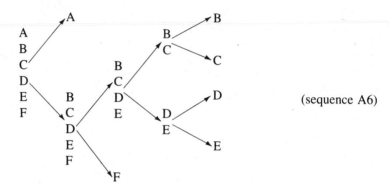

(sequence A6)

Sequence A7:  Arrange the splits favoring a balanced column
according to heuristic C2 (favor 50/50 splits)
Total annual cost = $1,153,000/yr (Seader and Westerberg, 1977; Nadgir and Liu, 1983)

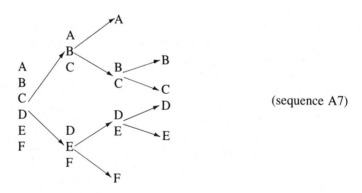

(sequence A7)

A comparison of sequences A6 and A7 suggests that *the dominating heuristic in multicomponent separation sequencing is heuristic C2 (favor 50/50 splits) and not heuristic S4 (perform easy separations first)*. This observation has also been confirmed by comparsions of initial sequences for many multicomponent separation problems (Nadgir and Liu, 1983).

Finally, we note that heuristic S4, based on the ranking of the coefficient of difficulty of separation (CDS) as described by Nath and Motard (1981), actually represents an attempt to incorporate heuristics D3 (favor direct sequence) and C2 (favor 50/50 split) into the CDS parameter.

## D.  Composition Heuristics

1. Heuristic C1 (remove most plentiful component first). A product composing a large fraction of the feed should be separated first, provided that the separation factor or relative volatility is reasonable for the separation (Nishimura and Hiraizumi, 1971; Rudd et al., 1973, pp. 167–169; King, 1980, p. 715).

2. Heuristic C2 (favor 50/50 split). If the component compositions do not vary widely, sequences that give a more nearly 50/50 or equimolal split of the feed between the distillate (D) and bottoms (B) products should be favored, provided that the separation factor or relative volatility is reasonable for the split (Harbert, 1957; Heaven, 1969; King, 1980, p. 715). If it is difficult to judge which split is closest to 50/50 and with a reasonable separation factor or relative volatility, perform the split with the highest value of the coefficient of ease of separation (CES) first.

Note that CES was defined previously in Equation (6.3), Section 6.2.B.1.


# APPENDIX 6.2.  SPECIFICATIONS OF STREAM AND DESIGN DATA FOR HEN SYNTHESIS PROBLEMS

The following material summarizes the specifications of process streams and design data for HEN synthesis problems that are included as illustrative examples in this chapter.


## A.  Stream Specifications

### 1.  Problem 3SP3 (Hohmann, 1971)

| Stream | Capacity Flow Rate (kW/°C) | Input Temperature (°C) | Output Temperature (°C) |
|---|---|---|---|
| $S_{h1}$ | 10.55 | 204.4 | 160.0 |
| $S_{c1}$ | 6.09 | 126.1 | 160.0 |
| $S_{c2}$ | 7.62 | 126.1 | 160.0 |

### 2.  Problem 4SP1 (Lee et al., 1970)

| Stream | Capacity Flow Rate (kW/°C) | Input Temperature (°C) | Output Temperature (°C) |
|---|---|---|---|
| $S_{c1}$ | 7.62 | 60.0 | 160.0 |
| $S_{c2}$ | 6.08 | 115.6 | 260.0 |
| $S_{h1}$ | 8.79 | 160.0 | 93.3 |
| $S_{h2}$ | 10.55 | 248.9 | 137.8 |

### 3. Problem 4SP2 (Ponton and Donaldson, 1974)

| Stream | Capacity Flow Rate (kW/°C) | Input Temperature (°C) | Output Temperature (°C) |
|---|---|---|---|
| $S_{c1}$ | 36.91 | −3.89 | 215.6 |
| $S_{h1}$ | 10.55 | 260.0 | 43.3 |
| $S_{h2}$ | 26.37 | 221.1 | 110.0 |
| $S_{h3}$ | 15.82 | 204.4 | 43.3 |

### 4. Problem 4TC2 (Linnhoff and Flower, 1978a)

| Stream | Capacity Flow Rate (kW/°C) | Input Temperature (°C) | Output Temperature (°C) |
|---|---|---|---|
| $S_{c1}$ | 3.0 | 60 | 180 |
| $S_{c2}$ | 0.4 | 30 | 130 |
| $S_{h1}$ | 2.0 | 180 | 40 |
| $S_{h2}$ | 4.0 | 150 | 40 |

### 5. Problem 5SP1 (Masso and Rudd, 1969)

| Stream | Capacity Flow Rate (kW/°C) | Inlet Temperature (°C) | Outlet Temperature (°C) |
|---|---|---|---|
| $S_{c1}$ | 11.39 | 37.8 | 204.4 |
| $S_{c2}$ | 12.92 | 65.6 | 182.2 |
| $S_{c3}$ | 13.03 | 93.3 | 204.4 |
| $S_{h1}$ | 16.62 | 248.9 | 121.1 |
| $S_{h2}$ | 13.29 | 204.4 | 65.6 |

### 6. Problem 6SP1 (Masso and Rudd, 1969)

| Stream | Capacity Flow Rate (kW/°C) | Inlet Temperature (°C) | Outlet Temperature (°C) |
|---|---|---|---|
| $S_{c1}$ | 8.44 | 37.8 | 221.1 |
| $S_{c2}$ | 17.28 | 82.2 | 176.7 |
| $S_{c3}$ | 13.90 | 93.3 | 204.4 |
| $S_{h1}$ | 14.77 | 226.7 | 65.6 |
| $S_{h2}$ | 12.55 | 271.1 | 148.9 |
| $S_{h3}$ | 17.72 | 198.9 | 65.6 |

## 7. Problem 7SP1 (Masso and Rudd, 1969)

| Stream | Capacity Flow Rate (kW/°C) | Inlet Temperature (°C) | Outlet Temperature (°C) |
|--------|----------------------------|------------------------|-------------------------|
| $S_{h1}$ | 14.77 | 226.7 | 65.6 |
| $S_{h2}$ | 12.55 | 271.1 | 148.9 |
| $S_{h3}$ | 17.72 | 198.9 | 65.6 |
| $S_{c1}$ | 8.44 | 37.8 | 221.1 |
| $S_{c2}$ | 17.28 | 82.2 | 176.7 |
| $S_{c3}$ | 13.90 | 93.3 | 204.4 |
| $S_{c4}$ | 9.94 | 176.7 | 210.0 |

## 8. Problem 10SP1 (Pho and Lapidus, 1973)

| Stream | Capacity Flow Rate (kW/°C) | Inlet Temperature (°C) | Outlet Temperature (°C) |
|--------|----------------------------|------------------------|-------------------------|
| $S_{c1}$ | 7.62 | 60.0 | 160.0 |
| $S_{c2}$ | 6.08 | 115.6 | 221.7 |
| $S_{c3}$ | 8.44 | 37.8 | 221.1 |
| $S_{c4}$ | 17.28 | 82.2 | 176.7 |
| $S_{c5}$ | 13.90 | 93.3 | 204.4 |
| $S_{h1}$ | 8.79 | 160.0 | 93.3 |
| $S_{h2}$ | 10.55 | 248.9 | 137.8 |
| $s_{h3}$ | 14.77 | 226.7 | 65.6 |
| $S_{h4}$ | 12.55 | 271.1 | 148.9 |
| $S_{h5}$ | 17.72 | 198.9 | 65.6 |

## B. Design Data

See Table A.1.

## ACKNOWLEDGMENTS

The author thanks the following individuals, from whom he has learned much about the field of process synthesis through their valuable publications and research discussion: Richard D. Colberg, California Institute of Technology; C. Judson King, University of California at Berkeley; Rudy L. Motard, Washington University; Vijay M. Nadgir, Du Pont Company; Naonori Nishida, Science University of Tokyo; Frederick A. Pehler, Shell Development Company; J. D. "Bob" Seader, University of Utah; D. William Tedder, Georgia Institute of Technology; and

**TABLE A.1  Specifications of Design Data for HEN Synthesis Problems**

|  | 4SP1 | 4TC2 |
|---|---|---|
| *Steam* | | |
| Pressure, kPA | 6635 | 4500 |
| Latent heat, kJ/kg | 1527.17 | 1676 |
| Temperature, °C | 282.22 | 258 |
| *Cooling water* | | |
| Temperature, °C | 37.8 | 30 |
| Heat capacity, kJ/kg-K | 4.1840 | 4.184 |
| Maximum water output temperature, °C | 82.2 | 80 |
| Maximum allowable approach temperature, °C: | | |
|   Heat exchanger | 11.1 | 10 |
|   Steam heater | 13.9 | 10 |
|   Water cooler | 11.1 | 10 |
| Overall heat-transfer coefficients, W/m$^2$-K: | | |
|   Heat exchanger | 851.71 | 1000 |
|   Steam heater | 1135.61 | 750 |
|   Water cooler | 851.71 | 1000 |
| Equipment down time (hr/yr) | 380 | 260 |
| Cost of heat-transfer area A(m$^2$) ($) | 1,456.3 A$^{0.6}$ | 3,000 A$^{0.6}$ |
| Annual rate of return, J | 0.1 | 0.1 |
| Cooling-water cost, $/kg | $1.1023 \times 10^{-4}$ | $1.5 \times 10^{-4}$ |
| Steam cost, $/kg | $2.2046 \times 10^{-3}$ | $6 \times 10^{-3}$ |

*Note:* Design data for all other problems are identical to those given for the 4SP1 problem, except that a saturated steam is available at 3102 kPa and 235.6°C with a latent heat of 1785.11 kJ/kg.

Arthur W. Westerberg, Carnegie–Mellon University. Special thanks are extended to Diane S. Cannaday for her patience while typing the manuscript for this chapter.

## NOMENCLATURE

| | |
|---|---|
| $A$ | Heat-transfer area of a single-pass countercurrent exchanger, heater, or cooler, Equation (6.14), m$^2$ or ft$^2$ |
| $A_{Ei}, A_{Hi}, A_{Ci}$ | Heat-transfer areas for the $i$-th heat exchanger, steam heater, and water cooler, respectively, Equation (6.5), m$^2$ or ft$^2$ |
| $A_N$ | Heat-transfer area for each shell pass of a multipass exchanger, heater, or cooler with $N$ shells in series. |
| $a,b$ | Constants in network investment-cost function, Equation (6.5), dimensionless |
| $B$ | molal flow rate of bottoms product, mol/hr |
| $C, C_i$ | Cooler or $i$th cooler in a given network |
| $C_{Ei}, C_{Hi}, C_{Ci}$ | Investment costs for the $i$th heat exchanger, stream heater, and water cooler, respectively, Equation (6.5), $ |

| | |
|---|---|
| $CDS$ | Coefficient of difficulty of separation defined by Equation (6.22), Appendix 6.1, dimensionless |
| $CES$ | Coefficient of ease of separation defined by Equation (6.3), dimensionless |
| $CW$ | Cooling water |
| $D$ | Molal flow rate of the overhead (distillate) product, mol/hr |
| $E, E_i$ | Exchanger or $i$th exchanger in a given network |
| $F$ | molal flow rate of the feed, mol/hr |
| $f$ | $D/B$ or $B/D$ such that $f < 1$, Equation (6.3), dimensionless |
| $F_N$ | Mean-temperature-difference (MTD) correction factor for a multipass exchanger, heater, or cooler of $N$ shells, Equation (6.14), dimensionless |
| $H, H_i$ | Heater or $i$th heater in a given network |
| $J$ | Total investment and utility-operating cost of the network, Equation (6.5), \$/yr |
| $L$ | Length of an exchanger, a heater, or a cooler, Figure (6.14), m or ft |
| $LMTD$ | Logarithmic-mean temperature difference between hot and cold streams entering and leaving an ideal single-pass, countercurrent exchanger, heater, or cooler, Equation (6.14), °C or °F |
| $MTD$ | Mean temperature difference between hot and cold streams entering and leaving an ideal single-pass, countercurrent exchanger, heater, or cooler, Equation (6.14), °C or °F |
| $N$ | Number of components in a mixture, Equations (6.1) and (6.2), dimensionless; number of shells in a multipass exchanger, heater, or cooler, Equations (6.16)–(6.18), dimensionless |
| $N_c, N_{cu}$ | Number of cold process and utility streams, respectively, Equation (6.7), dimensionless |
| $N_h, N_{hu}$ | Number of hot process and utility streams, respectively, Equation (6.7), dimensionless |
| $N_{min}$ | Most probable minimum number of units (exchangers, heaters, and coolers), Equation (6.7), dimensionless |
| $P$ | Thermal efficiency of an exchanger, a heater, or a cooler, Equation (6.15), dimensionless |
| $P_D$ | Optimal overhead pressure, Equation (6.4); MP or psia |
| $Q_{c,min}$ | Minimum cooling-utility requirement in a heat-exchanger network, kW or Btu/hr |
| $Q_{h,min}$ | Minimum heating-utility requirement in a heat-exchanger network, kW or Btu/hr |
| $R$ | Capacity ratio defined as the ratio of the heat-capacity flow rate of hot stream to that of cold stream, Equation (6.15), dimensionless |
| $R_{min}$ | Minimum reflux ratio of a distillation column, dimensionless |
| $S$ | Number of theoretically possible sequences for separating an $N$-component mixture into $N$ pure-component products by $T$ separation methods, Equation (6.2), dimensionless |

| $S_{hi}, S_{cj}$ | $i$th hot and $j$th cold process streams, respectively |
|---|---|
| $S_N$ | Number of theoretically possible sequences for separating an $N$-component mixture into $N$ pure-component products by one separation method only, Equation (6.1), dimensionless |
| $S_{ukl}$ | Amount of utility such as steam or water spent at the $l$th auxiliary equipment per year, Equation (6.5), kg/yr or lb/yr |
| $T$ | Number of separation methods in Equation (6.2), dimensionless; normal boiling point of a component in separation sequencing, or stream temperature in HEN synthesis, °C or °F, in all other places |
| $T_b$ | Boundary temperature of temperature intervals or subnetwork in a problem table, Table 6.4 |
| $T_{cj}, T_{cj}^*$ | Input and output temperatures of the $j$th cold stream, $S_{cj}$, respectively, °C or °F |
| $T_{hi}, T_{hi}^*$ | Input and output temperatures of the $i$th hot stream, $S_{hi}$, respectively, °C or °F |
| $T_{FB}$ | Normal bubble point of a feed stream, Equation (6.4), °C or °F |
| $U$ | Overall heat-transfer coefficient$_2$ for an exchanger, a heater, or a cooler, Equation (6.14), $\dot{W}/m^2$-°C or Btu/hr-ft$^2$-°F |
| $u_k$ | Operating cost of the utility $S_{uk}$ per year, Equation (6.5), \$/yr |
| $W_{hi}, W_{cj}$ | Heat-capacity flow rate (heat capacity multiplied by mass flow rate) of the $i$th hot and $j$th cold process streams, respectively, kW/°C or Btu/hr-°F |
| $x$ | Distance along the length of an exchanger, Figure (6.14), m |
| $y$ | Dimensionless distance along the length of an exchanger, $y = x/L$, Figure (6.14) |

## GREEK LETTERS

| $\alpha$ | Relative volatility of key components to be separated, dimensionless |
|---|---|
| $\delta$ | Annual rate of return on the investment cost, Equation (6.5), dimensionless |

## SUBSCRIPTS

| LK | Light key component |
|---|---|
| HK | Heavy key component |
| $i,j$ | $i$th and $j$th stream or exchanger (heater/cooler), respectively |
| min | Minimum value |
| sp | Split fraction of feed in a given product, Equation (6.22), Appendix 6.1, dimensionless |

## SYMBOLS

| Δ | $\Delta T$ or $(\alpha - 1) \times 100$ |
|---|---|

# REFERENCES

Bell, K. J., "Estimate S and T Exchanger Design Fast," *Oil Gas J.*, **59** (Dec. 1978).

Bell, K. J., "Thermal Design of Heat-Transfer Equipment," in *Chemical Engineers' Handbook*, 6th ed., R. H. Perry and D. Green (eds.), McGraw-Hill, New York, 1984, p. 10–24.

Cheng, S. H., "Systematic Synthesis of Sloppy Multicomponent Separation Sequences," Ph.D. dissertation, Virginia Polytechnic Institute and State University, Blacksburg, VA, 1987.

Colberg, R. D. and M. Morari, "Analysis and Synthesis of Resilient Heat Exchanger Networks," *Adv. Chem. Eng.*, in press (1987).

Colbert, R. W., "Industrial Heat Exchange Networks," *Chem. Eng. Prog.*, **78** (7), 47 (1982).

Elshout, R. V. and E. C. Hohmann, "The Heat Exchanger Network Simulator *(HENS),*" *Chem. Eng. Prog.*, **75** (3), 71 (1979), Elshout & Associates, Pasadena, CA 91106.

Floudas, C. A., A. R. Civic, and I. E. Grossmann, "Automatic Synthesis of Optimum Heat Exchanger Network Configurations," *AIChE J.*, **32,** 276 (1986). Use of *MAGNETS* (Mathematical Generation of Heat Exchanger Network Structures), Department of Chemical Engineering, Carnegie-Mellon University, Pittsburgh, PA 15213.

Floudas, C. A. and I. E. Grossmann, "Synthesis of Flexible Heat Exchanger Networks with Uncertain Flow Rates and Temperatures," paper presented at AIChE Annual Meeting, Miami, FL (Nov. 1986).

Gautnam, R. and J. A. Smith, "A Computer-Aided System for Process Synthesis and Optimization," paper No. 146, AIChE Meeting, Houston, TX (March 1985). Use of *ADVENT*. Union Carbide Corporation, P.O. Box 8361, South Charleston, WV 25303.

Gomez, M. A. and J. D. Seader, "Separation Sequence Synthesis by a Predictor-Based Ordered Search," *AIChE J.*, **22,** 970 (1976).

Gomez-Munoz, A. and J. D. Seader, "Synthesis of Distillation Trains by Thermodynamic Methods," *Comput. Chem. Eng.*, **9,** 311 (1985).

Harbert, W. D., "Which Tower Goes Where?" *Pet. Refiner,* **36** (3), 169 (1957).

Heaven, D. L., "Optimum Sequencing of Distillation Columns in Multicomponent Fractionation," M. S. thesis, University of California, Berkeley, 1969.

Hendry, J. E., "Computer Aided Synthesis of Optimal Multicomponent Separation Sequences," Ph.D. dissertation, University of Wisconsin, Madison, WI, 1972.

Hendry, J. E. and R. R. Hughes, "Generating Separation Process Flowsheets," *Chem. Eng. Prog.*, **68** (6), 71 (1972).

Henley, E. J. and J. D. Seader, *Equilibrium-Stage Separation Operations in Chemical Engineering*, Wiley, New York, 1981.

Hohmann, E. C., "Optimum Networks for Heat Exchange," Ph.D. dissertation, University of South California, 1971.

Hohmann, E. C. and F. J. Lockhart, "Optimum Heat Exchanger Network Synthesis," paper presented at AIChE 82nd National Meeting, Atlantic City, NJ (1976).

Huang, F. and R. Elshout, "Optimizing the Heat Recovery of Crude Units," *Chem. Eng. Prog.*, **72** (7), 64 (1976).

King, C. J., *Separation Processes*, 2nd ed., McGraw-Hill, New York, 1980.

Kotjabasakis, E. and B. Linnhoff, "Sensitivity Tables for the Design of Flexible Processes (1)—How Much Contingency in Heat Exchanger Networks Is Cost-Effective," *Trans. Inst. Chem. Eng.*, **64**, 197 (1986).

Lee, K. F., A. H. Masso, and D. F. Rudd, "Branch and Bound Synthesis of Integrated Process Designs," *Ind. Eng. Chem. Fund.*, **90**, 48 (1970).

Linnhoff, B. and J. R. Flower, "Synthesis of Heat Exchanger Networks, Part I. Systematic Generation of Energy Optimal Networks," *AIChE J.*, **24**, 633 (1978a).

Linnhoff, B. and J. R. Flower, "Synthesis of Heat Exchange Networks, Part II. Evolutionary Generation of Networks with Various Criteria of Optimality," *AIChE J.*, **24**, 642 (1978b).

Linnhoff, B. and E. Hindmarsh, "The Pinch Design Method of Heat Exchanger Networks," *Chem. Eng. Sci.*, **38**, 745 (1983).

Linnhoff, B. and E. Kotjabasakis, "Downstream Paths for Operable Process Design," *Chem. Eng. Prog.*, **82** (5), 23 (1986).

Linnhoff, B. and J. A. Turner, "Heat-Recovery Networks: New Insights Yield Big Savings," *Chem. Eng.*, 56 (Nov. 2, 1981).

Linnhoff, B., D. W. Townsend, D. Boland, G. F. Hewitt, B. E. A. Thomas, A. R. Guy, R. H. Marsland, J. R. Flower, J. C. Hill, J. A. Turner, and D. A. Reay, *User Guide on Process Integration for the Efficient Use of Energy*, The Institute of Chemical Engineers, London, 1982.

Liu, Y. A., "A Course in Process Synthesis," *Chem. Eng. Educ.*, **IX** (4), 184 (1980).

Liu, Y. A., "Recent Progress Towards the Systematic Multiobjective Synthesis of Heat Exchanger Networks," paper No. 20a, AIChE National Meeting, Orlando, FL (1982a).

Liu, Y. A., "A Practical Approach to the Multiobjective Synthesis and Optimizing Control of Resilient Heat Exchanger Networks," *Proceedings of the American Control Conference*, Vol. 3, Arlington, VA, Institute of Electrical and Electronic Engineers, Inc., New York, 1982b, pp. 1115–1126.

Liu, Y. A., F. A. Pehler, and D. R. Cahela, "Studies in Chemical Process Design and Synthesis: Part VII. Systematic Synthesis of Multipass Heat Exchanger Networks," *AIChE J.*, **31**, 487 (1985).

Lockhart, F. J., "Multi-Column Distillation of Natural Gasoline," *Pet. Refiner*, **26** (8), 105 (1947).

Marselle, D. F., M. Morari, and D. F. Rudd, "Design of Resilient Plants. II. Design and Control of Energy Management Systems," *Chem. Eng. Sci.*, **37**, 259 (1982).

Masso, A. H. and D. F. Rudd, "The Synthesis of System Designs: Heuristic Structuring," *AIChE J.*, **15**, 10 (1969).

McCabe, W. L. and J. C. Smith, *Unit Operations of Chemical Engineering*, 3rd ed., McGraw-Hill, New York, 1976.

Nadgir, V. M. and Y. A. Liu, "Studies in Chemical Process Design and Synthesis: Part V. A Simple Heuristic Method for Systematic Synthesis of Initial Sequences for Multicomponent Separations," *AIChE J.*, **29**, 926 (1983).

Nath, R., "Studies in the Synthesis of Separation Processes," Ph.D. dissertation, University of Houston, Houston, TX, 1977.

Nath, R. and R. L. Motard, "Evolutionary Synthesis of Separation Processes," *AIChE J.,* **27,** 578 (1981).

Nishida, N., S. Kobayashi, and A. Ichikawa, "Optimal Synthesis of Heat Exchange Systems—Necessary Conditions for Minimum Heat Transfer Area and Their Applications to System Synthesis," *Chem. Eng. Sci.,* **27,** 1408 (1971).

Nishida, N., Y. A. Liu, and L. Lapidus, "Studies in Chemical Process Design and Synthesis. Part III. A Simple and Practical Approach to the Optimal Synthesis of Heat Exchanger Networks," *AIChE J.,* **23,** 77 (1977).

Nishida, N., G. Stephenopoulos, and A. W. Westerberg, "A Review of Process Synthesis," *AIChE J.,* **27,** 321 (1981).

Nishimura, H. and Y. Hiraizumi, "Optimal System Pattern for Multicomponent Distillation Systems," *Int. Chem. Eng.,* **11,** 188 (1971).

O'Reilly, M., "Personal View," *The Chemical Engineer,* 46 (Jan. 1985).

Parkinson, A. R., J. S. Liedman, C. O. Pederson, and A. B. Templeman, "The Optimal Design of Resilient Heat Exchanger Networks," *AIChE Symp. Ser.,* **78** (214), 85 (1982).

Pehler, F. A. and Y. A. Liu, "Thermodynamic Availability Analysis in the Synthesis of Energy-Optimum and Minimum-Cost Heat Exchanger Networks," paper presented at AIChE National Meeting, Detroit, MI, August (1981); a shorter version appears in *ACS Symposium Series,* No. 235, *Efficiency and Costing: Second Law Analysis of Processes,* Richard A. Giaggioli (ed.), American Chemical Society, Washington, D.C., 1983, pp. 161–178.

Pehler, F. A. and Y. A. Liu, "Studies in Chemical Process Design and Synthesis: Part VI. A Thermoeconomic Approach to the Evolutionary Synthesis of Heat Exchanger Networks," *Chem. Eng. Comm.,* **25,** 295 (1984).

Peters, M. S. and K. D. Timmerhaus, *Plant Design and Economics for Chemical Engineers,* 3rd ed., McGraw-Hill, New York (1980).

Pho, T. K. and L. Lapidus, "Topics in Computer-Aided Design II. Synthesis of Optimal Heat Exchanger Networks by Tree Search Algorithms," *AIChE J.,* **19,** 1182 (1973).

Ponton, J. W. and R. A. B. Donaldson, "A Fast Method for the Synthesis of Optimal Heat Exchanger Networks," *Chem. Eng. Sci.,* **29,** 2375 (1974).

Rodrigo, B. F. R. and J. D. Seader, "Synthesis of Separation Sequences by Ordered Branch Search," *AIChE J.,* **21,** 785 (1975).

Rudd, D. F., G. J. Powers, and J. J. Siirola, *Process Synthesis,* Prentice-Hall, Englewood Cliffs, NJ (1973).

Saboo, A. K., M. Morari, and R. D. Colberg, "*RESHEX:* An Interactive Software Package for the Synthesis and Analysis of Heat Exchanger Networks. Part I. Program Description and Application, and Part II. Discussion of Area Targeting and Network Synthesis Algorithms," *Comput. Chem. Eng.,* **10,** 577 and 591 (1986). Department of Chemical Engineering, California Institute of Technology, Pasadena, CA 91125.

Seader, J. D. and A. W. Westerberg, "A Combined Heuristic and Evolutionary Strategy for the Synthesis of Simple Separation Sequences," *AIChE J.,* **23,** 951 (1977).

Simulation Sciences, Inc., *HEXTRAN Input Manual,* 1051 West Bastanchury Road, Fullerton, CA 92633, May (1982), pp. 2.1–2.4; *HEXTRAN New Features Guide,* March 1984 Release, Version 384, April (1984), pp. 1–2.

Souders, M., "The Countercurrent Separation Processes," *Chem. Eng. Prog.*, **62** (2), 75 (1964).

Stephanopoulos, G. and A. W. Westerberg, "Studies in Process Synthesis—II. Evolutionary Synthesis of Optimal Process Flowsheets," *Chem. Eng. Sci.*, **31**, 195 (1976).

Su, J. L. and R. L. Motard, "Evolutionary Synthesis of Heat Exchanger Networks," *Comput. Chem. Eng.*, **8** (2), 67 (1984).

Taborek, J., "Evolution of Heat Exchanger Design Techniques," *Heat Transfer Eng.*, **1**, 15 (July–Sept. 1979).

Tedder, D. W. and D. F. Rudd, "Parametric Studies in Industrial Distillation: Part II. Heuristic Optimization," *AIChE J.*, **24**, 316 (1978).

Thompson, R. W. and C. J. King, "Systematic Synthesis of Separation Schemes," *AIChE J.*, **18**, 941 (1972).

Townsend, D. W. and B. Linnhoff, "Heat and Power Networks in Process Integration: I. Criteria for the Placement of Heat Engines and Heat Pumps in Process Networks," *AIChE J.*, **29**, 742 (1983).

Umeda, T., J. Itoh, and K. Shiroko, "Heat-Exchange System Synthesis," *Chem. Eng. Prog.*, **74** (7), 70 (1978).

Van Winkle, M., *Distillation*, McGraw-Hill, New York, 1967.

Westerberg, A. W., "The Synthesis of Distillation-Based Separation Systems," *Comput. Chem. Eng.*, **9**, 421 (1985).

Westerberg, A. W. and G. Stephanopoulous, "Studies in Chemical Process Synthesis—I. Branch and Bound Strategy with List Techniques for the Synthesis of Separation Schemes," *Chem. Eng. Sci.*, **30**, 963 (1975).

Whistler, A. M., "Heat Exchangers as Money Makers," *Pet. Refiner*, **27**, 83 (1948).

Wood, R. M., R. J. Wilcox, and I. E. Grossmann, "A Note on the Minimum Number of Units for HEN Synthesis," DRC-06-75-84, Design Research Center, Carnegie-Mellon University, Pittsburgh, PA, Dec. (1984).

# 7

# COMPUTER-AIDED DESIGN (CAD): ADVANCES IN PROCESS FLOWSHEETING SYSTEMS

LAWRENCE B. EVANS

---

## 7.1. INTRODUCTION

The field of computer-aided process engineering has emerged in recent years as a major subdiscipline of chemical engineering. Computer-based models of process plants are used routinely as aids in process development and scale-up, and plant design and operation. The term *flowsheeting* has been coined to describe "the use of computer aids to perform steady-state heat and mass balancing, sizing, and costing calculations for a chemical process" (Westerberg et al., 1979). We use the terms "process flowsheeting" and "computer-aided process engineering" synonymously.

Today no major chemical or petrochemical plant would be built without first simulating the process on the computer. Commercially available software systems such as ASPEN PLUS (Aspen Technology, Inc., 1986), CONCEPT (Winter, 1982), DESIGN II (ChemShare Corporation, 1985), FLOWPACK II (Perris, 1982), FLOWTRAN (Rosen and Pauls, 1977), and PROCESS (Simulation Sciences, Inc., 1985) are widely used for this purpose and many companies also use their own proprietary in-house computer program.

The input and output to a typical flowsheeting program such as ASPEN PLUS are shown in Figure 7.1. The input consists of the information a process engineer would normally have from the process flow diagram. The output consists of the stream conditions (temperature, pressure, composition, and flow rate) of all the intermediate and product streams, the performance of the major unit operations, the sizes of major pieces of equipment, and the process economics. The steps in process simulation that produce these results are shown in Figure 7.2.

Every computer-aided engineering system is based on five major building blocks,

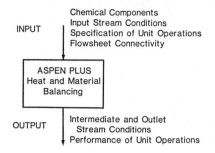

INPUT

Chemical Components
Input Stream Conditions
Specification of Unit Operations
Flowsheet Connectivity

ASPEN PLUS
Heat and Material
Balancing

OUTPUT

Intermediate and Outlet
Stream Conditions
Performance of Unit Operations

**FIGURE 7.1**  Input and output to ASPEN PLUS flowsheeting program.

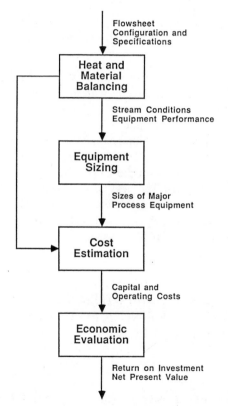

Flowsheet
Configuration and
Specifications

Heat and
Material
Balancing

Stream Conditions
Equipment Performance

Equipment
Sizing

Sizes of Major
Process Equipment

Cost
Estimation

Capital and
Operating Costs

Economic
Evaluation

Return on Investment
Net Present Value

**FIGURE 7.2**  Steps in process simulation.

as shown in Figure 7.3. The foundation of the system consists of the mathematical *models* of the engineering system to be designed (in this case a chemical process). These models are expressed in terms of equations or constraints that must be solved to predict the performance of the system. Because these nonlinear equations must be solved numerically, it is necessary to have good solution *algorithms*. The algorithms and the system itself, however, must be implemented in computer *soft-*

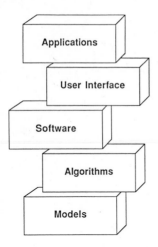

**FIGURE 7.3**  Building blocks of computer-aided engineering.

*ware*. It is also necessary to have a good *user interface* that allows the engineer to communicate easily with the system and to interact directly with the models. Finally, for a computer-aided engineering system to be successful, there must be good *applications* that justify its development and support.

This chapter assesses the state of the art in the development and use of process flowsheeting systems by first comparing the growth of traditional computer-aided design and manufacturing (CAD/CAM) systems with those for computer-aided process design and management (CAPD/CAPM). The next two sections discuss two current issues in the field: (1) What is the best architecture for steady-state process simulation? and (2) What is the right computer environment? Finally, the last two sections illustrate some current applications of flowsheeting and address future directions in the field.

## 7.2.  COMPARISON OF TRADITIONAL COMPUTER-AIDED DESIGN/COMPUTER-AIDED MANUFACTURING (CAD/CAM) SYSTEMS WITH THOSE FOR PROCESS DESIGN

### A.  Traditional CAD/CAM Systems

In the past decade, there has been an almost explosive growth in the use of CAD systems. As in shown in Figure 7.4, sales of CAD systems have grown at over 40% per year, exceeding $4 billion in 1985 (Daratech Associates, 1984). These systems are used by engineers to design everything from printing presses and bridges to household appliances and electronic circuits. It is estimated that 25% of all machined parts are designed today with CAD systems, and by 1990 the figure will reach 50%.

A breakdown of the major applications of CAD systems (Merrill Lynch, 1984) is shown in Table 7.1 with an indication also of the changes in use from 1980 to

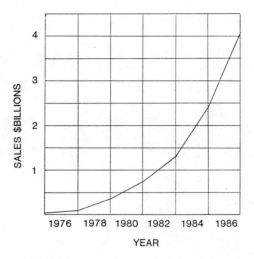

**FIGURE 7.4**  Growth in sales of CAD system.

**TABLE 7.1  Applications of Computer-Aided Design**

| Applications of CAD | 1980 (%) | 1984 (%) |
|---|---|---|
| Mechanical | 39 | 53 |
| Electrical | 29 | 22 |
| Architectural/Engineering | 15 | 18 |
| Mapping | 14 | 5 |
| Other | 3 | 2 |
| Total | 100 | 100 |

1984. As shown, the major applications are for the design of mechanical systems, with over 53% of the total in 1984, followed by electronic, architectural, and engineering systems. Chemical engineering applications are not even listed. The reason why such applications have lagged behind the other disciplines is of special interest to us and is addressed later in the chapter.

Figure 7.5 shows an engineer working at a CAD system workstation and Figure 7.6 shows the flow of information in the system. The engineer inputs a description of the device he is designing using a combination of graphics and keyboard input. In many systems, he can also perform modeling and analysis of the device. For a mechanical part, he might perform a stress calculation, determine the weight and cost of manufacture, or perform a heat-transfer analysis. The engineer could change

**FIGURE 7.5**  Engineer using a CAD workstation. (Photograph provided courtesy of Applicon, Ann Arbor, MI and reproduced with permission.)

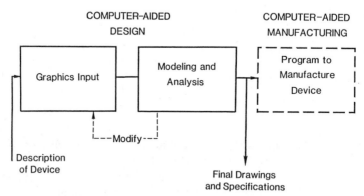

**FIGURE 7.6**  Flow of information in CAD system.

construction materials or other aspects of the design. Finally, when he is satisfied with the design, the CAD system can produce a complete set of drawings and specifications needed for manufacture.

As a further step in automation, the output of the CAD system can be used as the input to a computer-aided manufacturing (CAM) system. The CAM system, using numerically controlled machines and automated assembly processes, can manufacture at least a prototype of the device.

## B.  Systems for Computer-Aided Process Design and Management (CAPD/CAPM)

The analogous flow of information in a computer-aided process design (CAPD) system is shown in Figure 7.7. Here the input to the system is the process flow diagram, which constitutes a functional model of the process. The engineer can perform simulation and analysis of the system. Finally, the CAPD system can output the complete process design, including the process configuration, the sizes and performances of major pieces of equipment, a complete heat and material balance, and an estimate of the economic performance.

The analog to computer-aided manufacturing in process engineering is computer-aided process management. This involves use of the same model of the process that was employed in design as an aid in the management of the operations of the plant after it is on stream. We have coined the term CAPD/CAPM to describe the use of computer-based models of process plants to aid in their design and operation.

## C.  Differences Between CAD/CAM and CAPD/CAPM

A major distinction between traditional CAD/CAM and the process engineering applications of CAPD/CAPM is that CAD/CAM is concerned with the design of a product, while CAPD/CAPM is concerned with design of the process to make the product. The design of the products produced by the process industries has, in the past, been straightforward, particularly for bulk products such as gasoline, ethylene, ammonia, or other petrochemicals. It is the design of the manufacturing process that challenges the chemical engineer. This difference in application between CAD/CAM and CAPD/CAPM is responsible for the great differences in the nature of these systems.

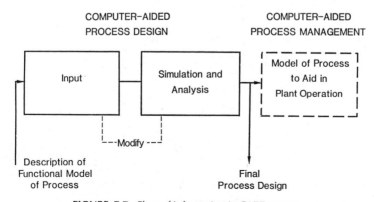

**FIGURE 7.7**  Flow of information in CAPD system.

In traditional CAD/CAM, the physical shape of the device being designed is of central importance. Drawings are the primary means of communication. Therefore, graphics are very important and most CAD systems have high-quality graphics workstations. Also, the types of analyses (e.g., stress calculations or heat-transfer analyses) involved in mechanical design are more routine and capable of being standardized than those in chemical-process systems.

In CAPD/CAPM, the structure of the process flowsheet is of central importance because it shows the interrelationship between flows of material and energy in the process. The process heat and material balances are the means of communication. Therefore, graphics are less important and need not be as detailed as the graphics required to show the physical shape of an object. Instead, they are used to produce process flow diagrams and graphs illustrating the relationship between process performance variables and independent design parameters.

Of course, traditional CAD/CAM systems are widely used in the process industries in the mechanical-design and plant-construction phases of a project. Figure 7.8 shows the stages in the life of a plant from process conception to plant operation. The roles of CAPD/CAPM are in the upstream steps of process design and the downstream stage of plant operations. Sandwiched in-between is the use of traditional CAD/CAM in the intermediate steps of mechanical design and plant construction. It is, therefore, of great current interest to be able to interface the CAPD systems for upstream process engineering with the CAD systems for downstream mechanical plant design.

The restructuring of the process industries that is now taking place in the industrialized countries will affect the requirements for chemical engineering CAD systems. There will be much more emphasis on the manufacture of specialty products as opposed to bulk commodity products. This means that in the future, chemical

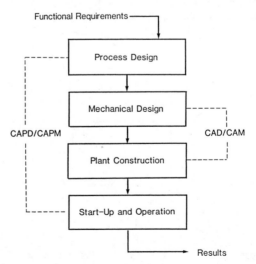

**FIGURE 7.8**  Role of CAD and CAPD in life-cycle of a chemical process plant.

engineers will be more concerned with product design than with process design. There will be a need for CAD systems to design specialty chemical products. In fact, some systems oriented toward molecular design already exist (Molecular Design, Ltd., 1985; Wipke, 1984).

In summary, the strength of traditional CAD/CAM systems has been their ability to use computer graphics to represent the physical shape of objects. In fact, they might better be called computer-aided drawing systems rather than computer-aided design systems. Their ability to do design calculations has been more primitive. However, the representation of physical shape does not play such an important role in chemical process design. Therefore, the application of traditional CAD systems to chemical process design has been very limited.

Computer-aided process design is itself a unique field and has its own requirements, which depend heavily upon the ability to do steady-state process simulation or flowsheeting calculations. In the next section, we discuss this application.

## 7.3. COMPUTING ARCHITECTURES FOR STEADY-STATE PROCESS SIMULATION

### A. Sequential-Modular vs. Equation-Solving Architecture

There has been a major controversy for many years over the best computational approach for solving flowsheeting problems (Rosen, 1980). Leading academic researchers have advocated use of the equation-solving approach, and prototype systems such as ASCEND (Locke et al., 1980) and SPEED-UP (Sargent, 1964) have been developed based on this approach. Stadtherr and Vegeais (1985) reviewed recent progress in equation-based process flowsheeting. Most commercial simulators in wide use today are based on the sequential-modular architecture. However, there is increased commercial interest in an equation-based approach (Barnard et al., 1986).

The starting point for flowsheet simulation is the process-flow diagram of the type shown in Figure 7.9. However, the engineer must next construct a model of the process by modeling each equipment item or processing step in the process with a modular unit-operation block from a set provided by the flowsheeting system. Each unit-operation block contains the equations that relate the outlet-stream and performance variables (the output variables) for the block to the inlet-steam variables and specified parameters (the input variables). The engineer must also specify the connectivity between the blocks.

A diagram of the model of the process is shown in Figure 7.10. There may be additional constraints which the engineer wishes to impose. The constraints, referred to as design specifications, free one of the independent variables (such as a feed-stream variable or an input parameter to one of the unit-operation blocks) and allow it to be adjusted to achieve a specified value of an output variable. In Figure 7.10, there is a constraint that the residence time in the reactor is to be adjusted to achieve a specified composition in outlet stream S4.

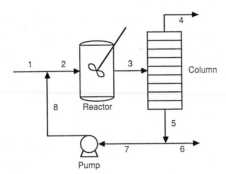

**FIGURE 7.9**  Typical process flow diagram.

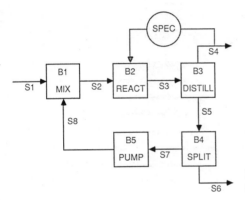

**FIGURE 7.10**  Diagram of simulation model of process.

Up to the point of the model diagram, there is no difference between the sequential-modular and equation-solving approaches. Both begin with the same model of the process. The difference is in how best to solve the equations that describe the process.

## B.  Sequential-Modular Approach

The sequential-modular method implements the unit-operation blocks as computer subroutines that calculate the output variables as functions of the input variables. To solve the problem, the system must partition the flowsheet, select tear streams, nest convergence of the tear streams, and determine the computational sequence.

Partitioning identifies those sets of blocks that must be solved together (referred to as a maximal cyclic subsystem). In Figure 7.10, all of the blocks in the model constitute such subsystems and must be solved together.

Tearing determines those streams or information flows that must be torn to render each subsystem acyclic. In Figure 7.10, Stream 8 and the information flow of the design specification represent a suitable tear set.

Nesting determines which tear streams are to be converged simultaneously and in which order collections of tear streams are to be converged. In the example, it is possible to nest the convergence of the design specification within the convergence of the tear stream or vice-versa, or to converge both tear streams simultaneously.

Most sequential-modular simulators have methods for performing all of these steps. Also, there are powerful numerical techniques for converging the tear streams using quasi-Newton methods.

## C. Equation-Solving Methods

The basic idea of equation-solving methods is simply to collect all of the equations describing the flowsheet and solve them as a large system of nonlinear algebraic equations.

Mathematically the problem can be stated as

$$\text{solve } f(x, u) = 0$$

$$\text{with } g(x, u) \leq 0$$

where    $x$ is the vector of state (dependent) variables
          $u$ is the vector of decision (independent) variables
          $f(x, u)$ is the set of process model equations
          $g(x, u)$ is the set of inequality or equality constraints

The decision variables include all of the input-block parameters and feed-stream variables; and the state variables include all of the intermediate and product-stream variables, internal variables within each unit operation block, and output-performance variables from each block.

Alternatively, the problem may be formulated mathematically as an optimization problem,

$$\text{minimize } h(x, u)$$

$$\text{with} \quad f(x, u) = 0$$

$$g(x, u) \leq 0$$

where $h(x, u)$ is the objective function and $f(x, u)$ and $g(x, u)$ have the same meaning as before. The equality constraints are the same set of equations described above, but rather than specify the decision variables arbitrarily, they are selected to minimize the objective function. It is widely held that the design problem is most naturally formulated as an optimization problem (Westerberg, 1981).

## D.   Advantages and Disadvantages of Each Approach

The advantages of the sequential-modular approach are that: (1) the approach is conceptually easy for process engineers to understand, (2) there is a large body of existing computer software organized in the form needed by the sequential-modular approach, (3) it is possible to include convergence heuristics developed by experience over the years (e.g., good initialization procedures), and (4) in case of an error, it is easy to give understandable error messages. A primary disadvantage is that for large, highly integrated problems, it may be difficult to find a satisfactory convergence scheme.

The advantages of the equation-solving approach are that: (1) it is the natural way to specify a problem since the design problem is by nature an optimization problem and the engineer does not have any other criterion for specifying many arbitrary variables, (2) it is easy to specify variables and constraints, and (3) it can handle highly integrated systems since all equations are solved simultaneously. Some disadvantages of the approach are that: (1) it may be more difficult to handle highly nonlinear and discontinuous relations required to represent physical properties, (2) it does not take advantage of the large investment of industry in unit-operation models based on the sequential-modular method, (3) it may be difficult for the user to diagnose a problem, and (4) the method requires good initial estimates of the variables.

Although the advocates of each method seem firm in their convictions, there appear to be some compromises in store. The equation-solving approach is seeing increased use within individual unit-operation blocks such as for distillation and heat-exchanger networks incorporated into sequential-modular simulators. And a new approach, the simultaneous-modular or two-tier algorithm, appears to combine many of the advantages of the sequential-modular and equation-solving approaches.

## E.   Two-Tier Simultaneous-Modular Approach

The two-tier approach was first proposed by Rosen (1962) and was called *simultaneous-modular* by Westerberg et al. (1979). The method has been further developed by Pierucci et al. (1982), and by our research at MIT (Jirapongphan, 1980; Jirapongphan et al., 1980; Mahalec et al., 1979; Trevino-Lozano, 1985; Kisala, 1985; Trevino-Lozano et al., 1985).

The two-tier approach employs two types of models: rigorous and simple. Rigorous models are the traditional unit-operation models now used in sequential-modular simulators. They are utilized, however, to determine parameters in the simple models which are represented algebraically. In the optimization formulation of the problem, the simple models constitute equality constraints for a reduced optimization problem.

As shown in Figure 7.11, the algorithm first makes a pass through the outside loop, executing the rigorous models for each unit operation to determine the parameters for the simple models. This sets up the reduced optimization problem.

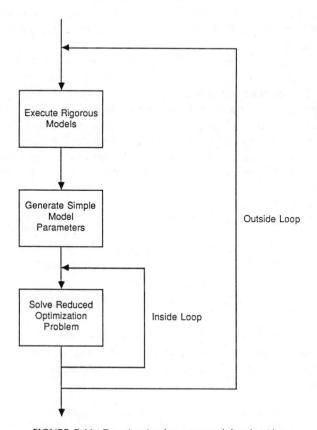

**FIGURE 7.11** Two-tier simultaneous-modular algorithm.

Then, in an inside loop, the reduced optimization problem is solved by an efficient equation-based nonlinear programming algorithm, such as the Han–Powell algorithm (Powell, 1977).

A key to the success of the simultaneous-modular algorithm is to have good simple models. These may be linear models whose coefficients are determined by numerical perturbation of the rigorous models. Alternatively, the simple models can be approximate engineering models, such as the Kremser (1930) model of an absorber or the Smith–Brinkley (1960) model of a distillation column. The simple models are in the form of equations, however. They have many fewer internal variables than the rigorous models. In a linear model, the only state variables are the outlet-stream variables. In an approximate engineering model, such as the Kremser model of an absorber, nonlinearities are represented more accurately than by a linear model. An approximate engineering model should represent the dominant nonlinearities as a smooth function with analytic partial derivatives. It should be very much simpler than the rigorous model and should avoid time-consuming calculation of thermophysical properties. The simple models, whether linear or nonlinear, are sometimes referred to as "reduced" models, because they involve many fewer variables.

Some illustrative results obtained with the simultaneous modular algorithm were published by Trevino-Lozano et al. (1985) for the hypothetical flowsheet shown in Figure 7.12. A feed of component $A$ reacts in a plug-flow reactor to obtain $P$ as the main product; $P$ further reacts to form $G$, an undesirable by-product. The stream leaving the reactor is introduced to a flash drum to separate the more volatile product from the leftover $p$. Part of the liquid stream leaving the flash drum is recycled back to the reactor. These $A$, $P$, and $G$ have the approximate physical properties (but not the chemical reactivity) of isobutyric acid, ethyl acetate, and $n$-butyric acid.

Five variations of this problem were solved to illustrate the efficiency of the algorithm on increasingly complex problems:

1. A simple recycle problem as shown in Figure 7.12 with a fixed reactor length.
2. Impose a design specification in which reactor length is adjusted to achieve 95% conversion of $A$ in effluent from flash.
3. Impose a design specification where split fraction at bleed stream is varied to obtain mole fraction of $G$ in feed to reactor of 0.02.
4. Impose both of the design specifications described above simultaneously.
5. Recast as optimization problem where reactor length is varied to maximize objective function $P$-$30G$ where $P$ and $G$ are flow rates of $P$ and $G$ in effluent from flash.

Each variation was solved two ways: using a simultaneous-modular algorithm implemented as a prototype in ASPEN PLUS, and using the default-convergence

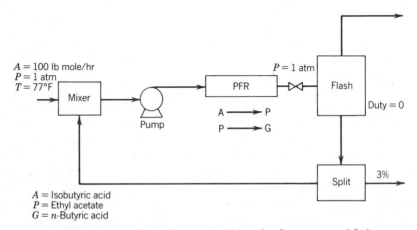

**FIGURE 7.12**  Hypothetical process involving plug-flow reactor and flash.

procedure generated by the sequential-modular version of ASPEN PLUS (bounded Wegstein for material recycles and Secant method for design specifications). To compare methodologies, the number of simulation-time equivalents was computed for each case. A simulation-time equivalent is defined as the ratio of the total CPU time necessary to solve the problem with the particular method to the time needed to converge the basic simulation problem (without design specifications) using the bounded Wegstein technique.

Table 7.2 shows the number of simulation-time equivalents needed to solve these problems using sequential-modular and nonlinear simultaneous-modular calculations.

The simultaneous-modular calculations were more efficient in every case than the standard sequential-modular techniques. Furthermore, the gap between the performance of the two approaches increased as the problem became more complex. A complete analysis of the performance of the algorithm is presented by Trevino-Lozano (1985). However, the general behavior shown above is characteristic of the method. The simultaneous-modular approach appears to offer the promise of being able to solve optimization problems with about the same computational effort as is now required to solve a standard simulation problem.

An important theoretical problem must be solved before a full-scale commercial implementation of the algorithm is feasible however. As pointed out by Trevino-Lozano (1985), Trevino-Lozano et al. (1985), and Biegler et al. (1985), the reduced optimization problem based on the nonlinear simple-model approximations may not have the same constrained optimum as do the rigorous models. This is because the Kuhn–Tucker conditions for the reduced problem are not the same as for the full-scale optimization problem. Even though the nonlinear reduced models give the same values for the state variables as do the rigorous models, they do not necessarily give the same values for the derivatives. (Note: This is not a problem with linear-reduced models, since in the limit at the optimum, they have the same gradients as the rigorous model.) Indeed, in practical implementations of the nonlinear simultaneous-modular method, we have observed that, while the optimum of the reduced problem is close to that of the rigorous model, it differs enough to be noticeable to users.

**TABLE 7.2   Simulation-Time Equivalents Required to Solve Example Problems by Sequential and Simultaneous Modular Algorithms**

| Example Problem Variation | Sequential Modular | Simultaneous Modular |
|---|---|---|
| Recycle problem | 1.0 | 0.80 |
| Specification on conversion | 1.9 | 0.85 |
| Specification on mole fraction | 4.8 | 0.84 |
| Two specifications | 6.4 | 0.73 |
| Optimization | 4.6 | 0.92 |

There are two proposed solutions to this difficulty. Both involve use of the nonlinear-reduced models with the simultaneous-modular algorithm to get close to the solution fast. Then there are two options to finish the algorithm and converge on the final solution: (1) continue with the simultaneous-modular algorithm, but switch to linear reduced models; (2) switch to an algorithm that uses the complete set of rigorous flowsheet models, such as the infeasible-path sequential-modular optimization technique (Biegler and Hughes, 1982; Kisala, 1985).

## 7.4. COMPUTING ENVIRONMENT FOR COMPUTER-AIDED PROCESS ENGINEERING

Another important issue facing the profession is the right computing environment for computer-aided process engineering. This dilemma is influenced by rapid changes in computer hardware and software. In just the past decade, we have evolved from extensive use of punched cards for input to the use of interactive computer terminals. Now there are two new developments: (1) the widespread availability of personal computer-based workstations on individual engineers' desks and (2) the desire of engineering organizations to coordinate the flow of information in their organizations through use of an engineering database.

The normal way of using a simulator in the stand-alone mode is shown in Figure 7.13. The engineer prepares an input file (usually at an interactive terminal with the aid of an editor) and inputs it to the simulator (ASPEN PLUS in this example). The simulator produces an output report and also saves a problem data file which can be used by the engineer to modify and rerun the problem.

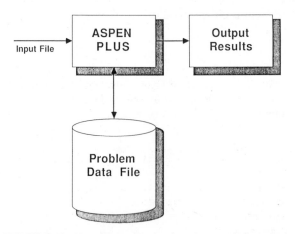

**FIGURE 7.13**   Use of a process simulator in a stand-alone mode.

## A.  Engineering Database Environment

A new type of environment is now available in the form of an engineering data-base. One of the most advanced of these systems is PRODABAS (Angus and Winter, 1985). The ASPEN PLUS simulator has been interfaced with PRODA-BAS and is used in this environment, as shown in Figure 7.14. The engineer does not prepare input directly from ASPEN PLUS. Instead he or she prepares input through an interactive terminal directly to the PRODABAS system. Convenient menu-driven "fill-in-the-blanks" input forms are available.

In the database environment, PRODABAS saves a complete description of the process in the process engineering database. It then calls on ASPEN PLUS as requested by the engineer to perform flowsheet calculations. ASPEN PLUS acts as the slave and PRODABAS is the master in this arrangement.

The main benefit of this arrangement is that it is possible to interface many different programs. For example, the output from flowsheet simulation can be used as the input to distillation-column or heat-exchanger design. ASPEN PLUS and PRODABAS form a complete system that can be used for integrated design, as shown in Figure 7.15.

Other database systems have been announced, such as DESIGNMASTER (Craft, 1985), and a few simulators in addition to ASPEN PLUS have been inter-faced with a database.

## B.  Workstation Environment

A problem with the database environment is that present implementations are too rigid for the needs of early-stage process engineering. In the early stages, the

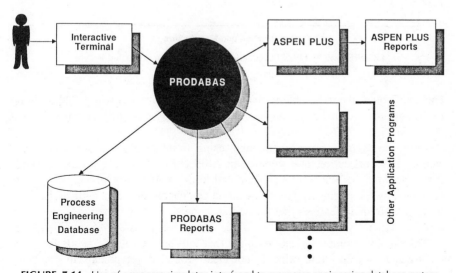

**FIGURE 7.14**  Use of a process simulator interfaced to a process engineering database system.

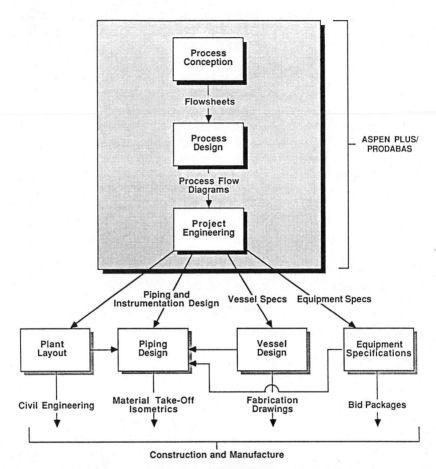

**FIGURE 7.15**   ASPEN PLUS and PRODABAS as an integrated system.

engineer would like to be able to make runs and try out ideas without committing the results to a formalized database. An interactive workstation environment is shown in Figure 7.16.

Here, the engineer prepares the input using an interactive, menu-driven workstation. The workstation then prepares the ASPEN PLUS input language and sends it to a mainframe. The results are returned to the workstation where they can be examined by the engineer. The workstation can check for errors and can have all of the user-friendly features that the engineers need. Finally, when the engineer is satisfied with a "work product," it can be sent to a process-engineering database.

The standard computing environment that will be used in the 1990s for computer-aided process engineering is not clear. Rapid developments in computer

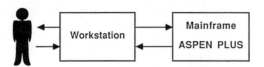

**FIGURE 7.16**   Interactive workstation environment.

graphics, display terminals, microcomputer-based workstations, operating systems, networking, databases, and expert systems will all have an important influence.

## 7.5.   APPLICATIONS OF FLOWSHEET SIMULATION

The ultimate success of any computer-aided engineering system rests on good applications that produce economic benefits for their users and can justify the large expense of developing and maintaining such systems. The use of computer-based models of process plants offers powerful benefits to the process industries by helping them to make better engineering and business decisions. This section describes some of the ways that flowsheet simulation has helped companies in process development, in plant design and retrofitting, and in plant operations. These examples are drawn from the author's experience and discussions with users of flowsheet simulation.

### A.   A New Way of Doing Process Development

A leading chemical manufacturer in the United States, interested in diversifying into new products, was developing a process to make a new polymer. At the earliest conceptual stages, the development team prepared a flowsheet simulation model of the process, including preliminary costing and economics. Pilot-plant experiments were used to determine parameters in the model. The model was then used to optimize the process and to assess its economics.

This represented a new way of doing development in that company. The pilot plant was not used to prove the workability of the process—that was done on the computer. Instead, the pilot-plant experimental program was used to get the data needed for the model, which involved experiments specifically designed for this purpose. Thus the flowsheet model drove the development program. Armed with the results of the flowsheet model, the development team was able to get management approval to go ahead with plant construction in record time.

The plant was constructed and successfully brought on stream with much fewer than the usual number of start-up problems. The leader of the development team was promoted to Director of Process Development and is now a Vice President with the company.

## B.   Speed Commercialization

In another example, a specialty chemical manufacturer was interested in bringing a new high-value product to market as soon as possible. They used flowsheet simulation to model the process and determine the preliminary process economics. The project was highly attractive.

Because of the confidence gained with the model, the company decided to bypass pilot-plant work and go directly from bench-scale experiments into commercial production. As a result, the product was brought to market 18 months ahead of schedule.

Because of the company's head start, a major competitor decided not to enter the market. As a result, the company enjoyed increased revenues measured in the millions of dollars and was able to establish a dominant position in the market.

## C.   Retrofitting of a Liquid-Extraction Process

A chemicals division of FMC Corporation (Gruber, 1985) was designing a process, shown in Figure 7.17, to extract an organic compound (compound A) from an aqueous solution. The solution also contained an organic impurity (compound B); compound A was mildly polar, and compound B was highly polar. Xylene was chosen as the extracting agent, because A is much more soluble in xylene than B and the goal was to extract A from the water without removing B.

The conventional wisdom was to use as little xylene in the extraction step as possible so that the highly soluble A would dissolve in the xylene and not B. This solution of A in xylene was used downstream in the process and since quite a lot of xylene was required, more would be added later in the process. Since B was undesired, any significant amount of B in the xylene would have to be removed.

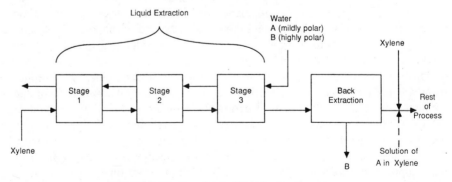

**FIGURE 7.17**  Flowsheet of liquid-extraction process.

To determine the optimum amount of solvent to use, the engineers developed a flowsheet model of the extraction process. They found that, contrary to expectations, the more xylene they used, the less B was extracted. Since xylene was added later in the process anyway, adding this xylene in the extraction to decrease the undesired B was not a problem.

The reason for the change in extraction strategy was that the simulation showed that the more concentrated solution of A in xylene was more polar; hence it was able to dissolve more B than the dilute solution could, which was more like xylene and therefore less attractive to B.

This finding had serious implications for the process. By using a large amount of xylene in the extraction, the engineers were able to keep the amount of B dissolved low, therefore eliminating a whole back-extraction process and the associated train of equipment.

The engineers verified the results of the flowsheet simulation with a quick pilot-plant study. Without the flowsheet simulation, the engineers would not even have thought to try this approach. The whole study required about 2 days of work and resulted in a savings of $250,000 in capital investment.

## D. Ammonia Plant Operations

The Chevron Chemicals Company (Moore et al., 1985) reported their experience of using a flowsheet simulation model of a 1500-ton-per-day ammonia plant located on the Gulf Coast of the United States. Plant engineers run the model on a regular basis to determine the effect of changing key operating parameters and to find the optimum operating conditions. These conditions change with variations in feedstock and ambient conditions. Chevron engineers report savings in excess of $1,000,000 per year from reduced operating costs as a result of using the model.

## E. Design of a Feed-Forward Control System

Another chemical company used a flowsheet simulation model to develop a feed-forward process control model of an energy-intensive distillation column. The goal of the process control system was to measure disturbances, such as changes in feed-stream composition, upstream of the process and then to adjust operating variables, such as the reflux rate, to correct for the effect of the disturbances before they had a chance to affect the process.

The control engineers made a series of flowsheet simulation runs to determine the effect of changes in input variables to the process on the output variables. With this information, they were able to implement a feed-forward control scheme that resulted in a 27% reduction in steam usage and improved product quality and consistency. The company had several columns of this type around the world and total savings were measured in the hundreds of thousands of dollars per year. The flowsheet model was also used to "debottleneck" the overall process.

## F.   Competitive Evaluation

A U.S. producer of commodity chemicals, which is the most efficient in the industry, developed a flowsheet model of their competitor's process. They use the model to estimate the competitor's manufacturing cost. This information enables the U.S. producer to set the price of their product at a level just low enough that the competitor's profit margin is unsatisfactory, but which results in good business for the U.S. company.

## G.   Reverse Engineering

A chemical manufacturer located in the Far East licenses a process from an engineering firm on a turnkey basis. The engineering firm designs the process, but does not provide technical details. They license the process on a "black box" or performance basis.

The company developed a flowsheet model of the process. They were able to "back-calculate" the important process parameters. With this model, the company was able to explore the effect of process modifications and changes in operating conditions.

Armed with this information, the company was better able to negotiate with the process licensor and was able to obtain a license for a plant expansion under much more favorable terms than they had before.

## H.   Raw-Material Bidding

An aluminum manufacturer purchases raw materials (bauxite) on the spot market. They have developed a flowsheet simulation model of the alumina refining process. They use their flowsheet model to determine what their manufacturing costs would be using raw materials with the specifications available. This tells the company how much they can afford to bid for the raw material.

## I.   General Comments

These examples are intended to provide a feeling for how models of processes are used on an everyday basis in the process industries to make better engineering and business decisions. Rigorous process models produce a powerful economic benefit for the companies that use them. These benefits are in terms of reduced manufacturing costs, improved plant throughput and efficiency, and reduced engineering risks.

## 7.6.   FUTURE DIRECTIONS IN COMPUTER-AIDED PROCESS ENGINEERING

The future of computer-aided process engineering is being driven by (1) the demands of industry and society, (2) advances in the field of computer-aided process design, and (3) developments in computer hardware and software.

## A.  Needs of Industry

The process industries are undergoing great changes now, as is evident from the news in 1986 of company mergers and plant closings in the industrialized countries. The manufacture of commodity products is becoming less attractive and is moving toward sources of inexpensive feedstocks. Companies in the industrialized countries are focusing on the manufacture of specialty or high-performance products in which there is a higher technology component. There is increased concern about toxic wastes, environmental pollution, and plant safety. This will affect the need for computer-aided engineering as follows:

- Because producers of commodity products must be more efficient, there will be increased use of flowsheet models to improve plant operations.
- Because there is less new plant construction, the use of models will be to study plant debottleknecking and retrofit designs.
- There is much more use of batch processing in the manufacture of specialty products. This will drive the development of improved tools for simulation and design of batch processes.
- There will be a need for new models of applications in biotechnology, toxic-waste cleanup, and specialty chemical manufacture.
- New computing tools will be needed to perform plant safety and reliability studies.
- There will be increased use of flowsheet simulation in industries that have traditionally made less use, such as metals and minerals, pulp and paper, and food.

## B.  Developments in Computer-Aided Process Design (CAPD)

There is extensive research underway in the field of computer-aided process design. This field is now an established branch of engineering. Some of the advances that will have an impact by the 1990s are:

- Process synthesis techniques will be used routinely in computer-aided engineering systems.
- Optimization techniques will be used routinely in flowsheet models and other modeling applications.
- Expert system techniques will be incorporated into computer-aided engineering systems and will be used to advise the engineer much as an experienced consultant would.
- Dynamic simulation will be used much more routinely, particularly in the study of batch processes and emergency operating conditions.
- There will be increased use of database technology to improve communications between engineers working on a project and to link conceptual process design with downstream project engineering.

- Plant simulation systems will be used much more routinely for operator training.

## C.  Developments in Computer Hardware and Software

There continue to be dramatic improvements in computer hardware and software that will also affect the practice of computer-aided process engineering. Some of these developments include the following:

- The continued decrease in cost of computing will put increasingly powerful workstation computers on the desks of engineers. These computers will eventually be able to do powerful calculations.
- Increased performance of computer-graphics terminals with color will lead to increased use.
- There will be increased use of distributed computer networks, so that the workstation computers can communicate with centralized mainframe computers containing large databases.
- The cost of computers with vector and parallel architectures will be reduced dramatically, making it feasible to use them for computationally intensive flowsheeting calculations, but requiring extensive changes in CAPD software. This is discussed by Stadtherr and Vegeais (1985).

## 7.7.  SUMMARY

In conclusion, this chapter has presented an overview of the state of the art of computer-aided design in chemical engineering.

We first showed the differences between traditional CAD systems used for mechanical, electrical, and civil engineering design and the CAPD systems used by chemical engineers. The focus of chemical engineers on design of a process rather than a product has led to unique requirements for CAPD systems.

The problem of flowsheet modeling and simulation is the central problem in CAPD. We discussed the pros and cons of sequential-modular and equation-solving architectures and showed how the new two-tier simultaneous-modular algorithms offer the promise of bridging the gap between the two.

There is an issue today concerning the proper computing environment for computer-aided engineering. Will engineers work under the control of a centralized engineering database, or will the proliferation of personal computers lead to a very decentralized environment? Again, there appears to be a compromise where engineers work in a decentralized mode on an engineering workstation, but send their work product when ready to a central database for improved communication.

Next, we discussed some applications of computer-aided engineering. Computer-based process models are used routinely for process development, plant design and retrofitting, and plant operations. The use of these models results in better engineering and business decisions and offers powerful benefits and economic savings.

Finally, we examined trends in the field that will affect the future of computer-aided process engineering. These trends are driven by the needs of industry, by developments in CAPD, and by advances in computer hardware and software.

## ACKNOWLEDGMENTS

Many of the examples and illustrations in this chapter were provided by Aspen Technology, Inc. (AspenTech). The assistance of my colleagues at AspenTech and many helpful discussions are gratefully acknowledged. We have drawn on ideas first presented (Evans, 1981) at the Conference on Foundations of Computer-Aided Chemical Process Design held in Henniker, New Hampshire in 1980.

## REFERENCES

Angus, C. J. and P. Winter, "An Engineering Database for Process Design," *Process Systems Engineering PSE '85, The Institution of Chemical Engineers Symposium Series*, No. 92, 1985, pp. 593–606.

Aspen Technology, Inc., *ASPEN PLUS Introductory Manual*, Aspen Technology, Inc., Cambridge, MA, 1986.

Barnard, W. L., D. R. Benjamin, D. L. Cummings, P. C. Piela, and J. L. Sills, "Three Issues in the Design of an Equation-Based Process Simulators," Paper No. 42d, Presented at AIChE Spring National Meetings, New Orleans, LA. (1986).

Biegler, L. T., I. E. Grossmann, and A. W. Westerberg, "A Note on Approximation Techniques Used for Process Optimization," *Comp. Chem. Eng.*, **9**, 201 (1985).

Biegler, L. T. and R. R. Hughes, "Infeasible Path Optimization with Sequential Modular Simulators," *AIChE J.*, **28**, 994 (1982).

ChemShare Corporation, *DESIGN II USER'S GUIDE*, ChemShare Corporation, Houston, TX, 1985.

Craft, J., "The Impact of CAD and Database Techniques in Process Engineering," *Process Systems Engineering PSE '85, The Institution of Chemical Engineers Symposium Series*, No. 92, 1985, pp. 565–580.

Daratech Associates, *CAD/CAM Handbook*, Daratech Associates, Cambridge, MA, 1984.

Evans, L. B., "Advances in Process Flowsheeting Systems," *Foundations of Computer-Aided Process Design*, Vol. 1, R. S. H. Mah and W. D. Seider (eds.), American Institute of Chemical Engineers, New York, 1981, pp. 425–469.

Gruber, G., personal communication, 1985.

Jirapongphan, S., "Simultaneous-Modular Convergence Concept in Process Flowsheet Optimization," Sc.D. Thesis, Department of Chemical Engineering, Massachusetts Institute of Technology, Cambridge, MA, 1980.

Jirapongphan, S., J. F. Boston, H. I. Britt, and L. B. Evans, "A Nonlinear, Simultaneous-Modular Algorithm for Process Flowsheet Optimization," paper presented at the Annual Meeting of the AIChE, Chicago, IL (November 1980).

Kisala, T. P., "Successive Quadratic Programming in Sequential-Modular Process Flowsheet Simulation and Optimization," Sc.D. Thesis, Department of Chemical Engineering, Massachusetts Institute of Technology, Cambridge, MA, 1985.

Kremser, A., *Natl. Pet. News*, **22** (21), 43 (May 21, 1930).

Locke, M. H., S. Kuru, P. A. Clark, and A. W. Westerberg, "ASCEND-II: An Advanced System for Chemical Engineering Design," 11th Annual Pittsburgh Conference on Modeling and Simulation, University of Pittsburgh, Pittsburgh, PA (May, 1980).

Mahalec, V., H. Kluzik, and L. B. Evans, "Simultaneous Modular Algorithm for Steady-State Flowsheet Simulation and Design," presented at the 125th Symposium on Computer Applications in Chemical Engineering, European Federation of Chemical Engineering, Montreux, Switzerland (April, 1979).

Merrill Lynch, *CAD/CAM Industry*, Merrill Lynch Capital Markets, New York, 1984.

Molecular Design, Ltd., *MACCS User Manual*, Molecular Design, Ltd., San Leandra, CA, 1985.

Moore, S. C., T. M. Piper, and C. C. Chen, "Computer Simulation of an Existing Ammonia Plant," presented at the 1985 Ammonia Symposium, AIChE, Seattle, WA (August, 1985).

Perris, F. A., "FLOWPACK II: A Third-Generation Flowsheeting System," *Computer-Aided Process and Plant Design*, M. E. Leesley (ed.), Gulf Publishing Company, Houston, TX, 1982, pp. 609–689.

Pierucci, S. J., E. M. Ranzi, and G. E. Biardi, "Solution of Recycle Problems in a Sequential Modular Approach," *AIChE J.*, **28,** 820 (1982).

Powell, M. J. D., "A Fast Algorithm for Nonlinearly Constrained Optimization Calculations," *Numerical Analysis, Dundee 1977, Lecture Notes in Mathematics No. 630*, G. A. Watson (ed.), Springer-Verlag, Berlin, Germany, 1977, pp. 144–157.

Rosen, E. M., "A Machine Computation Method for Performing Material Balances," *Chem. Eng. Progr.*, **58** (10), 69 (1962).

Rosen, E. M., "Steady-State Chemical Process Simulation—State-of-the-Art Review," *Computer Applications to Chemical Engineering, ACS Symposium Series No. 124*, R. G. Squires and G. V. Reklaitis (eds.), American Chemical Society, Washington, D.C., 1980, pp. 3–36.

Rosen, E. M. and A. C. Pauls, "Computer-Aided Chemical Process Design: The FLOWTRAN System," *Comp. Chem. Eng.*, **1**(1), 11(1977).

Sargent, R. W. H. and A. W. Westerberg, "SPEED-UP in Chemical Engineering Design," *Trans. Inst. Chem. Eng.*, **42,** 190 (1964).

Simulation Sciences, Inc., *PROCESS Input Manual*, Simulation Sciences, Inc., Fullerton, CA, 1985.

Smith, B. D. and W. K. Brinkley, "General Shortcut Equation for Equilibrium Stage Process," *AIChE J.*, **6,** 446(1960).

Stadtherr, M. A. and J. A. Vegeais, "Recent Progress in Equation-Based Process Flowsheeting," *Proceedings of 1985 Summer Simulation Conference*, Chicago, IL. (July, 1985).

Trevino-Lozano, R. A., "Simultaneous-Modular Concept in Chemical Process Simulation and Optimization," Ph.D. Thesis, Department of Chemical Engineering, Massachusetts Institute of Technology, Cambridge, MA, 1985.

Trevino-Lozano, R. A., L. B. Evans, H. I. Britt, and J. F. Boston, "Simultaneous-Modular Process Simulation and Optimization," *Process Systems Engineering PSE '85, The Institution of Chemical Engineers Symposium Series,* No. 92, 1985, pp. 25–36.

Westerberg, A. W., H. P. Hutchison, R. L. Motard, and P. Winter, *Process Flowsheeting,* Cambridge University Press, Cambridge, England, 1979.

Westerberg, A. W., "Optimization in Computer-Aided Design," *Foundations of Computer-Aided Design,* R. S. H. Mah and W. D. Seider (eds.), American Institute of Chemical Engineers, New York, 1981, pp. 149–183.

Winter, P., "CONCEPT: Interactive Process Flowsheeting," *Computer-Aided Process and Plant Design,* M. E. Leesley (ed.), Gulf Publishing Company, Houston, TX, 1982, pp. 512–535.

Wipke, W. T., *The Aster Guide to Computer Applications in the Pharmaceutical Industry,* David J. Fraade (ed.), Aster Publishing Corporation, Springfield, OR, 1984.

# 8

## INTEGRATION OF PROCESS DESIGN AND PROCESS CONTROL

### THOMAS J. McAVOY

## 8.1. INTRODUCTION

This chapter discusses techniques that can be used at the design stage of a process to assess process operability and controllability. Traditionally process control followed process design. Today, increased process integration is leading to tighter and tighter designs and forcing process designers to consider potential control problems early. The techniques that can be used by a designer to assess control and operability problems are split into three categories based upon the information they require. These categories are techniques that require steady-state, approximate dynamic, and detailed dynamic information. Techniques requiring only steady-state information appear to be the most promising for use at the design stage. In this chapter techniques in all three categories are discussed. Enough detail is given to allow the reader to gain an insight into the available methods. A number of current literature citations are presented so that the reader can further study techniques of interest.

## 8.2. HISTORICAL PERSPECTIVE

To discuss the current state of the integration of process design and control, it is worthwhile to consider the traditional relationship between these areas. Traditionally, and probably in many cases today, a plant is designed and *then* a control system is added. There was very little integration between design and control in the past. However, this situation is changing and one of the objectives of this chapter is to discuss the reasons for and the directions of these changes.

The early 1970s marked a key point in terms of both process design and control. In 1973 the oil embargo took place and this marked the end of cheap energy. When such energy was available, there was very little incentive either to design plants tightly or to apply advanced control concepts. Plants were typically over-designed and excess intermediate storage capacities were added to damp out the effects of upsets. As energy costs rose, it became economically attractive to design plants tighter, particularly with respect to energy integration. These modern, integrated designs in turn can require more advanced control systems for their operation than earlier designs. They also can require more advanced control systems for startup and shutdown, and during emergencies.

The early 1970s also marked the time when digital process control became an economic reality. Before that time, digital process controllers were both costly and unreliable. Today essentially all new plants and many old ones are run under digital control. In the 1980s, on-line expert systems have been introduced to help operators diagnose and solve operating problems. These expert systems consist of computer programs that contain knowledge about a process and its alarm conditions and a means to infer specific fault conditions. The programs are user-friendly and are designed to be run in real time. There are also several commercially available digital adaptive controllers that can adjust automatically to process changes and still achieve effective control performance. Given the tremendous advances taking place in digital technology, the future impact of digital devices on process control will clearly be substantial.

The economic implications of proper integration of process design and control can be illustrated by a recent industrial example. The $2 billion Syncrude Tar Sands Plant at Fort McMurray, Alberta, Canada has had serious operation/control problems with its fluid cokers for several years. These problems resulted in stream factors of less than 50%, very high maintenance costs, and severely depressed economic viability. In order to focus more clearly on the subject of integration of process design and control, several small examples are discussed. In considering these examples, keep the tar sands plant in mind as being indicative of the economic importance of the subject.

To put the past into perspective, it is useful to consider an actual industrial example taken from Shinskey (1979). This case, shown in Figure 8.1, involves a distillation tower with parallel reboilers. Three variables must be controlled, a tray temperature in the stripping section and the two reboiler levels. The three manipulated variables are the heat duty to each reboiler (manipulated via a BTU cascade controller) and the bottoms flow.

The system was designed in the 1960s. The reason for using parallel reboilers was not stated, but they caused great difficulty in the control of the tower. A single-input/single-output (SISO) strategy was used in which one manipulated variable was paired with one controlled variable. No matter how the manipulated and controlled variables were paired (e.g., bottoms flow controlling one level, heat duty controlling temperature, etc.), the system could never be made to work in actual practice. These operational problems can be traced to the designer's use of parallel reboilers. They do not occur in towers with a single reboiler. The operator's solution to this problem was to turn the control system off. Using excess energy or over-purifying the products was not considered significant enough to attempt to put the system back on automatic control.

The control solution to this parallel reboiler problem turns out to be very simple and it is discussed below. An important aspect of the solution is that it required

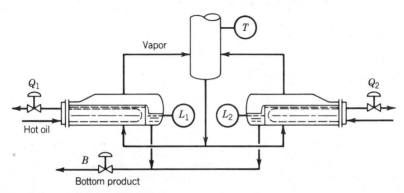

**FIGURE 8.1** A distillation tower with parallel reboilers (Shinsky, 1979). Reprinted with permission from McGraw-Hill Book Company.

the addition of a decoupler between the controller output and the manipulated variables. Since the original control system was pneumatic and inflexible, addition of this decoupler was not a trivial problem. Tubing had to be disconnected and then reconnected behind the control panel. In this example, the lack of integration between design and control created problems. Such problems would not have arisen if the designer had used a single reboiler, or if a more advanced control strategy was used from the beginning for the parallel reboilers. Additional interesting examples that illustrate the need to integrate process design and control are given by Shinskey (1983).

The control solution to the parallel reboiler system (Shinskey, 1979) is shown in Figure 8.2. It was developed by a simple and quick application of Bristol's (1966) relative-gain array (RGA) technique. (See further discussion of RGA in Section 8.5.A.3 below.) The application of the RGA to this problem did not require a detailed modeling study, but only knowledge of the process structure. The derivation of Shinskey's solution is given elsewhere (McAvoy, 1983). This solution treats the parallel reboiler system as if there were only a single reboiler. Thus the total accumulation of material in the reboilers, measured by the sum of the two levels, is controlled. Similarly the total heat input, given by the sum of the individual heat inputs, is manipulated to control the tray temperature. Finally, to complete the control structure, a differential heat input is used to control the difference in levels. When the controller shown in Figure 8.2 was implemented, all the operational difficulties disappeared and the system was left on control by the operators.

As can be seen in Figure 8.2, this revised control system requires adding and subtracting signals. In the days of pneumatic controllers, this task was not trivial.

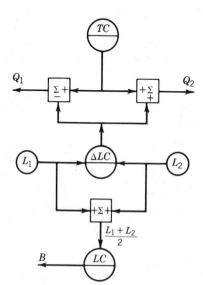

**FIGURE 8.2** Final control system for the tower shown in Figure 8.1 (Shinsky, 1979). Reprinted with permission from McGraw-Hill Book Company.

Today such calculations are done routinely. If this parallel reboiler system is encountered today, reconfiguring the control loops is much simpler than it actually was in the 1960s. If this is the case, one could argue that modern digital technology makes it easier for process control to be carried out after process design. If one control system doesn't work, it would appear to be a simple matter to reconfigure it. While such an approach can work in some cases, there are many important cases where it will fail. In these instances, the system is designed to have a structure such that good performance cannot be achieved using any type of control system. For these cases, it is imperative that the poor control performance be recognized at the design stage of the process. Examples of systems with inherently poor structure are discussed in the following section.

## 8.3. EXAMPLES OF SYSTEMS WITH POOR CONTROL STRUCTURE

### A.  Heat-Exchanger Network

A number of examples could be given to make the point that process design cannot be carried out today without considering the control and operational consequences of the design. Two papers are singled out for illustration. In discussing control system operability, resiliency, and flexibility, Grossmann and Morari (1984) presented a number of interesting examples. One involved the design of a heat-exchanger network, reproduced in Figure 8.3a, in which two hot streams are to be cooled and two cold streams are to be heated. A minimum-approach temperature of 10°C is desired for all exchangers. The stream flow rates and temperatures are given in Table 8.1. The heat-capacity flow rate (i.e., the average heat capacity multiplied by the flow rate) of stream H1 can range between 1 and 1.85 kW/°C. Figure 8.3a gives the base-case design for a heat-capacity flow of 1.0 kW/°C. Grossmann and Morari report that no other design with a smaller number of exchangers could be found. Figure 8.3b gives the results for a heat-capacity flow rate of 1.85 kW/°C. As can be seen for both extreme values, the network achieves the desired design performance in terms of approach and target temperatures. Since the network performs well at the extremes, one might be tempted to conclude that it will perform well for any intermediate value between 1 and 1.85 kW/°C. Figure 8.3c shows that for a heat-capacity flow rate of 1.359 kW/°C, the outlet temperature of H1 cannot be reduced below 71°C even if exchanger 1 had an infinite area. This temperature constitutes a 21° violation of the target temperature. Thus the structure given in Figure 8.3a is not capable of achieving the desired operating conditions regardless of the control system used. Clearly it is important that designers be aware of structural limitations to system performance such as those that occur in this example. By contrast to the parallel reboiler example, the structure problems in this example *cannot* be corrected after the fact by advanced control.

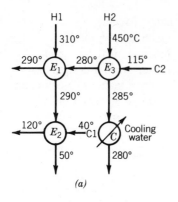

*(a)*

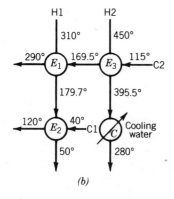

*(b)*

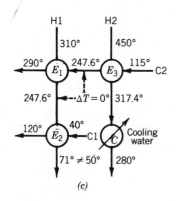

*(c)*

**FIGURE 8.3** (a) Base-case design of heat-exchanger network for stream data specified in Table 8.1 with the heat-capacity flow rate of H1 being equal to 1 kW/°C. (b) Extreme-case design of heat-exchanger network for stream data specified in Table 8.1, except the heat-capacity flow rate of H1 is 1.85 kW/°C. (c) Infeasible design of heat-exchanger network for stream data specified in Table 8.1, except the heat-capacity flow rate of H1 is 1.359 kW/°C. Note the temperature target violation by 21°C on H1, despite a zero approach temperature on exchanger E1 (infinite area). (Grossmann and Morari, 1984). Reprinted with permission from CACHE Corporation.

In the same paper, Grossmann and Morari discuss a distillation example where two different control strategies are used. The dynamic performance of both strategies, using a nominal plant model, is identical. However, for small changes in model parameters, the performance of one of the control systems deteriorates much more drastically than the other. This distillation example indicates the need for process designers to have tools that can be used to measure which systems are prone to operational problems due to parameter sensitivity.

**TABLE 8.1  Stream Data for Heat-Exchanger Networks of Figure 8.3[a]**

| Stream Number | Flow*Cp (kW / °C) | Initial Temperature (°C) | Target Temperature (°C) |
|---|---|---|---|
| H1 | 1 | 310 | 50 |
| H2 | 2 | 450 | 280 |
| C1 | 3 | 40 | 120 |
| C2 | 2 | 115 | 290 |

[a]From Grossmann and Morari (1984). Reprinted with permission from Pergamon Press.

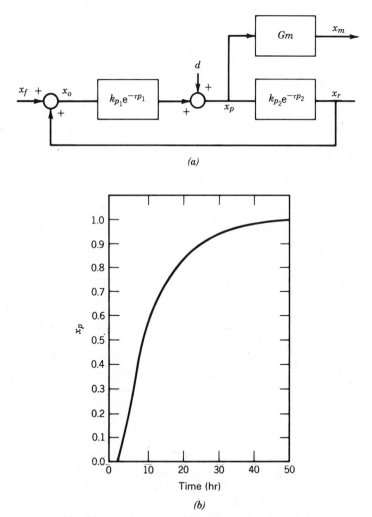

*(a)*

*(b)*

**FIGURE 8.4** (a) A simple recycle system block diagram (Rinard and Benjamin, 1982). (b) Transient response to step upset in $x_f$ for d = 0; $\tau_{P_1}$ = 0.5; $\tau_{P_2}$ = 1.0; $k_{p_1}$ = 1.0; $k_{P_2}$ = 0.9 (Rinard and Benjamin, 1982). Reprinted with permission from authors.

## B. Recycle Process Example

Another example of the effect of process structure on control-system performance was presented by Rinard and Benjamin (1982). They studied reactor systems in which there was a substantial recycle of unreacted material due to low conversion. They reported difficulty in controlling such reactors. These authors used a simple dead-time model for the reactor and separation systems, shown in Figure 8.4a, to illustrate their point. Since the process dead-times are 0.5 and 1 hr, one might expect that response to an upset would take about 1.5 hr. Figure 8.4b gives the response reported by Rinard and Benjamin. As shown, it takes 10–15 hr for the effect of upsets to die out. In this case the recycle structure of the process, coupled with the large recycle ratio, give rise to the unexpectedly long transients. Other authors (Gilliland et al., 1964; Attir and Denn, 1978; Denn, 1979; Denn and Lavie, 1982; and Kapoor et al., 1986a) noted similar effects. If there were no recycle, clearly the transients in the system would be of a much shorter duration. Even though this example involves very simple models, the essential point is clear. If a reactor system is designed for a large recycle of unreacted material, it may be impossible to achieve acceptable dynamic performance. The overall system performance may simply be too sluggish, regardless of the control system used.

## C. Summary

These examples demonstrate the need for designers to be aware of potential control problems at the design stage of a process. Some tools available today allow a designer to gain such insight. There is also a clear need for further research in this area. Before mentioning the tools that exist, a discussion of the unique nature of chemical process control is in order.

## 8.4. UNIQUE NATURE OF CHEMICAL PROCESS CONTROL

It is interesting to compare chemical process-control problems to control problems that arise in other disciplines. Take, for example, the case of designing a control system for a new jet-fighter aircraft. Typically the control-system designer starts with a dynamic model of the aircraft. The expense of developing and validating such a model is not a problem, since one cannot place a cost on defending the country. The cost producing a model is also mitigated by its application to the several hundred aircraft that will be built. Now consider the aircraft's pilot. He is a well-educated and technically competent individual who has the ability to comprehend advanced control systems. Hence, aircraft control designers are able to use some reasonably sophisticated approaches. It has been reported recently (Astrom and Wittenmark, 1984) that some modern fighter aircraft are open-loop unstable at low speeds, that is, on take-off and landing. Without advanced, sophisticated control systems, these aircraft would not be able to fly.

Now consider the case of chemical process control. First, almost all plants are custom-designed, because different contractors are used or technology changes occur over time. If a model is developed for one plant, there is no guarantee that it will apply to another plant. Even for individual units such as catalytic crackers and crude towers, there is no such thing as a completely general model. Each model has to be customized, and this requires considerable engineering time. It can be argued that increases in computer speed and performance will lead to more and more use of detailed dynamic simulations. However, if engineering manpower is the limiting cost factor in model development and simulation, expected advances due to computer speed may not be that substantial. Second, the most significant performance index in the process industries is *profit*. An advanced control system that is elegant will not be used unless it is profitable. By contrast with modern fighter aircraft, it is difficult to imagine producing chemical products with open-loop unstable reactors unless substantial profits would result. Third, chemical-plant operators tend to be relatively unsophisticated, compared with jet pilots. Control systems have to be understandable to these operators or they will simply shut the controllers off.

These unique aspects of chemical process control are illustrated by considering two of the fastest growing research areas in the field. These are adaptive control (Seborg et al., 1986; Astrom, 1986; Dumont, 1986) and model-predictive control (Garcia and Morari, 1982; Garcia and Prett, 1986; Popiel et al., 1986; Richalet and Froisey, 1986). Both are drawing considerable industrial interest. In the most common approach to adaptive control, one has to specify the form of the model and then use the process itself, through identification, to calculate model parameters. Alternatively, a "pattern recognition" adaptive controller, which requires essentially no modeling effort, has been marketed (Bristol and Kraus, 1984). In model-predictive control, numerical models, developed from experimental testing of actual systems, are used. In both adaptive and model-predictive control, the detailed and costly model development step is avoided or side-stepped. Both types of control, however, still have to be accepted by operators.

The unique aspects of chemical process control also have to be considered when addressing the integration of design and control. At the design stage, detailed dynamic models are available for only a small number of operations, such as distillation (Roat et al., 1986). *To be useful at the design stage, control techniques have to use only the information that is available or can be obtained consistent with process economics.* A number of control techniques only require steady-state or limited dynamic information. These techniques hold a great deal of promise for helping to integrate process design and process control. In the discussion below, techniques are split into three categories following the approach of Marlin et al. (1986). These are methods requiring steady-state and limited dynamic information, and a complete dynamic model.

## 8.5. TECHNIQUES FOR INTEGRATING PROCESS DESIGN AND PROCESS CONTROL

### A. Techniques Utilizing Only Steady-State Information

A number of techniques for analyzing control-system performance require only steady-state information. Since this type of information is readily available at the design stage of a process, these techniques are the most likely to be widely applied. However, neglect of process dynamics is clearly a weakness of the steady-state methods. A common theme that runs through these techniques is that they assume perfect steady-state control. This assumption allows one to focus on the process itself and not its controller to assess control-system behavior. Most of the techniques are based on linear theory, while real processes are nonlinear. In spite of this limitation, the steady-state methods hold great promise for integrating process design and process control.

## 1. Flexibility and Resiliency

Interesting results in the area of assessing the flexibility and resiliency of a given design have recently appeared. As defined by Grossmann and Morari (1984), resiliency refers to the ability of a plant to tolerate adverse conditions (e.g., parameter variations and disturbances) and to perform acceptably. Flexibility refers to the ability of a design to readily adjust to meet alternative, desirable operating conditions. As used in the literature, however, these two definitions sometimes overlap.

### a. FLEXIBILITY

In the last several years, a number of papers have appeared on designing plants for flexibility. In addition to the literature cited below, bibliographies can be found in Grossmann and Morari (1984), Morari (1983b), and Swaney and Grossmann (1985a). Design for flexibility represents a fundamental change in philosophy. The new approaches were developed from the necessity of *integrating* process design and control for today's tightly designed plants. In the past, plants were designed for nominal operating conditions and a number of extreme cases were checked. As the example shown in Figure 8.3 illustrates, this approach can fail when the worst operating conditions are not obvious.

Grossmann and co-workers (Grossmann and Halemane, 1982; Grossmann et al., 1983; Swaney and Grossmann, 1985a,b) developed a steady-state index to measure process flexibility. They treat a general plant described by a set of equality and inequality constraints:

$$h(d, m, x, \theta) = 0 \qquad (8.1)$$

$$g(d, m, x, \theta) \leq 0 \qquad (8.2)$$

where $d$ are the design variables, $m$ are the manipulated variables, $x$ are the state variables, and $\theta$ are the uncertain parameters (e.g., changes in process parameters, such as heat transfer coefficients, as well as disturbances). Thus the flexibility index proposed by Grossmann and co-workers measures both flexibility and resiliency as defined above. The state variables, $x$, are eliminated because they are controlled by $m$. The resulting problem involves a set of constraints:

$$f(d, m, \theta) \leq 0 \qquad (8.3)$$

that define the feasibility or infeasibility of operation for a given control, $m$, when $d$ and $\theta$ are given. To explain the flexibility index proposed by Grossmann and co-workers, it is convenient to refer to Figure 8.5a, which illustrates a two-dimensional example. Inside the enclosed cross-hatched region, Equation (8.3) is satisfied and the plant operation is feasible. To calculate the flexibility index of a design, the expected maximum and minimum changes in the uncertain parameters, $\theta$, have to be specified:

$$\theta^- \leq \theta \leq \theta^+ \tag{8.4}$$

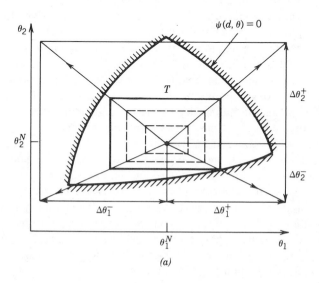

*(a)*

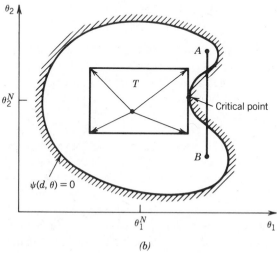

*(b)*

**FIGURE 8.5**  (a) Maximum scaled hyperrectangle $T$ inscribed within feasible region. (b) Region with nonconvex critical point (Swaney and Grossman 1985a). Reprinted with permission from the American Institute of Chemical Engineers.

The nominal values for $\theta$ are indicated as a superscript $N$ in Figure 8.5a and they can be anywhere in the range defined by Equation (8.4). The changes in $\theta$, normalized by their maximum changes, are all taken to be the same in both positive and negative directions and they are given by a single variable $\delta$. If some parameter variations are dependent on the variations of other parameters, these dependent parameters are eliminated before beginning the analysis. As $\delta$ increases, a series of "hyperrectangles" is inscribed inside the feasible region. The flexibility index for a given design is the value of $\delta$ that produces the maximum inscribed hyperrectangle in terms of area. By calculating $\delta$ for several competing designs, it is possible to determine a measure of the flexibility of each design. In this way, alternative designs can be compared with one another.

The calculation of $\delta$ is far from a trivial task. Determining $\delta$ is a semiinfinite programming problem that is difficult to solve. The problem is called semiinfinite because $\delta$ can take on any value from 0 to $+\infty$. In some cases, the optimization problem can be nonconvex; Figure 8.5b illustrates such a case. The critical point occurs in the midrange of the inscribed hyperrectangle rather than at a vertex, as is the case in Figure 8.5a. At the critical point, plant operation is feasible; however, any increase in $\delta$ beyond the critical value results in infeasible operation, meaning that Equations (8.3) and (8.4) have no solution. Swaney and Grossmann (1985a,b) determined conditions under which the critical point must occur at a vertex and they developed algorithms to solve these cases. They also give a simple example of designing a fluid-pumping system to illustrate the flexibility analysis. While the flexibility index represents a real advance in terms of evaluating a given design, an even more challenging problem involves the synthesis of flexible designs from the beginning. Work is being carried out in this area and no doubt interesting results will be published in the future.

### b. RESILIENCY

Morari and co-workers (Lenhoff and Morari, 1982; Marselle, et al., 1982; Morari, 1983a,b; Saboo and Morari, 1984; Holt and Morari, 1985a,b; Morari et al., 1985; Saboo et al., 1985) published a number of interesting papers on the subject of process resiliency. The basis for their analysis is the Internal Model Control (IMC) structure (Garcia and Morari, 1982) shown in Figure 8.6. In Figure 8.6, $G$ is the actual plant and $\tilde{G}$ is a model of $G$. If a perfect model is available, and if it contains no dead times or right half-plane zeroes, perfect control can be achieved by using the IMC structure and setting $G_c$ equal to $G^{-1}$. In general, for real systems, perfect control cannot be achieved. Morari and co-workers showed that three items limit such control and therefore limit process resiliency. These are (1) nonminimum-phase elements (dead times and right half-plane zeroes), (2) constraints on control action, and (3) plant–model mismatch. Item 1 requires a dynamic model, while items 2 and 3 can be calculated either from steady-state or dynamic models. In a number of cases, Morari and co-workers found that steady-state information alone was sufficient to determine resiliency and discriminate among various design alternatives. The resiliency analysis has two key aspects. First, the approach de-

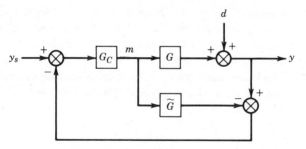

**FIGURE 8.6** An Internal Model Control structure (Morari, 1983a). Reprinted with permission from Pergamon Press.

pends only on the process model itself and is independent of the particular control system used. Second, the approach does not require a detailed knowledge of the disturbance affecting the system. Both of these characteristics greatly facilitate carrying out the resiliency analysis.

To determine the resiliency of a design, a model of the process is needed. This model is the open-loop transfer function of the plant. At steady state, the model is equal to the gain matrix. Using singular-value decomposition, SVD (Klema and Laub, 1980; see Section 8.5.A.2), the gain matrix $K$ is decomposed as

$$K = V \Sigma W^T \tag{8.5}$$

where

$\Sigma = \text{diag} \, (\sigma_1, \sigma_2 \ldots \sigma_n) > 0$

$\sigma_i = $ singular value of $K$

$V = $ matrix of left singular vectors

$W = $ matrix of right singular vectors

The resiliency analysis consists of analyzing the minimum singular value, $\sigma_m$, and the condition number (CN) of the system, defined as the ratio of the maximum ($\sigma^M$) to minimum singular values:

$$CN = \frac{\sigma^M}{\sigma_m} \tag{8.6}$$

The minimum singular value is proportional to the magnitude of the largest upset that can be handled without causing the manipulated variables to saturate. The condition number is a measure of how sensitive the system will be to errors in the process model. The most resilient system is the one with the smallest condition number and the largest minimum singular value.

As a simple illustration of the resiliency analysis, consider the system shown in Figure 8.7. It is desired to control the flow of material $F$ and the intermediate pressure $P_1$. The fluid is assumed to be incompressible and, to simplify the analysis, the flow through a valve is assumed to be proportional to the pressure drop across the valve. This last assumption can be relaxed and, for example, a square-root relationship can be used. Based on these assumptions, the following steady-state model can be written:

$$F = m_1(P_0 - P_1) \tag{8.7}$$

$$F = m_2(P_1 - P_2) \tag{8.8}$$

The manipulative variables are stem positions on the control valves, which are represented by $m_1$ and $m_2$ in Equations (8.7) and (8.8). The gain matrix for this system is given by

$$K = \begin{pmatrix} \left.\dfrac{\partial F}{\partial m_1}\right|_{m_2} & \left.\dfrac{\partial F}{\partial m_2}\right|_{m_1} \\[2mm] \left.\dfrac{\partial P_1}{\partial m_1}\right|_{m_2} & \left.\dfrac{\partial P_1}{\partial m_2}\right|_{m_1} \end{pmatrix} \tag{8.9}$$

The four elements in the $K$ matrix can be calculated by differentiating Equation (8.7) and (8.8) to give

$$K = \begin{pmatrix} \dfrac{(P_0 - P_1)^2}{(P_0 - P_2)} & \dfrac{(P_1 - P_2)^2}{(P_0 - P_2)} \\[3mm] \dfrac{(P_0 - P_2)^2(P_1 - P_2)}{F(P_0 - P_2)} & \dfrac{-(P_1 - P_2)^2(P_0 - P_1)}{(P_0 - P_2)F} \end{pmatrix} \tag{8.9}$$

For illustrative purposes, $F = 100$, $P_0 = 200$, and $P_2 = 100$. Two cases for $P_1$ are considered, namely, $P_1 = 199$ and $P_1 = 150$. Before examining the singular values for these two cases, it is helpful to consider the physical system first. When $P_1 = 199$, the physical system is unusual. One valve has a pressure drop of 1.0,

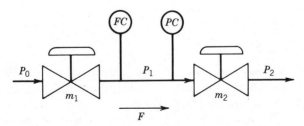

**FIGURE 8.7**  Flow- and pressure-control system.

while the other has a pressure drop of 99. Physically, the valve with the small pressure drop would have to be much larger than the valve with the large pressure drop, if the flow through each valve is the same. In fact, the small valve would have to be about 100 times smaller than the large valve.* One can anticipate operational problems in a system that uses a very small valve in series with a large valve. It is likely that the small valve will saturate when nominal process upsets occur. If valve saturation cannot be tolerated, one should look at redesigning the process for better operability. For the second case where $P_1 = 150$, both valves are the same size and neither is more prone to saturation than the other.

The SVD calculation supports the results based on physical insight. Using the expressions given by McAvoy (1983), the singular values for the two cases can be calculated as

*Case 1 ($P_1 = 199$)*

$$\sigma_m = 0.01 \tag{8.11}$$

$$\sigma^M = 98 \tag{8.12}$$

*Case 2 ($P_1 = 150$)*

$$\sigma_m = 17.7 \tag{8.13}$$

$$\sigma^M = 35.4 \tag{8.14}$$

The small $\sigma_m$ for Case 1 indicates that it will be prone to saturation problems. The large $\sigma_m$ for Case 2 suggests that it is a much better design than Case 1 in terms of its ability to deal with upsets. In this simple example, potential operational problems are obvious. For more complex physical processes, it can be very difficult to use physical insight to analyze potential operability problems. Downs and Moore (1981) presented examples involving azeotropic distillation that would be difficult to analyze without using SVD. With complex systems, SVD is particularly helpful for pointing out potential control-valve saturation problems.

Another aspect of analyzing process resiliency involves assessing the effect of errors in a process model. Figure 8.8 shows the results of applying the resiliency analysis to a model of an industrial reactor system (Morari et al., 1985). In Figure 8.8a, the nominal response is shown. Figure 8.8b gives the response of one possible control system design, while Figure 8.8c gives the response of another. For both b and c, identical errors are introduced into the nonlinear model. The steady-state resiliency analysis indicates that the system in Figure 8.8b is much more sensitive to plant–model mismatch than the system in Figure 8.8c. The detailed dynamic simulations bear out this fact. The deviation between Figure 8.8b and a is much more substantial than that between c and a.

---

*If $m_i$ is proportional to valve size, the size of the valves would differ by a factor of 99, the ratio of the two pressure drops.

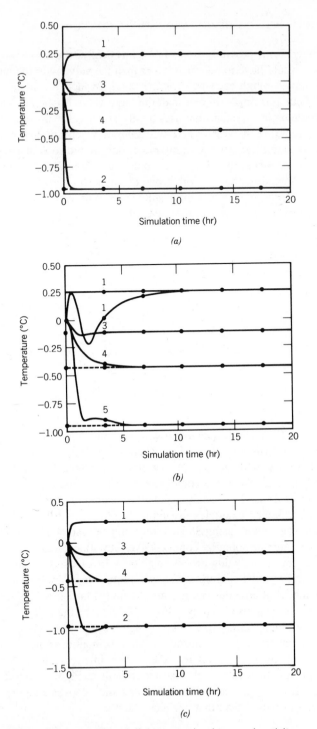

**FIGURE 8.8** (a) Closed-loop response of all designs in the absence of modeling error. (b) Closed-loop response of design 1B in the presence of modeling errors. (c) Closed-loop response of design 4B in the presence of modeling errors (Morari et al., 1985). Reprinted with permission from Pergamon Press.

While the resiliency analysis yields new, promising results, some aspects need further study. First, by not considering disturbances explicitly, it is possible that the method produces overly conservative estimates. Structured singular-value analysis (Doyle, 1982; Morari and Doyle, 1986; Grosdidier and Morari, 1986) may overcome this limitation, but only at the expense of requiring considerably more detailed knowledge about the expected disturbances. Second, the singular-value analysis is dependent upon how the system model is scaled and such scaling is arbitrary. If one changes scaling, the answers given by the method change. It has been proposed (Grossmann and Morari, 1984; Morari et al., 1985) that the appropriate method of scaling is that which minimizes the steady-state condition number. However, real control systems are not always scaled in this manner because of such variables as throughput changes. Additional work is needed to determine the best scaling approach for solving actual problems.

## 2. Singular-Value Decomposition (SVD)

Moore, Bruns, and co-workers (Smith et al., 1981; Downs and Moore, 1981; Bruns and Smith, 1982) were among the first to publish research on the use of steady-state singular-value decomposition to analyze chemical process-control problems at the design stage. They used $V$, $W$, and $\Sigma$ matrices to draw conclusions about various control schemes. They used the $W$ matrix to indicate the degree of interaction among manipulated variables. The $\Sigma$ matrix was utilized to determine how many degrees of freedom the system had from a practical point of view. They scaled the process-gain matrix $K$ based on how the control system was implemented (i.e., using actual valve and transducer sizes). If the singular values in the system were on the order of the noise in the system, they concluded that the system had lost one or more degrees of freedom and some control objectives had to be sacrificed (Downs, 1982). A loss of one or more degrees of freedom is equivalent to having small $\sigma_m$'s in the resiliency analysis discussed above. If one tries to implement too many loops when small singular values occur, valve saturation results because there are not enough degrees of freedom. Lastly, Downs and Moore used the $V$ matrix to determine the best location for process sensors. This point deserves some elaboration.

One of the advantages of singular-value decomposition is that it can be applied to systems that are not square. Such systems involve cases where there are more variables to be controlled than there are manipulative variables, and vice versa. In one example, Downs and Moore (1981) considered the control of a complex azeotropic distillation tower. They kept all tray temperatures as candidates for sensor location. By carefully numerically differentiating a steady-state simulation of the tower, they were able to determine process gains. Further, by applying singular-value decomposition, they were able to calculate the $V$ matrix. For the azeotropic tower, this matrix indicated what tray temperatures would be affected most by the action of the manipulated variables. The authors selected these tray temperatures and then verified their choice by running nonlinear dynamic simulations. Their approach holds much promise as a systematic method for selecting sensor locations based on steady-state information, and it has been applied to a number of industrial column-control designs (Roat et al., 1986).

## 3. Relative Gain Array (RGA)

The oldest technique for integrating process design and control is the relative gain array (RGA), which was introduced by Bristol (1966) and popularized by Shinskey (1979). A recent monograph (McAvoy, 1983) gives a thorough discussion of the RGA and its pertinent literature through 1983.

One carries out an RGA analysis by assuming that a conventional, multiloop PID control structure is used. With such a control system, one measurement is used to manipulate a single valve. One then focuses on a selected loop in the control system. Assume that the loop under consideration involves controlling $x_i$ by manipulating $m_j$. The RGA results from calculating the steady-state gain of this loop when all other loops are both off and perfect. Mathematically, the $ij$th element of the RGA is defined as

$$
\lambda_{ij} = \frac{\left. \dfrac{\partial x_i}{\partial m_j} \right|_{m_k} \quad k \neq j}{\left. \dfrac{\partial x_i}{\partial m_j} \right|_{x_k} \quad k \neq i} \tag{8.15}
$$

The various $\lambda_{ij}$'s are arranged in a matrix called the relative gain array. One practical method of calculating the numerator and denominator of $\lambda_{ij}$ is to perform numerical differentiation on standard steady-state design programs. However, care must be taken to avoid numerical errors (McAvoy, 1983). Since the RGA involves a ratio of gains, it is dimensionless and scale-independent. In fact, the RGA can be considered as a dimensionless number in process control. Just as the Reynolds number involves a ratio of viscous to inertial effects in fluid-dynamical systems, the RGA involves a ratio of open- to closed-loop effects in multiloop-control systems. Although the RGA is based on a linearized model, the ratio nature of the RGA tends to make it reasonably insensitive to process nonlinearity (Shinskey, 1979). Another important feature of the RGA that makes it valuable at the design stage of a process is that it is independent of controller design. Thus a designer does not have to carry out detailed control-system designs for processes that he only wants to screen for further study.

The most important uses of the RGA are to decide how loops should be paired in multiloop systems, whether or not loop interaction will be a problem, and if decoupling needs to be considered. In multiloop systems, one would like the closure of one loop to have a minimal effect on other loops in the system. If one pairs manipulative ($m_j$) and controlled ($x_i$) variables based on an RGA value ($\lambda_{ij}$) close to 1.0, loop interaction is minimized, at least at steady state. This point can be seen by noting that the numerator of the right-hand side of Equation (8.15) is the open-loop gain of $x_i$ to $m_j$. The denominator is the closed-loop gain of $x_i$ to $m_j$. Thus Equation (8.15) can be rewritten as

$$\text{Closed loop gain} = \frac{\text{Open loop gain}}{\lambda_{ij}} \qquad (8.16)$$

If $\lambda_{ij}$ is close to 1.0, closure of additional loops has little effect on the gain of the $x_i - m_j$ loop. However, one cannot always pair $x_i - m_j$ variables based on RGA values close to 1.0. Figure 8.9 gives the results for a $2 \times 2$ reactor-control system taken from Marino-Galarraga et al. (1985). In this case, the only stable way to pair manipulated and controlled variables is based on an RGA value of 13.7. As shown, the resulting control system behaves very poorly when a setpoint change is introduced into one loop. In general, one has to pair on positive RGA values to make possible good control-system performance. If a pairing is made on either a large RGA (relative to one) or a small fractional RGA element, the resulting control-system behavior is poor.

To illustrate the use of the RGA, the system shown in Figure 8.7 and modeled by Equations (8.7) and (8.8) is considered again. For such $2 \times 2$ systems, only one element of the RGA is independent (McAvoy, 1983). If $x_1 = F$ and $x_2 = P_1$, the RGA for the $F - m_1$ pairing can be calculated by differentiating Equations (8.7) and (8.8). First consider the numerator in equation (8.15). Differentiation of Equation (8.7) and (8.8), holding $m_2$ constant, gives

$$\frac{\partial F}{\partial m_1}\bigg|_{m_2} = P_0 - .P_1 - m_1 \frac{\partial P_1}{\partial m_1}\bigg|_{m_2} \qquad (8.17)$$

$$\frac{\partial F}{\partial m_1}\bigg|_{m_2} = m_2 \frac{\partial P_1}{\partial m_1}\bigg|_{m_2} \qquad (8.18)$$

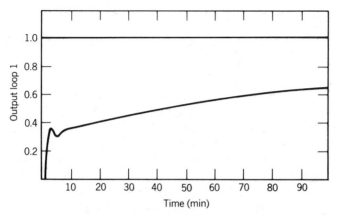

FIGURE 8.9 Response to step change in set point.

Note that in carrying out the differentiation, the upstream and downstream pressures, $P_0$ and $P_2$, are taken as constant. Elimination of $\partial P_1/\partial m_1 \big|_{m_2}$ between Equations (8.17) and (8.18) gives the numerator of Equation (8.15) as

$$\frac{\partial F}{\partial m_1}\bigg|_{m_2} = \frac{m_2}{m_1 + m_2}(P_0 - P_1) \tag{8.19}$$

In calculating the denominator of Equation (8.15), $P_1$ is taken as constant. Differentiation of Equation (8.7) gives

$$\frac{\partial F}{\partial m_1}\bigg|_{P_1} = (P_0 - P_1) \tag{8.20}$$

Substituting Equations (8.19) and (8.20) into Equation (8.15) gives

$$\lambda_{F-m_1} = \frac{\dfrac{\partial F}{\partial m_1}\bigg|_{m_2}}{\dfrac{\partial F}{\partial m_1}\bigg|_{P_1}} = \frac{m_1}{m_1 + m_2} \tag{8.21}$$

By solving for $m_1$ and $m_2$ using Equations (8.7) and (8.8), Equation (8.21) can be written in terms of pressure drops as

$$\lambda_{F-m_1} = \frac{P_0 - P_1}{P_0 - P_2} \tag{8.22}$$

Equation (8.22) shows that the RGA for the $F - m_1$ pairing is always a fraction, since the pressure drop across the first valve, $P_0 - P_1$, is always less than the total pressure drop, $P_0 - P_2$. Using the properties of the RGA (McAvoy, 1983), it is easy to show that the RGA for the $F - m_2$ pairing is

$$\lambda_{F-m_2} = \frac{P_1 - P_2}{P_0 - P_2} \tag{8.23}$$

In pairing manipulated and controlled variables using the RGA, one chooses loops that have $\lambda_{ij}$'s that are positive and as close to 1.0 as possible. Again consider the case where $P_0 = 200$ and $P_2 = 100$. If $P_1$ is equal to 120, $\lambda_{F-m_1}$ is 0.8. In this case, the $F - m_1$ pairing should be made. If $P_1$ is 180, $\lambda_{F-m_1}$ is 0.2 and the $F - m_1$ pairing should not be made. However, the RGA for the $F - m_2$ loop, $\lambda_{F-m_2}$, is 0.8 when $P_1 = 180$. In this case, $F$ should be paired with $m_2$. Once the $F$-loop pairing is made, the remaining manipulative variable is paired with $P_1$. In this example, the pairing rule resulting from an RGA analysis is that the valve having the largest pressure drop across it should be used to control the flow. The valve with the smallest pressure drop should be used to control intermediate pressure.

Pairing on negative RGA values deserves special consideration because of some recently published results on this subject (Grosdidier et al., 1984, 1985; Koppel, 1985). It can be shown rigorously that if one pairs on negative RGA values, the resulting system is either unstable or exhibits an inverse response. Further, Grosdidier et al. (1984, 1985) showed that if one pairs on a negative RGA element and the system is stable, and if the loop with the negative RGA element is lost (e.g., via a sensor failure), the resulting subsystem is unstable. These results are particularly powerful because (1) they are completely rigorous, (2) they require only steady-state information, and (3) they use an integral mode in the controller to ensure perfect control at steady state.

An important issue that the RGA does not address is how the control system is forced. The RGA is independent of forcing and this is advantageous because less information is required. However, being independent of forcing is also a disadvantage since the nature of loop forcing has a profound effect on the resulting transient response. To illustrate this point, Figure 8.10 shows the response of the same reactor-control system treated in Figure 8.9. In Figure 8.10, the response is shown for an upset occurring in one of the manipulative variables. Figure 8.9 shows the response of a step setpoint change in one loop. Although the RGA is the same for both figures, the nature of the transient responses is completely different. This example illustrates that one cannot judge control-system performance solely on the basis of the RGA.

Although the RGA is over 20 yr old, new research on its use and on alternatives continue to appear. Manousiouthakis et al. (1986) extended the RGA to cases where more than SISO controllers are considered. They call their approach "the block RGA." Jensen et al. (1986) questioned the use of the dynamic RGA and

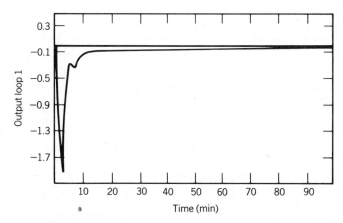

**FIGURE 8.10**   Response to step-load change (Marino-Galarraga et al., 1987b). Reprinted with permission from American Chemical Society.

proposed alternative dynamic interaction measures. One difficulty with their approach is that it requires that the controllers be designed and thus it appears to be of limited utility at the process-design stage. Grosdidier and Morari (1986) proposed using structured singular values as the best means of assessing dynamic interaction effects. From the recent rate of publication, it can be expected that new results on the RGA will be forthcoming for some time.

## 4. Relative Disturbance Gain (RDG)

To include disturbances in an analysis of classical multiloop control systems, Stanley et al. (1985) defined the relative disturbance gain (RDG). Several additional papers on the RDG have also appeared (Marino-Galarraga et al., 1985, 1987a,b; Marlin et al., 1986). The RDG can be applied to $n \times n$ systems. To explain the RDG, a $2 \times 2$ case, shown in Figure 8.11, is treated. As can be seen, a disturbance, $d$, enters the control system. The effect that $d$ has on the response of loop 1 is calculated for two cases for loop 2—when the control for loop 2 is both perfect and off. Mathematically, $\beta_1$ is defined as

$$
\beta_1 = \frac{\left.\dfrac{\partial m_1}{\partial d}\right|_{x_1, x_2}}{\left.\dfrac{\partial m_1}{\partial d}\right|_{x_1, m_2}}
\tag{8.24}
$$

The RDG is similar to the RGA in that it involves a dimensionless ratio of perfect closed-loop to open-loop effects. Since the RDG is a ratio, it possesses the same desirable properties as the RGA. The RDG is scale-independent and it should be

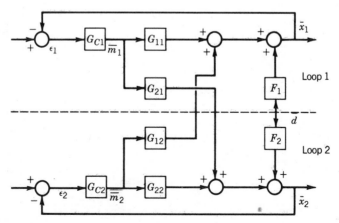

**FIGURE 8.11** A general $2 \times 2$ system (Marino-Galarraga et al., 1987b). Reprinted with permission from American Chemical Society.

reasonably insensitive to nonlinearities. The RDG can be calculated by numerically differentiating standard steady-state design programs. While the RGA is a fixed number for a particular control system, the RDG is different for each disturbance that affects a system. The RDG is also different for each loop in a multiloop system. The RDG for loop 2, $\beta_2$, can be calculated by interchanging the subscripts in Equation (8.24).

The most significant property of the RDG is its relationship to the area under the transient that results if $d$ is a step upset. Suppose that loop 2 is off and a step upset in $d$ occurs. The area under the resulting transient in the error in $x_1$ is designated as $A_{SISO}$. Suppose further that loop 2 is closed and the same step upset occurs in $d$. The error area for this case is designated as $A_{MV}$, where the subscript indicates multivariable control. It can be shown that $\beta_1$ is proportional to the ratio of the areas as

$$\beta_1 = \frac{A_{SISO}}{A_{MV}} \left( \frac{1}{f_1} \right) \tag{8.25}$$

The proportionality factor, $1/f_1$, is determined by how much the feedback controller in loop 1 has to be detuned as a result of interaction when loop 2 is closed. A simple method of estimating $f_1$ from steady-state data has been developed (Marino-Galarraga et al., 1987a). These authors also found that for many systems $f_1$ is in the range of 1 to 3.

To explain the significance of $\beta_1$, it is convenient to take $f_1$ as 1. In actual applications, $f_1$ should be estimated. If $f_1$ is 1.0, $\beta_1$ is equal to the area ratio for the single-loop and multiloop responses. Consider the response of the system in Figure 8.10. From a steady-state model, $\beta_1$ can be calculated as 1 in this case, indicating that $A_{SISO}$ equals $A_{MV}$. (A detailed example of how to calculate the RDG is given below.) Since the areas are equal, the SISO and multiloop transients should be roughly equal, indicating that for this disturbance, interaction between the loops does not cause a problem. Now consider the response shown in Figure 8.9. When only one setpoint in a multiloop system is changed, it can be shown that $\beta_1$ is equal numerically to the RGA, which is 13.7 in this case. Thus one expects that the area under the multiloop response will be 13.7 times as large as that under the SISO response. Because of the large response area, the resulting transient is poor, as Figure 8.9 shows. The results given in Figures 8.9 and 8.10 have implications in terms of on-line optimization, where control-loop setpoints are changed to track an economic optimum. While a particular control system might be fine for disturbance rejection (Figure 8.10), it may perform poorly when tracking a setpoint (Figure 8.9). The RDG can clearly point out these differences.

In essence, $\beta_1$ measures whether the closure of other loops in a multiloop system produces favorable interaction in terms of the performance of the loop under consideration. For a loop to have a good multiloop response, the absolute value of its RDG must be small. The size of the absolute value of the RDG can be used to decide if the loop under study needs to be decoupled from other loops in the system

(Marino-Galarraga et al., 1987a,b). One difficulty which arises in interpreting the RDG is that it involves net areas, and cancellation of positive and negative areas is possible. In spite of this limitation, the RDG gives a new dimension to the analysis of classical control schemes using only steady-state data. The RDG is not a substitute for the RGA, but rather it should be used in conjunction with the RGA. Both the RGA and the RDG are dimensionless numbers for control systems. An interesting question that arises is whether there are other dimensionless numbers that can be defined to help integrate process design and control.

As an example of calculating the RDG, the pressure-flow control system shown in Figure 8.7 is used. The steady-state pressures are $P_0 = 200$, $P_1 = 120$, and $P_2 = 100$. Based on an RGA analysis using these pressures, the recommended loop pairing is $F - m_1$ and $P_1 - m_2$, which is shown in Figure 8.12. Now suppose that the two significant variables which can change and upset the control system are $P_0$ and $P_2$. Further suppose that the flow loop, $F - m_1$, is of particular importance. One could ask: Which upset, $P_0$ or $P_2$, is more troublesome in terms of its effect on flow when both the $F$ and $P_1$ loops are closed? To answer this question, the RDG is calculated for each upset. First, consider $P_0$. The numerator of Equation (8.24) can be calculated by differentiating Equation (8.7) with respect to $P_0$, holding $F$ and $P_1$ constant, to give

$$\left.\frac{\partial m_1}{\partial P_0}\right|_{F_1,P_1} = \frac{-m_1}{P_0 - P_1} \tag{8.26}$$

The denominator of Equation (8.24) can be calculated by differentiating Equations (8.7) and (8.8) with respect to $P_0$ and holding $F$ and $m_2$ constant to give

$$0 = (P_0 - P_1)\left.\frac{\partial m_1}{\partial P_0}\right|_{F_1,m_2} + m_1\left(1 - \left.\frac{\partial P_1}{\partial P_0}\right|_{F_1,m_2}\right) \tag{8.27}$$

$$0 = m_2\left.\frac{\partial P_1}{\partial P_0}\right|_{F_1,m_2} \tag{8.28}$$

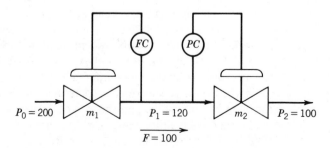

**FIGURE 8.12** Flow- and pressure-control system with loop pairings.

Elimination of $\partial P_1 / \partial P_0 |_{F_1, m_2}$ between Equations (8.27) and (8.28) yields

$$\left. \frac{\partial m_1}{\partial P_0} \right|_{F_1, m_2} = \frac{-m_1}{P_0 - P_1} \tag{8.29}$$

The resulting RDG can be determined by dividing Equation (8.26) by Equation (8.29):

$$\beta_{F, P_0} = 1.0 \tag{8.30}$$

By a similar technique, which is left to the reader as an exercise, the RDG for $P_2$ can be calculated as

$$\beta_{F, P_2} = 0 \tag{8.31}$$

Examination of Equations (8.30) and (8.31) shows that upsets in $P_0$ cause much more difficulty for the flow loop than upsets in $P_2$. In fact, the zero RDG value for $P_2$ indicates that no change in $m_1$ has to be made at steady state to counteract upsets in $P_2$. To illustrate this point, suppose that the $F - m_1$ flow controller is turned off when the $P_2$ upset occurs. The $P_1 - m_2$ controller will react to this upset and change the value of $m_2$. When steady state is achieved, $P_1$ will be back at its preupset value of 120. If the flow loop is then turned on, it is precisely at steady state and will remain there. In effect, the action of the $P_1 - m_2$ controller has not only restored $P_1$ back to steady state, but has helped to bring the flow back to its target value as well. Thus the $P_1 - m_2$ controller is helping the $F - m_1$ controller, and this is precisely what the RDG measures.

In the case of upsets in $P_0$, the reverse is true. Suppose that $P_0$ suddenly increases. The initial effect of this increase is to increase both $F$ and $P_1$. To decrease $P_1$ back to setpoint, the pressure controller will increase $m_2$, resulting in valve 2 being opened. However, opening valve 2 only serves to increase the flow rate even further. Thus, for $P_0$ upsets the action of the $P_1 - m_2$ controller does not help the $F - m_1$ controller. Given that $f_1$ in Equation (8.25) is typically greater than 1.0 and $\beta_{F, P_2} = 1$, one can anticipate that the area under the response curve when both loops are operational will be greater than that if only the $F - m_1$ loop is operational. This is equivalent to the fact that the $P_1 - m_2$ loop is not helpful for $P_0$ upsets.

As mentioned above, RDG can be used in conjunction with the RGA in analyzing control problems. To do so, one has to know which upsets and controlled variables are most important. One example of several possible uses of a combination of the RGA and RDG is given here. Again, assume that $F$ is more important to control than $P_1$. Further, at steady state, assume that the value of $P_1 = 150$, and that the $F - m_1$ and $P_1 - m_2$ pairings are used. In this case, the RGA is 0.5 and significant loop interaction can be expected. One way to eliminate the effect of interaction is to detune one of the loops, either $F - m_1$ or $P_1 - m_2$. If $P_0$ is the only significant upset, the $P_1 - m_2$ loop should be detuned since its action hurts the performance of the $F - m_1$ loop. If $P_2$ is the only significant upset, based on the RDG, the $F - m_1$ loop should be detuned. An alternative to detuning either loop is to use decoupling control. The RDG also provides insight into what structure should be used and what performance can be expected using this method (Marino-Galarraga et al., 1987a). This simple example has been presented to highlight the properties of the RDG. Additional information can be found in the references cited.

## 5. Expert Systems

Expert systems constitute a field of artificial intelligence that holds great promise for integrating process design and control. An expert system is a computer program that attempts to duplicate how an expert in a particular field solves a problem. The program consists of a knowledge base, embodying the knowledge of the expert, and an inference engine. This engine decides which parts of the knowledge base are appropriate to solve the task at hand. The interest in using expert systems to solve problems of process control is growing at an exponential rate.

Several authors (Shinskey, 1986; Niida and Umeda, 1986; Kirkwood et al., 1985) have discussed expert systems for process control that have implications for process design. A key feature of an expert system is that it captures the expertise of a skilled control-system designer and makes this knowledge readily accessible to a process-design engineer. In many cases, this design engineer may have only a rudimentary knowledge of process control. However, if a workable expert system is available, it can overcome this lack of knowledge. The expert system is similar to having an experienced control-system designer available at one's fingertips. As computer hardware becomes faster, it will be easier to design and even check control systems via expert systems. With this scenario, process control still follows process design. But the ease of designing control systems would be such that a process designer could iterate on many more designs if the control system proves to be inoperable.

Niida and Umeda (1986) discuss an expert system that can be used to synthesize and ultimately evaluate control systems. Following a proposed process design, their approach consists of six steps. These are: (1) determination of control objectives; (2) synthesis and selection of possible control loops for each unit; (3) analysis of control loops in each unit and coordination among control loops in the process system; (4) detailed design of each control loop; (5) confirmation of control-loop performance by using dynamic simulators; and (6) confirmation of validity of control loops in real plants.

Shinskey (1986) discussed an expert system for analysis and design of distillation control systems. Distillation is by far the most important separation technique used in the process industries. The distillation expert system selects a control system and presents its results both graphically and numerically. Both multicomponent and two-product sidestream towers are treated. It may be noted that both Shinskey's expert system and that of Niida and Umeda make use of the RGA.

Douglas and co-workers (Douglas, 1982, 1985; Fisher et al., 1985a,b) studied the use of short-cut models for both process design and control. Their approach has recently been integrated into an expert-system package (Kirkwood et al., 1985). A key aspect of their approach is that it is hierarchical and focuses on the key economic tradeoffs that have to be made. By using short-cut models, they are able to screen a large number of process alternatives very quickly and at a time when process changes can be made easily. As a result, designs that have inherently poor operability or controllability can be avoided.

As part of their design methodology, Douglas and co-workers consider steady-state control (Fisher et al., 1985a) as a first step in a hierarchical approach to control system synthesis. Subsequent steps involve normal dynamic operation, abnormal operation, and implementation. Three sources of poor operation which they identify are (1) insufficient number of manipulated variables, (2) insufficient overdesign, and (3) neglect of process interactions, particularly those due to recycles. To improve operability, they use their short-cut package to study optimum steady-state control. The optimum steady-state control problem which they formulate differs from the design problem in that only operating costs, and not capital costs, are minimized. For various disturbances, they determine the sensitivity of the optimum operating policy to available manipulative variables. Those variables that significantly affect the optimum are retained in the control structure. This procedure is carried out at each of the hierarchical stages of process design. If the process is not controllable, several options can be followed. These include addition of new manipulated variables, redesign to avoid constraints, and reformulation of the problem to eliminate the controlled variables that are economically the least important. Stated differently, their goal is to guarantee that the process has a large flexibility index, $\delta$ (Swaney and Grossmann, 1985a). It is the speed with which process and control alternatives can be examined that makes this short-cut approach attractive as a screening tool.

## B. Techniques Utilizing Dynamic Information

In this section, techniques are classified into those requiring approximate dynamic information and those requiring detailed dynamic models. Clearly, this last category is the most costly and time-consuming owing to the work involved in developing complete models.

## 1. Approximate Dynamic Information

Very few studies have been carried out that either developed accurate approximate dynamic process models or made use of only approximate information to analyze control systems. This area appears to be a fertile ground for further research. Silverstein and Shinnar (1982) presented an excellent example of just what can be accomplished using the limited dynamic information typically available at the design stage. They studied the control of a fixed-bed, exothermic catalytic reactor with feed preheat via recycle. Between the feed preheater and reactor, a furnace was used to add additional heat. Their system had characteristics similar to many industrial reactors (e.g., a hydrocracker).

It is well known that such a recycle-reactor configuration can be unstable even if the reactor itself is stable. The instability results from the positive feedback nature of the recycle system. Silverstein and Shinnar assumed a stable reactor and used linear models for the furnace, heater, and reactor. They concluded that the reactor is the key dynamic element which determines the operability of the system. Stability of the entire system can be guaranteed if the loop gain is less than 1.0 for all frequencies. Although this criterion appears conservative, Silverstein and Shinnar demonstrated that it is not, given that system parameters are likely to vary during actual operation. The loop-gain criterion allowed them to design the heater and furnace and their related control systems (e.g., a bypass around the heater) to guarantee stable operation for a given reactor. They were able to determine the key dynamic reactor information needed, and this information was surprisingly little. In some cases it could be obtained from similar operating units, while in others it could be determined from pilot-plant data. Silverstein and Shinnar extended their stability analysis to include important nonlinearities, especially in the reactor. Finally, they presented simple models and an approach to designing recycle reactors for controllability. Included was a discussion of the effects of scale-up and design changes on system stability.

This paper is excellent because it shows how the control/design problem can be simplified and how only a minimum amount of information is needed. As the authors point out, it is especially important to consider design and control together, since advanced control has only a small effect on improving design-generated problems. The report by Silverstein and Shinnar is specific to recycle reactors. It would be interesting if their approach could be generalized to apply to other systems. However, it is not obvious how such a generalization can be carried out.

Marlin et al. (1986) discuss an integrated, short-cut methodology for process control and operability analysis. This methodology makes use of approximate dynamic models when they are available. Two important cases where such models have been developed are distillation towers (Kapoor et al., 1986b) and steam systems (Bertrand, 1986). For the distillation-tower models, the method of moments (Gibilaro and Lees, 1969) is used to reduce the complex tower transfer-functions to first order with dead-time expressions. The time constant, gain, and dead time are given as functions of tower flows, relative volatility, and tower holdups. For the steam system, an approximate linear-dynamic model of the pressure in the high-pressure header was developed from a simplified physical model. This model compared very well with detailed nonlinear simulations and it was used effectively to analyze how the pressure loop should be tuned (Bertrand and McAvoy, 1986). Marlin and co-workers propose that these short-cut models be employed in a user-friendly simulation to facilitate control-system analysis. They also describe the characteristics of such a simulation package. As part of the integrated methodology, the short-cut models are used to estimate single-loop tuning parameters for use in an RDG analysis.

Another paper discussing short-cut dynamic analysis was published by Asbjorn-sen and Loe (1985). They studied a backup ethylene vaporizer that was only used if an ethylene plant had to be shut down. The key question to be answered was whether the vaporizer system could respond fast enough to avoid shutting down the downstream polyethylene and vinyl-chloride plants as well. By using a short-cut dynamic model, Asbjornsen and Loe showed that the proposed control system for the vaporizer could not recover from a compressor trip, and the downstream consumers would have to be shut down. Further, their analysis suggested how the control-system performance could be improved via simple process modifications. This particular problem was one where significant economic benefits resulted from the short-cut analysis.

Much research remains to be done in the area of short-cut dynamic modeling. If such research is successful, it will have a significant impact on process design.

## 2. Complete Dynamic Models

For some systems, a complete dynamic model is available and dynamic simulations can be carried out. Two of the earliest reports on synthesizing both process flowsheets and control structures in the face of uncertain parameters were published by Nishida et al. (1975, 1976). These authors assume that a dynamic model is known. They pose the synthesis problem as a variable-structure dynamic optimization problem with the structure parameters being the splitting factors of the process streams connecting the various units. They solve a minimax problem in which the effect of the worst variation in the uncertain parameters is minimized. They use a sum of integral performance indices for all individual processing units. This performance index is very general and can, for example, represent operating costs, profit, and so on. Theoretical as well as computational issues are addressed and several illustrative examples are given.

In carrying out dynamic simulations, the goal should be to maximize the insight that is gained into potential control problems. Rinard and Slaby (1984) surveyed both steady-state and dynamic simulation packages and concluded that no one readily available package exists that facilitates analyzing process operability and control at the design stage. While excellent individual packages exist, these only contain a part of what Rinard and Slaby conclude is necessary for a working package to study operability and control. They propose that such a simulation package must be efficient, consistent, and have an operability package interfaced to it. Efficiency avoids having an engineer cycle between several programs to carry out his analysis. Consistency is insured by having a common physical property database and common numerical routines. As a minimum, Rinard and Slaby conclude that the associated operability package should be able to calculate frequency responses directly from the simulation models. Rinard and Slaby also developed a prototype simulation/operability package and information about it can be found in the references cited and in Slaby's thesis (Slaby, 1985). As computer hardware and software develop, efficient packages of the type proposed by Rinard and Slaby will become available. These packages should help in integrating process design and control.

For some unit operations, it is economically attractive to develop general-purpose dynamic simulation packages. Distillation is one case where this approach is feasible. Roat et al. (1986) describe how rigorous simulation can be integrated with various short-cut methods to synthesize distillation tower control systems in a timely manner. These authors use the RGA, SVD, and Niederlinski (1971) operability indices to screen out undesirable control-system alternatives. Then they simulate the remaining candidates to arrive at a final control-system design. They point out that in the case of distillation, thermodynamics and tray hydraulics have advanced to the point where rigorous simulation is feasible. In carrying out the simulation, they take advantage of an efficient numerical integration package for stiff systems (Downs and Vogel, 1985) and the speed of modern computers. Since distillation is by far the most common separation process, the cost of developing a general-purpose simulation can be prorated over the many case studies that use it.

As noted, the resiliency approach of Morari and co-workers can be carried out using either steady-state or dynamic information. In the dynamic case, results are generated in the frequency domain. In addition, Arkun (1986) recently provided a summary of his work and that of others on assessing dynamic process operability. As Arkun points out, the goals of the various approaches are to quantify dynamic operability, to pinpoint plant bottlenecks, and to systematically improve the process design in terms of its operability. One of the advantages of these dynamic tools is that they can help a designer to select what simulations should be run. It should also be noted that they can be applied to approximate dynamic models.

## 8.6. SUMMARY

Integration of process design and process control is discussed in this chapter. To place this subject into context, an historical perspective is given. The need to integrate design and control in today's tightly integrated processes is illustrated by a heat-exchanger network and a recycle-reactor example. The unique nature of process control, in terms of what information is available at the design stage, is used to separate techniques for design/control integration into three categories. These are methods requiring steady-state, approximate-dynamic, and detailed-dynamic information. The most promising category is that involving only steady-state information. Detailed discussions of techniques in each of the three areas are presented. It is concluded that a number of promising methods have been developed and that the future should see much more integration of process design and control.

## ACKNOWLEDGMENTS

The author acknowledges the many useful discussions that he has had on process operability with Dr. Tom Marlin of Stone and Webster Corporation. I also recognize the excellent suggestions given by the two reviewers, Dr. Jim Doss and Dr. Pierre Latour, and by the volume editor, Dr. Y. A. Liu.

## NOMENCLATURE

| | |
|---|---|
| $A$ | Net error area |
| CN | Condition number defined by Equation 8.6 |
| $d$ | Design variables |
| $f$ | Function defined by Equation 8.3 |
| $f_i$ | Detuning factor for loop $i$ |
| $F$ | Flow rate |
| $g$ | Inequality constraints |
| $G$ | Process transfer function matrix |
| $\tilde{G}$ | Model of $G$ |
| $h$ | Equality constraints |
| $K$ | Process gain matrix |
| $m$ | Manipulated variables |
| $P$ | Pressure |
| $V$ | Matrix of left eigenvectors of $K^T K$ |
| $W$ | Matrix of right eigenvectors of $K^T K$ |
| $x$ | State variables |

## Greek Letters

| | |
|---|---|
| $\beta_i$ | Relative disturbance gain for loop $i$ |
| $\delta$ | Flexibility index |
| $\lambda_{ij}$ | Relative gain element |
| $\sigma$ | Singular values (the positive square roots of the eigenvalues of $K^T K$) |
| $\Sigma$ | Matrix of singular values |
| $\theta$ | Parameters that can change |

## Subscripts

| | |
|---|---|
| 0 | Upstream |
| 1 | Intermediate |
| 2 | Downstream |
| $i$ | Index |
| $j$ | Index |
| $m$ | Minimum |
| MV | Multivariable |
| SISO | Single-input, single-output |

## Superscripts

| | |
|---|---|
| $M$ | Maximum |
| $T$ | Transpose |
| $+$ | Positive deviation |
| $-$ | Negative deviation |

## REFERENCES

Arkun, Y., "Dynamic Process Operability. Important Problems, Recent Results and New Challenges," in *Chemical Process Control—III*, M. Morari and T. J. McAvoy (eds.), Elsevier, New York, 1986.

Asbjornsen, O. A. and I. Loe, "Transient Analysis Improves Reliability of an Integrated Plant Complex," *Proceedings of the American Control Conference*, Session WA10, Boston, MA, June 1985.

Astrom, K., "Adaptation, Auto-Tuning, and Smart Controls," in *Chemical Process Control—III*, M. Morari and T. J. McAvoy (eds.), Elsevier, New York, 1986.

Astrom, K. J. and B. Wittenmark, *Computer Controlled Systems*, Prentice-Hall, Englewood Cliffs, NJ, 1984, p. 159.

Attir, U. and M. M. Denn, "Dynamics and Control of an Activated Sludge Process," *AIChE J.*, **24,** 693 (1978).

Bertrand, C., *A Study on the Dynamic Simulation and Control of Steam Systems*, M.S. Thesis, University of Maryland, College Park, MD, 1986.

Bertrand, C. and T. J. McAvoy, "Short-Cut Analysis of Pressure, in Steam Headers," *Proceedings of the American Control Conference*, Seattle, WA, June 1986.

Bristol, E. F., "On a New Measure of Interaction for Multivariable Process Control," *IEEE Trans. Autom. Control*, **AC-11,** 133 (1966).

Bristol, E. F. and T. Kraus, "Life with Pattern Recognition," *Proceedings of the American Control Conference*, Session TP3, 888, San Diego, CA, June 1984.

Bruns, D. D. and C. R. Smith, "Singular Value Analysis: A Geometric Structure for Multivariable Processes," AIChE National Meeting, Orlando, FL, February 1982.

Denn, M. M., "Modeling for Process Control," in *Control and Dynamic Systems*, Vol. 15, C. T. Leondes (ed.), Academic, NY, 1979, p. 148.

Denn, M. M. and R. Lavie, "Dynamics of Plants With Recycle," *Chem. Eng. J.*, **24,** 55 (1982).

Douglas, J. M., "Process Operability and Control of Preliminary Designs," in *Chemical Process Control—II*, T. Edgar and D. E. Seborg (eds.), Engineering Foundation, New York, 1982.

Douglas, J. M., "A Hierarchical Decision Procedure for Process Synthesis," *AIChE J.*, **31,** 353 (1985).

Downs, J. J., "The Control of Azeotropic Distillation Columns," Ph.D. Dissertation, University of Tennessee, 1982.

Downs, J. J. and C. F. Moore, "Steady State Gain Analysis for Azeotropic Distillation," *Proceedings of the Joint Automatic Control Conference*, Session WP-7, Charlottesville, VA (June, 1981).

Downs, J. J. and E. F. Vogel, "An Interactive Dynamic Distillation Simulator for Analysis and Control Systems Development," *Proceedings of the Summer Computer Simulation Conference*, Chicago, IL, 1985, p. 337.

Doyle, J., "Analysis of Feedback Systems with Structured Uncertainties," *IEE Proc.*, **129,** D(6), 242 (1982).

Dumont, G., "A Survey of the Applications of Adaptive Control in the Process Industries," in *Chemical Process Control—III*, M. Morari and T. McAvoy (eds.), Elsevier, New York, 1986.

Fisher, W., M. F. Doherty, and J. M. Douglas, "Steady-State Control As a Prelude to Dynamic Control," *Proceedings of the PSE '85 Conference, The Institution of Chemical Engineers Symposium Series*, No. 92, 1985a, pp. 643–650.

Fisher, W., M. F. Doherty, and J. M. Douglas, "Operating Heuristics for the Control of Complete Chemical Plants," *Proceedings of the American Control Conference*, Session WA10, Boston, MA, 1985b.

Garcia, C. and M. Morari, "Internal Model Control I. A Unifying Review and Some New Results," *Ind. Eng. Chem. Process Des. Dev.*, **21**, 308 (1982).

Garcia, D. and D. M. Prett, "Advances in Model Predictive Control," in *Chemical Process Control—III*, M. Morari and T. J. McAvoy (eds.), Elsevier, New York, 1986.

Gibilaro, L. and F. Lees, "Reduction of Complex Transfer Function Models to Simple Models Using the Method of Moments," *Chem. Eng. Sci.*, **24**, 85 (1969).

Gilliland, E. R., L. A. Gould, and T. J. Boyle, "Dynamic Effects of Material Recycle," *Proceedings of Joint Automatic Control Conference*, Stanford University, Palo Alto, CA, 1964, p. 140.

Grosdidier, P. and M. Morari, "Interaction Measures for Systems Under Decentralized Control," *Automatica*, **22**, 309 (1986).

Grosdidier, P., M. Morari, and B. R. Holt, "Integral Controllability, Failure Tolerance, Robustness, and the Relative Gain Array," *Proceedings of the American Control Conference*, Session FA4, San Diego, CA, 1984, p. 1290.

Grosdidier, P., M. Morari, and B. R. Holt, "Closed-Loop Properties from Steady-State Gain Information," *Ind. Eng. Chem. Fund.*, **24**, 221 (1985).

Grossmann, I. and K. Halemane, "Decomposition Strategy for Designing Flexible Chemical Plants," *AIChE J.*, **28**, 686 (1982).

Grossmann, I., K. Halemane, and R. Swaney, "Optimization Strategies for Flexible Chemical Processes," *Comput. Chem. Eng.*, **7**, 439 (1983).

Grossmann, I. and M. Morari, "Operability, Resiliency, and Flexibility—Process Design Objectives for a Changing World," *Proceedings of the 2nd International Conference on Foundations of Computer Design*, A. W. Westerberg and H. H. Chien (eds.), CACHE Corporation, Austin, TX, 1984, pp. 931–1010.

Holt, B. R. and M. Morari, "Design of Resilient Processing Plants V, The Effect of Deadtime on Dynamic Resilience," *Chem. Eng. Sci.*, **40**, 1229 (1985a).

Holt, B. R. and M. Morari, "Design of Resilient Processing Plants VI. The Effect of Right Half Plane Zeros on Dynamic Resilience," *Chem. Eng. Sci.*, **40**, 59 (1985b).

Jensen, N., D. G. Fisher, and S. L. Shah, "Interaction Analysis in Multivariable Control Systems," *AIChE J.*, **32**, 959 (1986).

Kapoor, N., T. J. McAvoy, and T. E. Marlin, "Effect of Recycle Structure on Distillation Tower Time Constants," *AIChE J.*, **32**, 411 (1986a).

Kapoor, N., T. J. McAvoy, and T. E. Marlin, "An Approximate Analytical Approach to Dynamic Modeling of Distillation Towers," DYCORD 86, IFAC Symposium on Dynamics and Control of Chemical Reactors and Distillation Towers, Bournemouth, UK, December 1986b.

Kirkwood, M., M. Locke, and J. M. Douglas, "An Expert System for Synthesizing Flow-sheets and Optimum Designs," Annual AIChE Meeting, Paper 70f, Chicago, IL, November 1985.

Klema, V. C. and A. J. Laub, "The Singular Value Decomposition: Its Computation and Some Applications," *IEEE Trans. Autom. Control*, **AC-25**, 164 (1980).

Koppel, L. B., "Conditions Imposed by Process Statics on Multivariable Process Dynamics," *AIChE J.*, **31**, 70 (1985).

Lenhoff, A. and M. Morari, "Design of Resilient Processing Plants I. Process Design Under Considerations of Dynamic Aspects," *Chem. Eng. Sci.*, **37**, 245 (1982).

Manousiouthakis, V., R. Savage, and Y. Akrun, "Synthesis of Decentralized Process Control Structures Using the Concept of Block Relative Gains," *AIChE J.*, **32**, 991 (1986).

Marino-Galarraga, M., T. E. Marlin, and T. J. McAvoy, "Using the Relative Disturbance Gain to Analyze Process Operability," *Proceedings of the American Control Conference*, Boston, MA, June 1985.

Marino-Galarraga, M., T. J. McAvoy, and T. E. Marlin, "Short-Cut Operability Analysis II. Estimation of $f_i$ Detuning Parameter for Classical Control Systems," *Ind. Eng. Chem. Res.*, **26**, 511 (1987a).

Marino-Galarraga, M., T. J. McAvoy, and T. E. Marlin, "Short-Cut Operability Analysis III. Short-Cut Methodology for the Assessment of Process Control Designs," *Ind. Eng. Chem. Res.*, **26**, 521 (1987b).

Marlin, T. E., T. J. McAvoy, N. Marino-Galarraga, and N. Kapoor, "A Short-Cut Method for Process Control and Operability Analysis," in *Chemical Process Control—III*, M. Morari and T. J. McAvoy (eds.), Elsevier, New York, 1986.

Marselle, D. F., M. Morari, and D. F. Rudd, "Design of Resilient Processing Plants II: Design and Control of Energy Management Systems," *Chem. Eng. Sci.*, **37**, 259 (1982).

McAvoy, T. J., *Interaction Analysis: Principles and Applications*, Instrument Society of America, Research Triangle Park, NC, 1983.

Morari, M., "Design of Resilient Processing Plants III. A General Framework for the Assessment of Dynamic Resilience," *Chem. Eng. Sci.*, **38**, 1881 (1983a).

Morari, M., "Flexibility and Resiliency of Process Systems," *Comput. Chem. Eng.*, **7**, 423 (1983b).

Morari, M. and J. D. Doyle, "A Unifying Framework for Control System Design Under Uncertainty and its Implications for Chemical Process Control," in *Chemical Process Control—III*, M. Morari and T. J. McAvoy (eds.), Elsevier, New York, 1986.

Morari, M., W. Grimm, M. Oglesby, and I. Prosser, "Design of Resilient Processing Plants VII. Design of Energy Management System for Unstable Reactors—New Insights," *Chem. Eng. Sci.*, **40**, 187 (1985).

Niederlinski, A., "A Heuristic Approach to the Design of Linear Multivariable Interacting Control Systems," *Automatica*, **7**, 691 (1971).

Niida, K. and T. Umeda, "Process Control System Synthesis by an Expert System," in

*Chemical Process Control—III*, M. Morari and T. McAvoy (eds.), Elsevier, New York, 1986.

Nishida, N., Y. A. Liu, and A. Ichikawa, "Studies in Chemical Process Design and Synthesis: I. Optimal Synthesis of Dynamic Process Systems by a Gradient Method," Paper No. 57d (Microfiche No. 4), AIChE National Meeting, Boston, MA (September, 1975).

Nishida, N., Y. A. Liu, and A. Ichikawa, "Studies in Chemical Process Design and Synthesis: II. Optimal Synthesis of Dynamic Process Systems with Uncertainty," *AIChE J.*, **22**, 539 (1976).

Popiel, L., T. Matako, and C. Brosilow, "Coordinated Control," in *Chemical Process Control—III*, M. Morari and T. J. McAvoy (eds.), Elsevier, New York, 1986.

Richalet, J. and B. Froisey, "Industrial Applications of IDCOM," in *Chemical Process Control—III*, M. Morari and T. J. McAvoy (eds.), Elsevier, New York, 1986.

Rinard, I. H. and B. Benjamin, "Control of Recycle Systems," American Control Conference, Session WA5, Arlington, VA, 1982.

Rinard, I. H. and J. Slaby, "A Prorotype Multipurpose Simulator for Process Operability Analysis," AIChE National Meeting, Anaheim, CA, 1984.

Roat, S. D., J. J. Downs, E. F. Vogel, and J. E. Doss, "The Integration of Rigorous Dynamic Modeling and Control System Synthesis for Distillation Columns: An Industrial Approach," in *Chemical Process Control—III*, M. Morari and T. J. McAvoy (eds.), Elsevier, New York, 1986.

Saboo, A. and M. Morari, "Design of Resilient Processing Plants—IV. Some New Results on Heat Exchanger Network Synthesis," *Chem. Eng. Sci.*, **39**, 597 (1984).

Saboo, A., M. Morari, and D. Woodcock, "Design of Resilient Processing Plants VII. A Resilience Index for Heat Exchanger Networks," *Chem. Eng. Sci.*, **40**, 1553 (1985).

Seborg, D. E., T. F. Edgar, and S. Shah, "Adaptive Control Strategies for Process Control: A Survey," *AIChE J.*, **32**, 881 (1986).

Shinskey, F. G., *Process Control Systems*, 2nd ed., McGraw-Hill, New York, 1979, pp. 213–214.

Shinskey, F. G., "Uncontrollable Processes and What To Do About Them," *Hydrocarbon Process.*, **62**, (November 1983).

Shinskey, F. G., "An Expert System for the Design of Distillation Controls," in *Chemical Process Control—III*, M. Morari and T. J. McAvoy (eds.), Elsevier, New York, 1986.

Silverstein, J. and R. Shinnar, "Effect of Design on the Stability and Control of Fixed Bed Catalytic Reactors with Heat Feedback. 1. Concepts," *Ind. Eng. Chem. Process Design Dev.*, **21**, 241 (1982).

Slaby, J., *The Multipurpose Simulator as a Tool for Process Operability Analysis*, Ph.D. Dissertation, Polytechnic Institute of New York, Brooklyn, NY, 1985.

Slaby, J. and I. H. Rinard, "Decomposition in Distillation Systems," AIChE National Meeting, Houston, TX, March 1985.

Slaby, J. and I. H. Rinard, "A Complete Interpretation of the Dynamic Relative Gain Array," AIChE Annual Meeting, Miami, FL, November 1986.

Smith, C. R., C. F. Moore, and D. D. Bruns, "A Structural Framework for Multivariable Control Applications," Joint Automatic Control Conference, Charlottesville, VA, June 1981.

Stanley, G., M. Marino-Galarraga, and T. J. McAvoy, "Short-Cut Operability Analysis. 1. The Relative Disturbance Gain," *Ind. Eng. Chem. Process Des. Dev.*, **24,** 1181 (1985).

Swaney, R. and I. Grossmann, "An Index for Operational Flexibility in Chemical Process Design. Part I: Formulation and Theory," *AIChE J.*, **31,** 621 (1985a).

Swaney, R. and I. Grossmann, "An Index for Operational Flexibility in Chemical Process Design. Part II: Computational Algorithms," *AIChE J.*, **31,** 631 (1985b).

# 9

# SAFETY IN PROCESS
# AND PLANT DESIGN

## GEOFFREY L. WELLS

## 9.1.  INTRODUCTION

As technologies develop and the scale of production becomes larger, in order to reduce overall costs per unit of production, potential danger is generally increased and safety assumes a major role in deciding what type of plant to build and how to build it.

New attitudes and methodologies have developed with this intensified awareness of safety. Legislation and regulations bring all companies in a country to a minimum accepted level of safety. However, despite the evaluation of fatal accident rates and the probability of accidents, plant safety remains one of the chemical industry's major problems. Strictly speaking, the chemical industry is one of the safest industries. But the nature of the incidents, such as occasional catastrophic failures and multiple deaths, induces more public reaction than the mundane accidents of the building trade. The public is concerned because of the general level of ignorance about chemistry, and justifiably so, because they are subjected to risk without choice.

This chapter considers the general way in which safety improvements are being implemented, rather than the equivalent improvements in design calculations and equipment performance. The emphasis is placed on the need to make any changes necessary to improve safety at an early stage of the process synthesis and design to reduce the danger of a major hazardous incident. The need to carry out the design according to good practice, using appropriate codes and standards, is emphasized. These are subsequently backed by some form of risk analysis as a final check against potential error.

The chapter is a survey of methods in use and does not provide detailed information on any technique. This must be obtained from the literature cited.

## 9.2.  LEGISLATIVE ASPECTS OF SAFETY IN PROCESS AND PLANT DESIGN

Legislation on safety represents the reaction of government to circumstances and to public opinion as modified by consultation with some of the interested parties. It is backed by appropriate bodies whose task is inspection, enforcement, advice, and policy development. Major legislation relating to industrial safety is indicated in Table 9.1.

The control of major hazards in Europe was accelerated by the major accidents at Pernis, Holland in 1968, at Beck, Holland and Flixborough, United Kingdom

TABLE 9.1 A Sample of Legislation Relating to Industrial Safety

*UK Legislation*

| | |
|---|---|
| 1802 | The Health and Morals of Apprentices Act |
| 1831 | The Factories Act |
| 1906 | The Alkali Act |
| 1927 | The Explosive Substances Act |
| 1951 | The Rivers, Prevention of Pollution Act |
| 1956 | The Clean Air Act |
| 1960 | The Noise Abatement Act |
| 1961 | The Factories Act |
| 1974 | The Health and Safety at Work Act |
| 1974 | The Control of Pollution Act |
| 1984 | The Control of Industrial Major Accidents Regulations |

*USA Legislation*

| | |
|---|---|
| 1908 | Explosives Transportation Act |
| 1956 | Federal Water Pollution Control Act |
| 1968 | Natural Gas Pipeline Safety Act |
| 1970 | Hazardous Materials Transportation Act |
| 1970 | Clean Air Act |
| 1970 | Occupational Safety and Health Act |
| 1972 | Noise Control Act |
| 1972 | Deposit of Poisonous Wastes Act |
| 1976 | Toxic Substances Control Act |

in 1974, and at Seveso, Italy in 1976. These events were eventually followed by the European Economic Community's "Seveso" directive of 1982 (82/501/EEC) and in the United Kingdom by *The Notification of Installations Handling Hazardous Substance Regulations* by Her Majesty's Stationary Office in 1982 (HMSO, 1982). More recently in the United Kingdom, HMSO has published *The Guide to the Control of Industrial Major Accident Hazards (CIMAH)* (HMSO, 1984). Substances named in this guide and present in quantities greater than stated inventory levels are subject to these regulations. It is then necessary for the industry to demonstrate on request by the governmental health and safety organizations that major hazards have been identified, preventive steps taken, and on-site information, training, and equipment provided to an appropriate standard. Already this has meant, in a few cases, the reduction of inventory levels and the elimination of specific chemicals from a number of sites. It will no doubt influence new designs to keep below the inventory criterion. The regulations provide a helpful framework for the discussion of site safety. There is a requirement to produce both on- and off-site emergency plans to local governmental authorities and safety information to the public concerned. The latter requires delicate handling, particularly because property values may decline in affected areas.

## 9.3. THE MAJOR HAZARDS

Table 9.2 gives a selection of accidents that caused the largest property losses and fatalities in the hydrocarbon–chemical industry. This table is based on Marsh and

**TABLE 9.2  Some Major Accidents Over Three Decades in the Hydrocarbon–Chemical Industries**

| Date | Location | Brief Description |
|---|---|---|
| 8/27/55 | Whiting, IN | Detonation in a fluid hydroformer fragmented the reactor and the resulting projectiles caused a tank-farm fire. |
| 5/22/58 | Signal Hill, CA | A large tank of hot oil frothed over and this escalated into a large major fire. |
| 10/4/60 | Kingsport, TN | A nitrobenzene plant explosion occurred in the processing equipment with a power equivalent to 6 tons of TNT. |
| 6/16/64 | Niigata, Japan | An earthquake of magnitude 7.7 resulted in two major fires in an oil refinery. |
| 8/25/65 | Louisville, KY | Localized heating in moving metal parts initiated decomposition and subsequent explosion in a neoprene process. |
| 1/4/66 | Feyzin, France | An uncontrolled release of propane through a 50-mm connection on a storage sphere ignited and the tank exploded. |
| 8/8/67 | Lake Charles, LA | Valve failure released isobutane, which gave rise to an unconfined vapor-cloud explosion. |
| 12/19/67 | El Segundo, CA | A fuel oil reservoir was ignited by lightning and burned for 12 days. |
| 1/20/68 | Pernis, Holland | Mixing hot oil and water caused violent vapor release and boil-over, which resulted in a fire covering 30 acres. |
| 1/10/69 | Escombreras, Spain | Propane vapors were released from bullets and resulted in an unconfined vapor-cloud explosion. |
| 12/5/70 | Linden, NJ | A reactor on the hydrocracking unit operating at 175 bar failed explosively because of localized overheating. |
| 9/8/71 | Bound Brook, NY | A chemical plant was among the damaged property when a dam burst following a 28-mm (an 11-in.) rain storm. |
| 8/4/72 | Trieste, Italy | Terrorists destroyed crude storage tanks by charges on the discharge lines and the ensuing fires spread. |
| 2/10/73 | Staten Island, NY | A fire broke out in a prestressed concrete liquid-natural-gas tank when it was repaired nearly a year after being taken off-stream. |
| 7/8/73 | Tokuyama, Japan | An exothermic reaction, upon accumulation of hydrogen in an acetylene hydrogenation column, caused leakage and explosion. |

**TABLE 9.2** *(Continued)*

| Date | Location | Brief Description |
|---|---|---|
| 6/1/74 | Flixborough, UK | Massive failure of a 0.5-m bypass around a cyclohexane oxidation reactor gave rise to an unconfined vapor-cloud explosion of cyclohexane followed by fire. |
| 11/29/74 | Beaumont, TX | A leak in a suction line gave rise to a vapor-cloud explosion of hydrocarbons and further fires. |
| 2/10/75 | Antwerp, Belgium | A leak of high-pressure ethylene following fatigue failure on a vent connection caused widespread blast and fire damage. |
| 7/11/75 | Beek, Holland | A leak of propylene following cold-brittle fracture of a 40-mm connection resulted in a vapor-cloud explosion. |
| 7/10/76 | Seveso, Italy | Operating errors compounded an unexpected exothermic decomposition which led to a release of dioxin-containing material to the atmosphere via a ruptured disk. |
| 4/3/77 | Umm Said, Qatar | A large tank containing refrigerated propane failed massively on a repaired tank weld causing a major fire. |
| 5/11/77 | Abqaiq, Saudi Arabia | Failure of a 0.76-m crude-oil line operating at 64 bar resulted in a major fire. |
| 7/8/77 | Near Fairbanks, AK | A leak during maintenance on a pump strainer allowed crude oil at 17 bar to release and ignite. |
| 10/18/77 | Palmyra, MO | A nitration plant was at start-up when the detonation of its reactor caused extensive blast and fire damage. |
| 12/8/77 | Brindisi, Italy | A major gas release on an ethylene plant ignited and caused severe blast and fire damage. |
| 4/15/78 | Abqaiq, Saudi Arabia | Failure of a pipeline caused fire and explosion to a tank and a gas-separation plant. |
| 5/30/78 | Texas City, TX | An unidentified failure led to the release of light hydrocarbons. An intense fire developed in the tank farm giving rise to spectacular fireballs and missiles. |
| 10/3/78 | Pitesti, Rumania | An overhead vapor line fractured, releasing propane and propylene which gave rise to a vapor-cloud explosion. |
| 1/8/79 | Bantry Bay, Ireland | A tanker caught fire after unloading its first parcel of Arabian crude at a deep-water jetty. |

**329**

**TABLE 9.2** (*Continued*)

| Date | Location | Brief Description |
|---|---|---|
| 4/19/79 | Port Neches, TX | A tanker explosion occurred after it was struck by lightning when taking on ballast water following off-loading crude oil. |
| 7/21/79 | Texas City, TX | The failure of a 180-mm elbow on a depropanizer released gaseous and liquid hydrocarbons resulting in an explosion. |
| 8/1/79 | Deer Park, TX | Nearly simultaneous explosions occurred during a severe storm igniting a tanker and an ethanol tank. |
| 2/26/80 | Brooks, AA | The largest property damage loss sustained by the natural-gas pipeline industry occurred upon rupture of a buried line. |
| 10/21/80 | New Castle, DE | Improper maintenance resulted in removal of the valve instead of only the motor operator of a plug valve. The release at 10 bar gave rise to a vapor-cloud explosion. |
| 8/20/81 | Shaiba, Kuwait | A fire occurred while pumping naphtha and spread to other tanks in the plant. |
| 2/9/82 | Philadelphia, PA | Cumene hydroperoxide was being heated in a holding tank when vented cumene ignited explosively and ruptured the tank. |
| 4/18/82 | Edmonton, Canada | Ethylene was released from a 3-mm tube which failed as a result of fatigue caused by vibration from a compressor. |
| 1/7/83 | Newark, NJ | Gasoline overflowed while filling a large floating roof tank and ignited 300 m away, causing blast and fire damage. |
| 4/7/83 | Avon, CA | A 0.3-m line on a catalytic cracker ruptured and a slurry ignited. Firefighting was hampered by failure of a steam main. |
| 2/14/83 | Bontang, Indonesia | A cryogenic heat exchanger ruptured when relief valves failed to discharge correctly due to a closed valve. |
| 5/26/83 | Prudhoe Bay, AK | A low-pressure liquefied-natural-gas tank ruptured violently when high-pressure gas from downstream backed up past valves. |
| 2/24/84 | Cubatao, Brazil | A gasoline pipeline ruptured causing a huge fire which killed 79 people. |

**TABLE 9.2** (*Continued*)

| Date | Location | Brief Description |
|------|----------|-------------------|
| 7/23/84 | Romesville, IL | An methylamine column ruptured causing an explosion followed by a second explosion in the alkylation unit. |
| 11/19/84 | Mexico City, Mexico | A series of liquefied-petroleum-gas explosions at a gas distribution plant resulted in more than 500 people being killed. |
| 12/3/85 | Bhopal, India | The ingress of water initiated reactions which caused the release of methyl isocyanate and possibly hydrogen cyanide, killing over 2000 people. The incident reached the proportions it did because some safety measures were inoperative. |

McLennan (1986) and the Loss Prevention Bulletins of the Institution of Chemical Engineers, United Kingdom. The table shows little change in the causes of accidents, and the time-adjusted costs of accidents per year continue to escalate despite all the attention of recent years to loss prevention. This is discussed further by Kletz (1985a). Basically, the problem arises because of the increased scale of production, the inherent nature of the danger, particularly through fire, explosions, and exposure to toxic materials, and the wide international variation in the quality of operators, management, and the like. However, there has been an improvement in safety as any detailed study of industrial records would show.

## 9.4. LOSS PREVENTION AT AN EARLY STAGE OF THE DESIGN

There has been a growing awareness of the need for loss-prevention considerations to be implemented early in the design of the plant (I. Chem. E., 1976, 1977, 1985a,b; Wells, 1980; Kletz, 1985b; AIChE, 1985; Wells and Rose, 1986). The aim is to complete the first stage of the design with reasonable confidence that major expenditure to avoid process hazards will not arise at a later stage of the project. To achieve this objective, it is essential that process engineers are aware of the consequences of their decisions.

The key to safety studies and related training is to cultivate an awareness of the hazard. The nature of any danger may be assessed from information on process materials. There are many such sources, particularly governmental agencies. Often the nature of the hazard is obvious from the inherently hazardous elements.

Data on chemical hazards can be obtained from:

1. The manufacturer of a purchased chemical and associations of industrial chemical manufacturers such as the Chemical Manufacturers Association, Washington, D.C. and the Chemical Industries Association, London.

2. Classic sources such as *The Merck Index* (Windholy, 1978), *The Handbook of Chemistry and Physics* (Chemical Rubber Company, 1985), Sax (1979), Bretherick (1985), and Verscheuten (1977).

3. Principal governmental sources include the U.S. Department of Health and Human Services, Washington, D.C. and the Health and Safety Executive, London, United Kingdom. Other sources include the U.S. National Fire Protection Association, Boston, the Fire Prevention and Publication Center, London, and many professional institutions and societies.

4. Data banks are particularly useful for physical properties and less useful for threshold limit values and similar dangerous properties.

5. Chemical abstracts and patent search. If a literature survey does not provide the necessary information, a testing program is required.

Table 9.3 indicates some potentially, very dangerous types of processes. This list is not new, yet these processes continue to give rise to catastrophic incidents. Of all these processes, it is arguably the identification of exothermic reactions that requires the most care. This is largely because of the need to identify all probable reactions which take place, including those in unplanned locations and side-reactions with impurities from feed chemicals and common contaminants such as water. A check with the manufacturer of the chemical can suggest such reactions and possible impurities in the feedstock. Also, a literature survey of dangers related to the process and its chemicals often gives access to reported incidents (NFPA, 1985). The developments in the field of reaction-path synthesis (Govind and Powers, 1981) may yield useful information on side reactions. There is a need for available, extensive data banks which permit the assessment of the free energy of the reaction of interest.

**TABLE 9.3   Some Potentially Dangerous Types of Processes**

Processes subject to explosive reaction of detonation.
Processes that react energetically with water or common contaminants.
Processes subjected to spontaneous polymerization or heating.
Processes that are exothermic.
Processes containing flammables that are liquid existing above the atmospheric boiling point as a result of pressure or refrigeration.
Processes in which intrinsically unstable compounds are present.
Processes operating in or near the explosive range of materials.
Processes involving highly toxic materials, radiation, or dangerous bacteria.
Processes subjected to a dust or mist explosion hazard.
Processes with a large inventory of stored pressure energy.

Clearly, it is important to know if a process has any record of or reputation for being particularly dangerous. This affects planning permission and can make continuity of manufacture a problem, if a major accident occurs with the same chemical in exceptional cases even when the process is different. Also, it should be appreciated that the danger of a process is modified according to whether the company has experience with the type and scale of this or similar processes, and has skilled and experienced personnel available at all company levels.

The use of hazard indices such as the DOW (AIChE, 1981) or MOND index (Lewis, 1977; Tyler, 1985) and those used for insurance purposes (Redmond, 1984) can help because the ranking list gives an indication of the relative danger of the process. However, the assessment of inventory, operating conditions, and type of chemical serve much the same purpose. The selection of an appropriate method to use is largely dictated by the pressure of outside sources or company policy.

Checklists are also a great help in readily identifying hazards and some good examples can be found in Balemans (1974), Wells (1980), and AIChE (1985). It is true that many checklists suffer by not being comprehensive. Nevertheless, they can be applied rapidly and cheaply to reduce the risk.

## 9.5. IMPROVING PLANT SAFETY

The process engineer must take action as the design and project proceeds to improve plant safety. The following simple guide shows how such action should proceed.

For each part of the process and in each operating mode of the plant, carry out studies as follows:

First, identify the hazardous element and the hazardous condition, usually by common-sense methods.

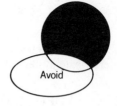

Take action to AVOID the hazard: further keywords— SUBSTITUTE, INTRODUCE, ELIMINATE.

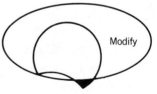

Take action to MODIFY the danger: further keywords—GOOD PRACTICE, ATTENUATE, MODERATE, IMPROVE.

To preserve this position, it is necessary to CORRECT and MAINTAIN the system. This is a function not only of control and maintenance systems, but also of the skill of the operator and the effectiveness of the loss-prevention program. The keywords here are IMPROVE and MODIFY.

Take action to MITIGATE the danger: further keywords—DETECT, ATTEN-UATE, ISOLATE, FIGHT.

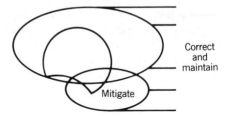

Further action is required to implement additional corrective and contingency action. At least TWO SAFEGUARDS must be provided against every contingency.

However, even then complete safety is not assured. For example, all the hazards may increase such that an incident occurs, and design errors may have been made or equipment damaged.

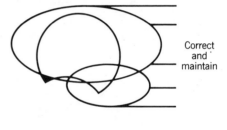

A modifying constraint may break down, causing or increasing the extent of an incident, or plant may deteriorate in service.

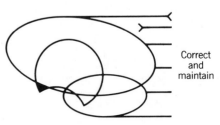

Further work is necessary to identify and reduce risks, although developing a completely safe system is infeasible.

## 9.6. SAMPLE USE OF KEYWORDS

Examples involving the use of the keywords include:

1. *Avoid* by substituting less hazardous process materials.
2. *Avoid* by introducing material that prevents the occurrence of a hazardous reaction.

3. *Modify* the use of highly dangerous reactions or operating conditions.

4. *Moderate* an operation by incorporating a catalyst that permits reaction at lower temperatures.

5. *Modify* the sequence, concentration, method, and operating conditions.

6. *Modify* by decrease (or rarely by increase) of the plant inventories.

7. *Attenuate* the size of available openings, particularly of valves and pipe openings.

8. *Attenuate* the maximum size of a fire.

9. *Modify* or eliminate some equipment, plant items or both, particularly prime movers and fired heater.

10. *Eliminate* a source of ignition.

11. *Avoid* by combining a reaction and a separation, or a quench and a separation.

12. *Improve* the system reliability and availability.

13. *Improve* the containment of hazardous materials.

14. *Eliminate* some crossovers and cross-connections between streams.

## 9.7. GOOD PRACTICE AND DESIGN

Good practice is the main suppressor of hazards. The underlying safety basis of all engineering design of equipment structures is that properly designed, constructed, operated, and maintained equipment will not fail catastrophically if its design conditions of pressure and temperature are not exceeded, and if the properties of the equipment are not allowed to deteriorate through corrosion, embrittlement, or other mechanisms. The method of achieving this aim is by good specification of design criteria and adherence to codes, standards, and regulations. However, the perfect identification of every hazard that might give rise to a catastrophic failure may not be achieved. For example, a lack of knowledge when the standard is written can lead to oversights. After all, few people knew about the zinc embrittlement of hot stainless steel before the Flixborough incident in the United Kingdom and its subsequent investigation (HMSO, 1978).

Exxon Chemicals use codes and practices as the standard against which the acceptability of a design is evaluated (Solomon, 1983). Thus, in seeking the potential hazards in a process, the question is not "What happens if it breaks?" but rather, "What conditions can arise that can cause the pressure or temperature to exceed the design parameters and then cause failure?"

This can be the basis for both the original design and testing of the process against company standards. Such a method permits the lessons learned over many years of experience to be implemented. Its satisfactory use depends on maintaining an up-to-date set of design and basic practices and giving training in their use.

One can only be envious about access to such a system. Many smaller companies attempt to produce their own standards, sometimes with poor results. There is also the problem of different standards applying in different countries, such that at the commencement of a project, it is necessary to determine which standards are being used.

A further requirement of good practice is the development of good control and maintenance coupled with a reliable plant (Sims, 1980). A feature of the last decade has been the progress made in this area. For example, the methodology of maintenance has improved and condition monitoring has become widely accepted. There has also been a greater involvement of maintenance engineers in project activities.

There has been considerable growth in the field of advanced process control with increased use of logic diagrams and alarm analysis. Shaw (1985) warns that many of the more advanced process-control strategies and algorithms can cause incorrect operator actions which may negate the advantages of the advanced control. For example, this problem arose at Three Mile Island (Nuclear Regulatory Commission, 1978; Perrow, 1984). The last decade has also seen even further improvement in the design, reliability, and maintenance of protective systems and their equipment. Symposiums on loss prevention cover this field regularly.

There is much emphasis on the calculation of plant reliability in the literature (Freeman, 1985; Henley and Kumamoto, 1981). There is less emphasis on how to get plant reliability right the first time. Articles on plant integrity emphasize the need for further testing and inspections (Wicks, 1983), but it is appreciated that this is expensive. Of course, reliability engineering is much more than guarding against release of materials. The major breakthrough in this field as far as the small chemical company is concerned is still awaited. This will only come from increased sharing and pooling of equipment reliability data and information on undesired events. Inhibitions regarding company confidentiality must be broken down.

## 9.8. RISK ANALYSIS

The studies outlined above have aimed at identifying the major hazards and putting together the process. The next stage is risk analysis. This activity can be summarized by three questions.

1. What can go wrong?
2. What are the effects and consequences?
3. How often will it happen?

A preliminary study is used to identify the major hazards of the process plant and storage installations and their possible effects. This preliminary screening should always be carried out before application for funds and can be helpful in comparing the safety risks of different routes.

A quantified risk analysis is performed to identify hazards and how they are realized. A rough estimate is made to identify whether any events are serious enough to justify the need for a quantitative estimate, in which the likelihood of events occurring are evaluated and the consequences placed in context to assess the overall level of risk of the project.

The general approach used for risk analysis is illustrated in Figure 9.1. The main difference between methods centers around the cause-and-effect section of the procedure. This can be illustrated by a simple example, which considers the following scenario of events.

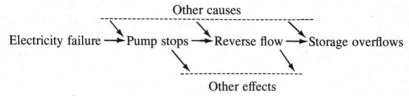

Each event is part of a cause-and-effect chain. It is convenient to identify the event used to set off a search for causes and effects as the deviation. The chain shown above then becomes one of

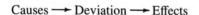

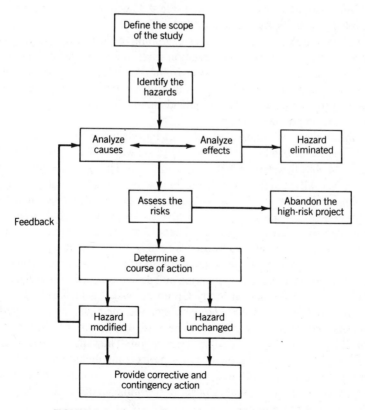

**FIGURE 9.1** The overall procedure used in a risk analysis.

If the event "pump stops" is considered to be the deviation, the study starts at this point to look back at causes and forward to effects and consequences. This particular deviation would tend to stem most readily from a failure mode and effect analysis (see Section 9.8.B). Alternatively, "reverse flow" might be postulated as the deviation. Such a deviation is most likely to be used as the starting point in operability studies (see Section 9.8.A). Another option is to start with what is the final event and work backward. In this case, it is helpful to know the typical causes of leaks, spills, and normal emissions. Such an approach might stem from the use of a check list or in developing a "fault tree" (see Section 9.8.C).

The process engineer must be able to start work at any point in the chain. This promotes a wider range of ideas and the engineer is arguably less constrained by custom and procedure.

The main methods used in risk analysis are described below and are illustrated in Appendix 9.1.

## A. Hazard and Operability Analysis (HAZOP)

The main method of risk analysis employed in the chemical industry on new designs starts with the hazard and operability study (HAZOP). A brief resume of the relevant information needed is given in Table 9.4, which is based on the work of the Chemical Industries Association (1976).

A HAZOP study essentially involves examining each line on a piping and instrumentation diagram. A possible deviation leads to a consequence and the cause of the deviation can readily be assessed from a knowledge of typical problems. HAZOP can be applied to all designs, as the system identifies all manner of problems that can prevent efficient operation and should generate savings in operating costs as well as improving safety.

As with all good techniques, people start introducing refinements such as additional keywords or combining checklists and keywords. This may well add rigor to the method, but this author feels it is a mistake. The important criterion is to have an analyst skilled in the use of the technique present as a member of the team.

Preliminary HAZOP studies can be carried out on flowsheets before a detailed design is initiated. Such a preliminary analysis, however conducted, does not remove the need for a full HAZOP study later (Kletz, 1986). Note that the preliminary hazard analysis as outlined by James (1969) is easier to use.

Alternatively, it may be more suitable for an organization to carry out a comparison with well-documented engineering and design practices as outlined by Solomon (1983). The International Study Group on Risk Analysis (ISGRA), as reported by Holden et al. (1985), stresses that the methods, or combination of methods, selected for risk analysis should be those that best fit into the other design and control activities of the particular organization. They specifically prefer HAZOP, or a similar fundamental method, for projects involving a new technology.

For further guidance in this area, see the AIChE (1985) publication on hazard-evaluation procedures which makes positive recommendations on application and implementation of procedures.

**TABLE 9.4 Some Information for a HAZOP Study**

- Define the scope of the study.
- Organize the team and data for the study.
- Examine each piping and instrumentation diagram (P&ID) section by section, and line by line.
- For each line, select a deviation (e.g., LESS FLOW). Determine if there is a realistic cause which can give rise to this deviation. If the deviation is meaningful, examine its consequences and decide if these are hazardous to people or equipment or work against an efficient operation. Determine whether the operator will know about the change and decide whether changes in the plant or methods will prevent this deviation from occurring, make it less likely to occur, or protect against the consequences. Decide if the cost of the change is worthwhile on grounds of plant safety, plant economics, or both.
- Agree to changes, responsibility for action, and subsequent follow-up.
- Guide words and their meaning:

| | |
|---|---|
| NO, NOT | The complete negation of the designer's intention as to how this particular part of the process should operate. |
| REVERSE | The logical opposite of the designer's intentions. |
| MORE, HIGHER | Quantitative increase. |
| LESS, LOWER | Quantitative decrease. |
| PART OF | A qualitative decrease with only some of the designer's intentions being realized. |
| AS WELL AS | A qualitative increase in which all the designer's intentions are realized together with some additional activity, for example, impurities or extra phases are present. |

All guide words are applied on a line-by-line basis, and it is frequently recommended that only "OTHER THAN" be applied to equipment. However, changes inside the vessel are important, such as HIGHER or LOWER level. Other guide words suggested include WHERE ELSE THAN, SOONER THAN, LATER THAN, STARTING, STOPPING, CONTROLLING, FLUCTUATING, DECONTAMINATING, INGRESS, and ISOLATION.

- Property words or parameters used with the guide words to make up the deviation include FLOW, TEMPERATURE, PRESSURE, LEVEL, TRANSFER, REACTION, PURGE, CONCENTRATION, VISCOSITY, ONE PHASE, OPERATION, and other specific stream properties. A typical deviation is LESS FLOW, which is caused by partial blockage, leaks, pump failure, poor suction head, and the like.
- The study is reported under the headings: guide word, duration, possible causes, possible consequences, action required.

## B. Failure Mode and Effect Analysis (FMEA)

The reliability of equipment affects plant availability and is directly related to plant safety and performance. In a failure mode and effect analysis (FMEA), the cause of the hazard is evaluated from a knowledge of equipment failure and error modes.

The effects of this failure are then further examined. Table 9.5 gives information on the method based on the work of Lambert (1973) and AIChE (1985). FMEA works upward from equipment failures to a top event of interest. It starts with specific types of equipment-component failure and assesses the consequences thereof. CONCAWE (1982) suggests the use of the technique, particularly in an existing plant. It is especially effective in educational use, because it identifies trigger events clearly and simply for inexperienced students. It tends not to identify dangers that arise when the equipment functions correctly.

Human errors are generally not examined by FMEA as often they manifest themselves as component failures. However, a knowledge of them assists in assessing the cause, frequency, and consequences of equipment failure. Human errors include:

---

**TABLE 9.5   Some Information on Failure Mode and Effect Analysis (FMEA)**

A formal FMEA involves:

- Determining the scope of the analysis.
- Identifying the plant/system to be analyzed, the physical system's boundary conditions, and the operating conditions at the interfaces, and assembling the relevant information needed for the study.
- Each piece of equipment is then examined, item by item.
- For each item, identify under appropriate headings on a chart:
  1. The component under analysis (include identifiers off the piping and instrumentation diagram).
  2. The failure or error mode. These include:
     a. Failure to open/start/stop/continue operation.
     b. Spurious failure.
     c. Degradation of performance.
     d. Erratic operation.
     e. Scheduled service/replacement/test.
  3. The immediate effect of the failure on other equipment/system.
  4. The effect of the failure on the whole system.
  5. Give either qualitative hazard classification and assessement of failure frequency (as per Table 9.7), or give a criticality ranking as follows (AIChE, 1985):
     a. No effects of failure on the system.
     b. Minor process upset, with small hazard to facilities and personnel and not causing a plant shutdown.
     c. Major process upset, with significant hazard to facilities and personnel and requiring the orderly shutdown of plant.
     d. Immediate hazard to personnel and equipment requiring the emergency shutdown of the plant.
  6. Indicate how the failure is detected by the operator.
  7. Note contingency provisions and any remarks.
- Decide on appropriate changes, actions, and follow-up.

Failure to perform part of a task.

Incorrect performance of a task.

Performance out of sequence.

The introduction of a task or step that should not be performed.

Failure to perform the task or sequence in the allotted time.

## C. Fault-Tree Analysis (FTA)

Fault-tree analysis (FTA) enables the systematic construction of a logic diagram of event sequences that lead to a specified failure, the TOP event of interest. The analyst starts at this TOP event and works down level by level. Quite complicated combinations of faults can be investigated and the significant cause-and-event sequences are identified and ranked in order. Table 9.6 indicates some of the steps involved. For a more detailed calculation method, see Doelp et al. (1984).

Fault-tree analysis requires access to considerable resources. The choice of probability factors is difficult and the chain of events is subject to the bias of the analyzer. In fault diagnosis, it is receiving a considerable boost from the development of expert-system programming/interpreting tools, which have made the development of suitable programs much easier and accessible (Andow, 1985). In these, the ''user'' supplies the knowledge for the knowledge-intensive program, which requires human expertise in a specific language that is interpreted and acted on by the supplied software programming/interpreting tools.

Data on equipment failures rates are difficult to obtain for use in specific processing environments. However, the prediction of human error, especially in abnormal and stressful circumstances, still cannot be made with confidence. Techniques used in human-failure analysis (Swain and Guttman, 1983) assist in generating rough data.

The use of fault-tree analysis is the basis for a structural approach to examine common-mode failures. Such studies require extensive resources and are normally only used in specific high-risk situations. It is very much an area of activity for the specialist.

## D. Hazard Analysis (HAZAN)

Hazard analysis makes wide use of fault trees. These are often greatly simplified, such that it is more appropriate to state that hazard analysis uses a tree structure. Such trees trace the undesirable event back to its routes. In reading such a tree, it is normal to reverse the process of construction such that the primary events are developed forward. Also, the consequences of a top event are further developed. Thus the analyst ends up with either an event tree or a cause–consequence diagram.

The problem in using the HAZAN technique lies in assessing the consequences of the release of toxic and flammable materials. The prediction of explosion or

**TABLE 9.6   The Stages in Formal Fault Evaluation**

*Problem Definition*

- The TOP EVENT is the accident or undesired event that is the subject of the fault-tree analysis. It must be described precisely indicating "what" happens, "where" it happens, and "when" it occurs, in terms of the operational configuration of the plant.
- The scope of the analysis is defined with particular emphasis on the bounds of the analysis, including events allowed and unallowed and the level of detail to be incorporated in the study.
- The system and its initial state must be defined, indicating the physical boundaries of the system and the operational state of each item of equipment.

*Fault-Tree Construction*

- Commence at the TOP EVENT and proceed by completing each "gate," level by level, until the BASIC EVENTS have been reached. Describe the "what, where, and when" for each fault that arises in the event statement. If the fault consists of an equipment failure, it is a "state-of-equipment" fault, and an OR gate can be added on the fault tree; the analyst then looks for the primary, secondary, and command failures that can result in the event. Otherwise, it is a "state-of-system" fault and the analyst looks for the immediate, necessary, and sufficient causes of the fault event (AIChE, 1985). The inputs to each gate must be fault events and no gate may be connected directly to another gate.

*Fault-Tree Evaluation*

The following diagram will be analyzed:

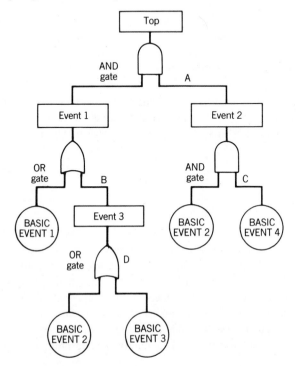

**TABLE 9.6** (*Continued*)

- Identify all gates and basic events as shown in the diagram. For convenience, we have omitted descriptions of events.
- Enter the TOP EVENT as follows:

A

- The first input event is an AND gate. The rule on analysis is for an AND gate (or IN-HIBIT or DELAY gate)—replace the gate identifier with the inputs to the AND gate, inserting one input per column.

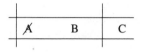

| A̶ | B | C |
|---|---|---|

- For an OR gate, replace the gate identifier by one input to the OR gate. Repeat for each input to the gate, one input per row.

| B̶1 | C |
|---|---|
| B̶D | C |

- Continue until only the basic events remain by replacing first gate C

| 1 | C̶2 | 4 |
|---|---|---|
| D | C̶2 | 4 |

and then gate D.

| 1 | 2 | 4 |
|---|---|---|
| D̶2 | 2 | 4 |
| D̶3 | 2 | 4 |

- Remove any duplicate events that appear in a row.

| 1 | 2 | 4 |
|---|---|---|
| 2 | 4 | |
| 3 | 2 | 4 |

**TABLE 9.6**   *(Continued)*

- Delete all supersets. In this example, it is necessary for sequence 2, 4 to occur for both 1, 2, 4 and 3, 2, 4 to arise. Hence we are left with the set

- The sequence above is the "minimal cut set," which represents all the combinations of equipment failure that can result in the top event. Simple inspection of the fault tree shows this to be the correct solution.
- Finally, the probability of the occurrence of each event chain and of the TOP event can be assessed from the individual probability of each event.

flash fire and the extent of dispersion are extremely difficult to evaluate, even for the expert. Many large uncertainties and assumptions are included in the calculated values. A number of references are incorporated in the bibliography to cover these points. But it does remain an unsatisfactory area which needs extensive work to identify the best procedures from many often conflicting techniques and formulas. Apart from the obvious uncertainty caused by meteorological conditions and elevation, and obstructions to the path, there is concern that few large-scale experiments since Katan (1951) have taken place on the dispersion of heavy gases (McQuaid, 1985). Hence it is difficult to place much accuracy on calculation methods. For an account of problem areas identified by the European International Study Group (ISGR), see Holden et al. (1985). Headlines on the release of toxic/reactive materials are unfortunately common. Also, much remains to be done to allocate responsibility for emitting chemicals presenting a long-term hazard back to the companies concerned.

Quantification of hazards is to be encouraged. However, great care should be made in comparing results as they are usually accurate to vastly different confidence limits. Quantified values related to the safety of the plant are statistical in nature, yet are generally presented without reference to statistical accuracy.

There are many ways of assessing risks. These include fatal accident rates (FAR), which are less often quoted these days, particularly by publicists in the nuclear industry. The guide given in Table 9.7 appears to be along the right lines for scientific judgments only. There can never by an absolute criterion which is universally acceptable to the general public, but at the same time the public has the right to expect companies to work toward an ever-decreasing target.

## 9.9   EMERGENCY PLANS AND THE LOSS-PREVENTION PROGRAM

### A.   Emergency Plans

It seems appropriate to repeat at this stage that contingency plans include all actions to be taken against fire and fire-spread explosions, poisoning, and so on. The

TABLE 9.7  Hazard Classification and Consequences

| Area at risk | Description of Risk | Hazard Classification | | | | |
|---|---|---|---|---|---|---|
| | | Minor | Appreciable | Major | Severe | Total |
| Plant | Damage | Minor < $1000 | Appreciable < $10,000 | Major < $100,000 | Severe < $1,000,000 | Total destruction |
| | Effect on personnel | Minor injury only | Injuries | 1 in 10 chance of a fatality | Fatality | Multiple fatalities |
| Works | Damage | None | None | Minor | Appreciable | Serious |
| Business | Business loss | None | None | Minor | Severe | Total loss of business |
| Public | Damage | None | Minor | Minor | Appreciable | Widespread |
| | Effect on people | None | Smell/nuisance | Some hospitalization | 1 in 10 chance of a fatality | Fatality |
| | Reaction | None | Minor local outcry | Considerable press reaction | Severe press reaction | Severe pressure to stop business |
| Guide values of publically acceptable frequencies | | Once per year | Once per decade | Once per century | Once per thousand years | Once per ten thousand years |

areas of major hazard must be assessed by qualified staff, using methods indicated above. Procedures must be formulated for dealing with the worst foreseeable contingency in each case. These procedures must be promulgated and tested. Prugh (1985) describes the methods to mitigate a major hazard. A study on a vapor-cloud hazard requires a full and formal examination of the following features, not necessarily in the given order.

Toxicity hazards
Blast effects
Leakage fractions
Evaporation from boiling pools
Atmospheric conditions
Cloud length, width, area, and volume
Number of people endangered
Detection of vapor releases
Evacuation prior to cloud arrival
Escape from a cloud
Protection offered by havens
Effectiveness of medical treatment
Estimated frequency of releases
Limiting the release quantity
Countermeasures for releases
Acceptability criteria

All these features do not apply should fireballs develop and explosions arise. Such a study yields much useful data, which should be made available in a suitable form for quick reference by the emergency control center. The emergency plans vary in extent according to the size of the incident. Key considerations involve the use of a control center and appropriate communications, the assignment of key personnel, the establishment of operating procedures to be followed, the media interface, and the vital liaison with outside authorities and services. These all require extensive discussion with the operating staff, public emergency services, and neighboring firms that may be affected in the event of a major accident. Full cooperation with these services is essential to providing appropriate warning for all employees and the general public who may be at risk. In the United Kingdom, the Health and Safety Executive must receive written emergency procedures for scrutiny, if the installation handles particular quantities of listed hazardous substances. They must also be informed if an accident occurs. Appropriate further information is awaited from the Health and Safety Executive, and this is particularly useful when it gives information about which quantifying methods to use.

Every attempt must be made to reduce hazards, and the preparation and periodic updating of emergency plans may well suggest improvements which should be made on the plant. Appropriate feedback to the design on all aspects related to the mitigation of the hazard is essential.

## B. The Loss-Prevention Program

It is emphasized that loss prevention requires and receives more attention than indicated herein. The process engineer is involved during detailed design in many reviews and related documentation directly related to plant safety. These cover items as diverse as the basic design philosophy and concepts, design specifications, isolation of equipment for maintenance, flare blowdown and relief systems, review of piping and instrumentation diagrams (P&ID), fire fighting, and operating instructions.

Management must establish monitoring and control systems that are augmented by training in design safety and in techniques for hazard analysis with the objective of ensuring that potential hazards are recognized and controlled during process design. Furthermore, the aims of the designer must be implemented. Safety devices must be operable at all times and the production personnel must be aware of the reasons why such devices and safeguards are necessary and must not operate the equipment outside of the specified operational constraints. However, it must be accepted that change is taking place in every activity on a chemical plant, including process feedstock and materials, operating conditions, the state of plant materials of construction and of personnel, the operating status of the plant, and plant modifications. A fresh hazard may be introduced into a plant by items as diverse as a maintenance change or an impurity in the process materials which gives rise to a dangerous reaction. Also, the modifying constraints on the plant may change. All these can result in an accident taking place. Process changes must be subjected to appropriate safety design checks by qualified staff.

A detailed analysis of the key accidents in process plants suggests that the main causes of accidents are:

1. Deficiencies in plant operation.
2. Design faults, equipment failures, construction and modification errors.
3. Inadequate maintenance and inspection.

The loss-prevention program must emphasize the need to pay attention to these key areas. A company should produce a statement of its overall safety policy, and establish an organizational structure and control system to ensure that the many safety aspects of operation and maintenance are given proper attention and priority throughout the life of the plant. Furthermore, it is essential that the experience so gained within the organization is used to feed information back to design practices and to engineering standards.

## APPENDIX 9.1.  AN ILLUSTRATION OF APPLICATION OF RISK-ANALYSIS TECHNIQUES TO A PLANT

The various safety studies described in the text are applied to a small section of a plant. No attempt is made to provide a comprehensive analysis. Instead, the aim is to illustrate various ways in which events might be identified as the design is developed.

### A.  The Process

The process under study involves the hydrodeaklylation of toluene by the exothermic reaction

$$C_6H_5CH_3 + H_2 \rightleftharpoons C_6H_6 + CH_4$$

which takes place at average conditions of 640°C and 2000 kPa or 20 bar over a fixed bed of catalyst. A hydrogen-to-aromatics ratio of $4:1$ is needed at the outlet of the reactor to safeguard against cracking reactions, causing carbon formation in the reactor.

A section of this process is shown in Figure 9.2. The process feed including recycles is heated from 300 to 600°C in a heater, fired by fuel gas which is largely a by-product of the process.

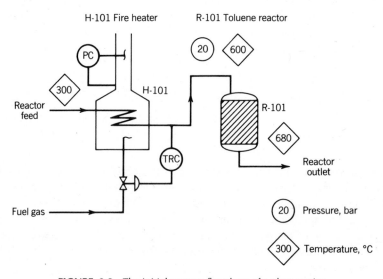

**FIGURE 9.2**  The initial process flowsheet of a plant section.

## B. Preliminary Hazard Analysis

A preliminary analysis of the fired-heater section of the process might proceed as follows at an early stage of the design.

| | |
|---|---|
| 1. Subsystem under investigation: | Fired heater, H-101. |
| 2. Its operating mode: | Normal operation. |
| 3. Trigger event causing hazardous condition: | Momentary loss of fuel and fuel valve open. |
| 4. Hazardous condition: | Fuel enters heater and forms explosive mixture. |
| 5. Trigger event causing accident: | Ignites explosively from hot walls. |
| 6. Potential accident: | Explosion and fire. |
| 7. Possible effects: | Damage to equipment. |
| 8. Contingency provisions and measures to prevent occurrence: | Provide flame-failure alarm and automatic shutdown system. |

This is repeated for other dangerous situations and shows that a feed failure in the fired heater can lead to fire damage of the coil, with potential to release hydrocarbon, causing fire or explosion. Also a "TRC" control failure can lead to a high temperature at the inlet to the reactor, R-101, in which the exothermic reaction is taking place.

These dangers are noted for further study. The need for a temperature alarm at the inlet to the reactor and smothering steam for the furnace are deemed to be essential, in addition to the flame-failure alarm and automatic shutdown system mentioned earlier. It is stressed that these changes would be incorporated without this analysis by the application of good practice.

## C. Failure Mode and Effect Analysis (FMEA)

A failure mode and effect analysis is carried out on the system as shown in Figure 9.3, which represents a later stage of the design. The diagram is greatly simplified for ease in this presentation.

| | |
|---|---|
| 1. Item/description: | Control valve, CV-1 on fuel gas in normal operating mode. |
| Failure mode A: | Fails to open or fails to close when required. |
| Effects: | Process-side coil outlet temperature rises and alarm, TA-1, sounds at 650°C. Further reaction occurs in the exothermic reactor, R-101, increasing temperatures to 750°C. |
| Actions: | Automatic shutdown is desirable if the inlet temperature reaches 700°C. This is accomplished by having the high-temperature cut-out close the fuel gas and shut |

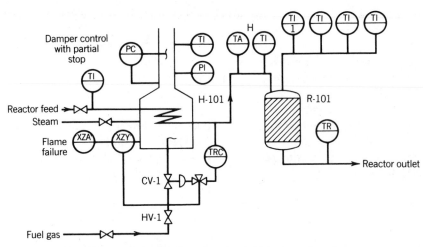

**FIGURE 9.3**   A simplified version of an initial process line diagram.

|                   |                                                                                                                                                                                                                                          |
|-------------------|------------------------------------------------------------------------------------------------------------------------------------------------------------------------------------------------------------------------------------------|
|                   | down the toluene pumps. A gas quench of recycle gas to the center of the reactor reduces the chance of reactor runaway and offers other advantages in the reactor. It is adopted.                                                         |
| Failure mode B:   | Leak to outside environment or rupture.                                                                                                                                                                                                 |
| Effects:          | Release of fuel gas followed by external fire with some damage of equipment.                                                                                                                                                            |
| Actions:          | Ensure the manual valve, HV-1, is accessible at least 16 m away from the fired heater. In addition, the application of good practice in plant layout locates this unit 16 m away from other units. A low-pressure alarm on the fuel gas may warn of the danger; similarly hydrocarbon detection devices may be helpful but are not essential or justified. |
| Failure mode D:   | Spurious failure caused by flame-failure shutdown system, closing the valve due to a command fault.                                                                                                                                     |
| Effects:          | Process reactor temperature falls, giving low conversion.                                                                                                                                                                               |
| Actions:          | None; a more reliable shutdown system was considered.                                                                                                                                                                                   |

2. Item/description:  Fired heater, H-101, in normal operating mode.
   Failure mode A:    Coil leaks.
   Effects:           Flammable material burns in firebox and may increase the temperature of the feed to the reactor.
   Action:            A low-flow alarm on the fuel gas was considered but rejected. The alarm, TA-1, warns the operator if there is a danger of reactor runaway.
   Failure mode B:    Coil fractures.

| | |
|---|---|
| Effects: | A major fire in the firebox. |
| Action: | A check valve on the coil outlet reduces the risk of reverse flow. The manual valves for smothering steam and feed are to be located 16 m from heater. Explosion hatches on the firebox were rejected. |
| Failure mode C: | Coil blockage. |
| Effects: | No flow to reactor. The pressure increases upstream with pressure alarm. The furnace temperature increases and the coil may be damaged. |
| Action: | Provide a low-flow alarm and cut-out on the feed to furnace. The cut-out would shut off the fuel to furnace. Depending on the chemistry in the reactor R-101, it may also be desirable to cut out the recycle hydrogen flow if feed stops. |

The study goes on to consider the firebox side of the furnace, making decisions about pilot lights, control, and so on.

## D. An Operability Study

This operability study is greatly abbreviated to avoid repetition of all the points made above. Only the more meaningful deviations are mentioned. Two lines are considered. They are the fuel gas to the fired heater, and the line from the fired heater to the reactor. The greatly simplified diagram, Figure 9.4, represents the plant at this stage of the study.

Details of this study are given in Table 9.8. As expected, the study picks up additional problems not noted earlier. However, none of these involves any delay on the project. Neither does their cost place the project in jeopardy.

## E. A Fault Tree

A fault tree is developed for a top event of interest. This is a high temperature, above 700°C, at the inlet to the reactor.

The fault tree is given in Figure 9.5. It was developed by working downward from the top event and doing one level at a time. The coil leak specified is not so great as to cause back-flow from the reactor.

The minimal cut sets for this fault tree are made up of the following basic events:

$$1,4,2; \quad 1,4,3; \quad 1,5,2; \quad 1,5,3$$

This seems a sensible solution of the problem and the likelihood of each of these sequences arising should now be quantified.

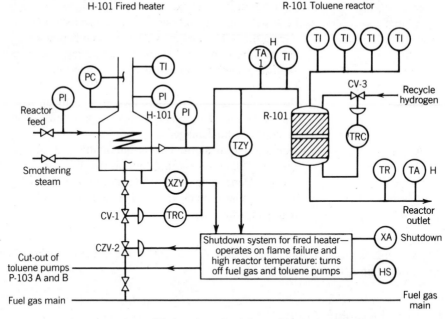

**FIGURE 9.4** A simplified version of a piping and instrument line diagram.

## F.  Hazard Analysis (HAZAN)

Only an event tree is developed for this section of the process. This is illustrated in Figure 9.6. The probability of each event arising can be assessed and the probability of each sequence occurring can be evaluated by multiplying probabilities as shown on the diagram.

This diagram was developed as essentially a follow-up on the other studies. But does it give an accurate picture of what is occurring? The sequences are believed to be correct. However, the analyst has yet to study the additional fire which is going on in the firebox of the heater. The normal emergency shutdown procedures must be modified, possibly to incorporate the need to depressurize and purge the plant. Also, the plant leak may escalate into a major fire, should the coil rupture.

An event tree that reflects these hazards might include the following safety functions, as shown in Figure 9.6:

**B.** Plant shutdown by operators or by automatic shutdown system.

**C.** Plant depressurized and purged by operators.

**D.** Fire prevented from escalating by operators' actions and local firefighting.

Such an event tree will doubtless lead to action to discuss the actions required under safety function D. The event tree may even be extended to consider the possible escalation of this fire to other plant areas.

**TABLE 9.8  Operability Study (HAZOP) of the Plant in Figure 9.4[a]**

| Guide Word and Deviation | Possible Causes | Possible Consequences | Actions |
|---|---|---|---|
| 1. LINE NUMBER: Fuel gas line LESS flow | Coil leaking. | TA-1 sounds; reactor runaway stopped by operator or shut-down system. | Hazard analysis requested as both safeguards may fail. |
| LESS pressure | Loss/reduction of fuel-gas supply from process knock-out drum D-103. | Low temperature at inlet to reactor with operator being unaware of reason. | Low-pressure alarm on fuel gas. |
| MORE pressure | Failure of pressure control valve on D-103. | High-pressure gas can enter heater. Relief valve set at 5 bar. | Discuss with manufacturers. High-pressure alarm on fuel gas. |
| AS WELL AS extra phase | Carry-over of aromatics from D-103 and/or condensation in fuel-gas line. | Spitting in heater and possible flame failure. | Additional knock-out pot in the fuel-gas main fed from this plant. |
| 2. LINE NUMBER: Reactor inlet line PART OF flow | Reduction in flow of hydrogen to reactor R101. | Production of carbon in the reactor. | Low-flow alarm on hydrogen feed to the reactor preheat circuit. Hydrogen low-flow cut-out added to fired heater shutdown system. |
| MORE temperature | Coil leak or failure of control valve to control disturbances. | TA-1 sounds as noted above under LESS flow. This is included because it shows how a further check is made. | |

[a]This study only notes meaningful deviations and considers only deviations that can be resolved from the information given about this process herein.

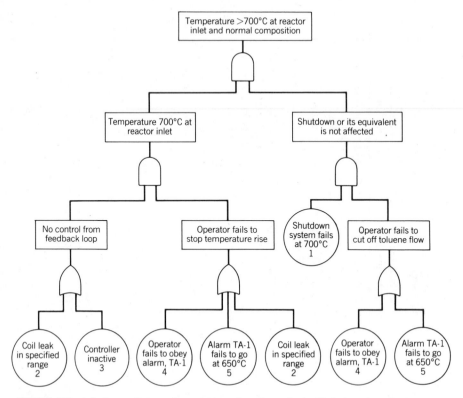

**FIGURE 9.5** A fault-tree diagram for a section of a toluene hydroalkylation plant. Basic event 4 requires that the operator first ensures that the fuel is cut off to the burner; if this does not stop the temperature rise, he then cuts off the toluene flow. ⌓ represents an AND gate and indicates that the output event occurs only when all the input events occur. △ represents an OR gate and indicates that the output event only occurs if any of the input events occur.

## G.  Other Notes

In addition to the actions described above, the designers carrying out the detailed design of the plant will implement various checklists and company standards. The process is fairly safe in the areas being studied. On a similar plant, this particular heater has been damaged badly by fire on one occasion in the last 20 yr. It has also served as a source of ignition on other occasions. So is it necessary? Actually, the heat-exchanger network can be so arranged for this particular process that this heater is only used for start-up purposes. The preliminary safety studies on the plant should generate such an option at the start of the study. A good checklist will always produce suggestions, such as eliminating all sources of ignition.

For companies having basic design practices and/or checklists, it would be expected that items such as loss of fuel gas or feed to a furnace would be anticipated and suitable protection provided via alarms, cut-out, manual valves, and snuffing steam, without going through all the steps listed above. Similarly, we would expect

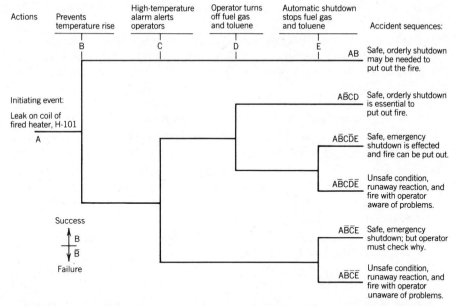

**FIGURE 9.6** An event tree examining the effects on the reactor leak of a leak on the coil of a fired heater.

loss of flow to a reactor, or a reactor temperature excursion, to be anticipated and suitable protection provided to warn of the occurrence and cut-out flow of the hydrogen reactant. The AIChE (1985) document has an excellent set of equipment-oriented checklists given in its Appendix A.

## REFERENCES

AIChE, *DOW's Fire and Explosion Index: Hazard and Classification Guide*, 5th ed., Technical Manual, American Institute of Chemical Engineers, New York, 1981.

AIChE, *Guidelines of Hazard Evaluation Procedures*, American Institute of Chemical Engineers, New York, 1985.

Andow, P. K., "Fault Diagnosis Using Intelligent Knowledge-Based Systems," *Process Systems Engineering PSE '85: The Use of Computers in Chemical Engineering*, pp. 145–156, Institute of Chemical Engineers Symposium Series No. 92, London, United Kingdom (1985).

Balemans, A. W. M., "Check-lists: Guidelines for Safe Design of Process Plants," in *Loss Prevention and Safety Promotion in the Process Industries*, C. H. Bushman (ed.), Elsevier, The Hague, Holland, 1974.

Bretherick, L., *Handbook of Reactive Chemical Hazards: An Indexed Guide to Published Data*, 3rd ed., Butterworths, Stoneham, MA, 1985.

Chemical Industries Association (CIA), *A Guide to Hazard and Operability Studies (HAZOP)*, Chemical Industries Association, London, UK, 1976.

Chemical Rubber Company, *Handbook of Chemistry and Physics*, Chemical Rubber Co., Cleveland, OH, 1985.

CONCAWE, "Methodologies for Hazard Analysis and Risk Assessment in the Petroleum Refining and Storage Industry," Den Haag, Holland (1982). CONCAWE is the Oil Companies' International Study Group for Conservation of Clean Air and Water–Europe.

Doelp, L. C., G. K. Lee, R. E. Linney, and R. W. Oronsby, "Quantitative Fault-Tree Analysis: Gate-by-Gate Method," *Plant/Operations Progress, 3*, 227 (1984).

Freeman, R. A., "The Use of Risk Assessment in the Chemical Industry," *Plant/Operations Progress, 4,* 85 (April, 1985).

Govind, R. and G. J. Powers, "Studies in Reaction Path Synthesis," *AIChE J., 27,* 429 (1981).

Henley, E. J. and Kumamoto, H., *Reliability Engineering and Risk Assessment*, Prentice-Hall, Englewood Cliffs, NJ, 1981.

Her Majesty's Stationary Office (HMSO), "The Flixborough Disaster," Report of the Court of Inquiry, HMSO, London, UK, 1978.

Her Majesty's Stationary Office (HMSO), *Hazardous Installations (Notifications and Survey) Regulations*, London, UK, 1982.

Her Majesty's Stationary Office (HMSO), *A Guide to the Control of Industrial Major Accident Hazards (CIMAH) Regulations*, London, UK, 1984.

Holden, P. L., D. R. T. Lowe, and G. Opschoor, "Risk Analysis in the Process Industries: An ISGRA Update," *Plant/Operations Progress, 4,* 63 (1985).

I. Chem. E., *Flowsheeting for Safety*, Institution of Chemical Engineers, London, UK, 1976.

I. Chem. E., *A First Guide to Loss Prevention*, Institution of Chemical Engineers, London, UK, 1977.

I. Chem. E., *"The Assessment and Control of Major Hazards,"* Institute of Chemical Engineers Symposium, Series No. 93, Institution of Chemical Engineers, London, UK, 1985a.

I. Chem. E., *"Risk Analysis in the Process Industries,"* E.F.C.E. Publication Series, No. 45, European Federation of Chemical Engineering, Institution of Chemical Engineers, London, UK, 1985b.

James, C. R. (Ed.), *Systems Safety Analytical Technology*, Publication No. DZ-113073-1, Boeing Co., Seattle, WA, 1969.

Katan, L. L., "The Fire Hazard on Fueling Aircraft in the Open," Fire Research Technical Paper No. 1, Fire Research Association, London, UK, 1951.

Kletz, T. A., *What Went Wrong*, Gulf Publishing Co., Houston, TX, 1985a.

Kletz, T. A., "An Atlas of Safety Thinking," *The Chemical Engineer,* 47 (May, 1985b).

Kletz, T. A., *HAZOP and HAZAN: Notes on the Identification and Assessment of Hazards*, 2nd ed., Institution of Chemical Engineers, London, UK, 1986.

Lambert, H. E., "Systems Safety Analysis and Fault-Tree Analysis," Report No. UCID-16238, University of California, Livermore, CA, 1973.

Lewis, D. J., "The Mond Fire Explosion and Toxicity Index," *AIChE Loss Prevention Series*, No. 11, American Institute of Chemical Engineers, New York, 1977, p. 20.

Marsh and McLennan, *One Hundred Largest Losses: A Thirty-Year Review of Property Damage Losses in the Hydrocarbon–Chemical Industries*, (9th ed.), M&M Protection Consultants, Chicago, IL, April 1986.

McQuaid, J., "Trials on Dispersion of Heavy Gas Clouds," *Plant/Operations Prog.*, **4**, 58 (1985).

National Fire Protection Association (NFPA), *Manual of Hazardous Chemical Reactions*, NFPA Manual No. 491M, Washington, D.C., 1985.

Nuclear Regulatory Commission, "Three Mile Island—Two Lessons Learned," Task Force Final Report, United States Nuclear Regulatory Commission, Report No. NUREG-0585, Washington, D.C., October, 1978.

Perrow, C., *Normal Accidents*, Basic Books, New York, 1984.

Prugh, R. W., "Mitigation of Vapor Cloud Hazards," *Plant/Operations Prog.*, **4**, 95 (1985).

Sax, N. I., *Dangerous Properties of Industrial Materials*, 5th ed., Van Nostrand Reinhold Company, New York, 1979.

Shaw, J. A., "Human Factor Aspects of Advanced Process Control," *Plant/Operations Prog.*, **4**, 111 (1985).

Solomon, C. H., "The Exxon Chemicals Method for Identifying Potential Process Hazards," Institution of Chemical Engineers, Loss Prevention Bulletin No. 52, London, UK, August 1983.

Swain, A. D. and H. E. Guttman, *Handbook of Human Reliability Analysis with Emphasis on Nuclear Plant Applications*, NUREG/CR-1278, U.S. Nuclear Regulatory Commission, Washington, D.C., 1983.

Tyler, B. J., "Using the Mond Index to Measure Inherent Hazards," *Plant/Operations Prog.*, **4**, 172 (1985).

Verscheuten, K., *Handbook of Environmental Data on Organic Chemicals*, Reinhold, New York, 1977.

Wells, G. L., *Safety in Process Design*, George Godwin/Institution of Chemical Engineers, London, UK, 1980.

Wells, G. L. and L. M. Rose, *The Art of Chemical Process Design*, Elsevier, Amsterdam, Holland, 1986.

Wicks, K. M., "Inherent Safety by Choice of Chemistry," *AIChE Loss Prevention Series*, No. 52, American Institute of Chemical Engineers, New York, 1983.

Windholy, M. (ed.), *The Merck Index: An Encyclopedia of Chemicals and Drugs*, Merck and Company, Rahway, NJ, 1978.

## FURTHER READING

The following bibliography gives a selection from the most recent literature in this field not cited in the text of this chapter. These in turn indicate further reading and in particular the originators of some of the methods here described.

The general references were selected as providing an overall feel for the topic. They start with the book by Armistead (1959), which arguably was the first major effort in this field. The two-volume text by Lees (1980) gives comprehensive coverage of the topic. Wells (1980) is more concerned with creating an awareness of danger and ways for its reduction.

The section on risk-assessment methods and hazard indices lists some of the extensive publications that assist in gaining a full grasp of these techniques. The AIChE (1985) review is a good starting point for further study as it deals with all organizational features of carrying out the different types of analysis. It is remarkable how little conflict arises in techniques described by various authors. The section on reliability analysis and related topics reflects the different styles and mathematical treatments.

The remaining sections are a mix of papers relating to hazard, corrective, protective, and contingency actions together with some selected reports. These tend to refer to specific and detailed topics covered in the literature.

## A.  General References

Armistead, G., *Safety in Petroleum Refining and Related Industries*, John Simmonds, New York, 1959.

Handley, W. (ed.), *Industrial Safety Handbook*, McGraw-Hill, New York, 1969.

Jain, R. K., L. V. Urban, and G. S. Stacey, *Environmental Impact Analysis*, Van Nostrand Reinhold, New York, 1977.

Jones, A. L., *Occupational Hygiene: An Introduction Guide*, Crom Helm, London, UK, 1981.

Kletz, T. A., "Eliminating Potential Process Hazards," *The Chemical Engineer*, **92,** 49 (1985).

Lees, F. A., *Loss Prevention in the Process Industries*, Vols. 1 and 2, Butterworths, London, UK, 1980.

Lees, F. A., "The Hazard Warning Structure of Major Hazards," *Trans. Inst. Chem. Eng.,* **60** (4), 211 (1982).

Shabica, A. C., "Evaluating the Hazards in Chemical Plants," *Chem. Eng. Prog.,* **59** (9), 57 (1963).

Spiegelman, A., "Risk Evaluation of Chemical Plants," *AIChE Loss Prevent. Ser.* **6,** 1 (1969).

Walker, A., *Law of Industrial Pollution Control*, George Godwin, London, UK, 1979.

## B.  Risk Assessment Methods and Hazard Indices

Cizek, J. S., "Diamond Shamrock Loss Prevention Review Program," Canadian Society of Chemical Engineers Conference, 1982.

Department of Navy, "Procedures for Performing a Risk Analysis," Department of Navy, Report No. MLSTD-1629A, Washington, D.C., 1977.

Gibson, S. B., "The Design of New Plants Using Hazard Analysis," *Inst. Chem. Eng. Symp. Ser.*, No. 47, 135 (1976).

Harris, N. C. and A. M. Moses, "The Use of Acute Toxicity Data in the Risk Assessment of the Effects of Accidental Releases of Toxic Gases," *I. Chem. Eng. Symp. Ser.,* **80,** 136 (1983).

Insurance Technical Bureau, "Process Factor Algorithm for Oil and Petrochemical Plants," Insurance Technical Bureau, London, UK, 1983.

Knowlton, R. E., "Hazard and Operability Studies," Chemetics International Ltd., Vancouver, B.C., Canada, 1981.

Knowlton, R. D., "An Introduction to Guide Word Hazard and Operability Studies," Canadian Society of Chemical Engineers Conference, 1982.

Lambert, H. E., "Failure Mode and Effect Analysis," NATO Advanced Study Institute, Urbino, Italy, July 1978.

Lawley, H. G., "Operability Studies and Hazard Analysis," *AIChE Loss Prevent. Ser.*, **8**, 1 (1974).

Munday, G., "Instantaneous Fractional Annual Loss," 3rd International Symposium on Loss Prevention and Safety Promotion in the Process Industries, Basel, Belgium, September 15, 1980.

Roach, J. P. and F. P. Lees, "Some Features and Activities in Hazard and Operability Studies," *The Chemical Engineer,* 456 (October, 1981).

Taylor, J. R., "Cause-Consequence Diagrams," NATO Advanced Study Institute, Urbino, Italy, July 1978.

## C. Reliability Analysis and Related Topics

ANSI/ISA-S5.2-1986, "Binary Logic Diagram for Process Operations," American National Standard Institute and Instrument Society of America, Research Triangle Park, NC, 1981.

Barlow, R. E., J. B. Fussell, and N. D. Singpurwalla (eds.), *Reliability and Fault Tree Analysis*, Society of Industrial and Applied Mathematics (SIAM), Philadelphia, PA, 1975.

Bell, B. J. and A. D. Swain, "A Procedure for Conducting a Human Reliability Analysis for Nuclear Power Plants," Report No. NUREG/CR-2254, U.S. Nuclear Regulatory Commission, Washington, D.C., May 1983.

Brown, D. M. and P. W. Ball, "A Simple Method for the Approximate Evaluations of Fault Trees," 3rd International Symposium on Loss Prevention, Basel, Belgium, September 15–19, 1980.

Fussell, J. B. and W. E. Vesely, "A New Methodology for Obtaining Cut Sets for Fault Trees," *Trans. Am. Nuc. Soc.,* **15**, 262 (1972).

Green, A. E. and J. R. Bowne, *Reliability Technology*, Wiley, New York, 1972.

Henley, E. J. and H. Kumamoto, *Designing for Reliability and Safety Control*, Prentice-Hall, Englewood Cliffs, NJ, 1985.

Himmelblau, D. M., *Fault Detection and Diagnosis in Chemical and Petrochemical Processes*, Elsevier, Amsterdam, Holland, 1978.

Kumamoto, H., K. Ikenchi, K. Inoue, and E. J. Henley, "Application of Expert Systems to Fault Diagnosis," *Chem. Eng. J.,* **29,** 1 (1984).

Lieberman, N. P., *Process Design for Reliable Operations*, Gulf Publishing Co., Houston, TX, 1982.

Nielson, D. S., "Use of Cause-Consequence Charts in Practical Systems Analysis," Report No. RISO-M-1743, RISO National Laboratory, Roskilde, Denmark, 1984.

Powers, G. J. and S. A. Lapp, "Computer-Aided Fault Tree Synthesis," *Chem. Eng. Prog.*, **72** (4), 89 (1976).

U. S. Air Force, "Nonelectronic Reliability Notebook," Rome Air Development Center, Griffiss Air Force Base, Bawe, New York, 1975.

Vesely, W. E. and R. E. Narum, "PREP AND KITT—Computer Codes for the Automatic Evaluation of Fault Tree," Idaho Nuclear Corp., Scientific and Technical Report, Contract Number AT(10-1)-1230, U.S. Atomic Energy Commission, Washington, D.C., 1977.

Vesely, W. E., *Fault Tree Handbook*, Nuclear Regulatory Commission, NUREG-0492, Washington, D.C., 1981.

Voller, V. and B. Knight, "Expert Systems," *Chem. Eng.*, 93 (June 10, 1985).

Woodson, E. W., *Human Factors Design Handbook*, McGraw-Hill, New York, 1981.

## D.  Hazards, Leaks, and Spills

Briller, R. E. and R. F. Griffiths, *Dense Gas Dispersion*, Elsevier, Amsterdam, Holland, 1982.

British Standards Institute, BS CP 5908, "Code of Practice for Fire Precautions in Chemical Plant," London, UK, 1980.

Burgess, D., "Volume of Flammable Mixture Resulting from the Atmospheric Dispersion of a Leak or Spill," 15th International Symposium on Combustion, Tokyo, Japan, 1974.

Buschmann, C. H., "Methods for the Calculation of the Physical Effects of the Escape of Dangerous Materials," Netherlands Organization for Applied Research, TNO, The Netherlands, 1980.

Clancey, V. J., "The Evaporation and Dispersion of Flammable Liquid Spillages," *I. Chem. E. Symp. Ser.*, No. 39a, 80 (1974).

Craven, A. D., "Fire and Explosion Hazards Associated with Small-Scale Unconfined Spillages," *I. Chem. E. Symp. Ser.*, No. 47, 37 (1976).

Dutch Chemical Industry Association, *Handling Chemicals Safely*, Holland, 1980.

Gugan, K., *Unconfined Vapor Cloud Explosions*, Institutions of Chemical Engineers, Godwin, London, UK, 1979.

Her Majesty's Stationary Office (HMSO), "The Packaging and Labelling of Dangerous Substances Regulations," London, UK, 1978.

International Oil Insurers, "The Evaluation of Estimated Maximum Loss from Fire and Explosion in Oil, Gas and Petrochemical Industries with Reference to Percussive Unconfined Vapor Cloud Explosions," London, UK, 1983.

Klein, L., *River Pollution, Vol. 1, Chemical Analysis*, Butterworths, London, UK, 1959.

Klein, L., *River Pollution, Vol. 2, Causes and Effects*, Butterworths, London, UK, 1962.

Klein, L., *River Pollution, Vol. 3, Control*, Butterworths, London, UK, 1966.

Kusnetz, H. L., "Industrial Hygiene Factors in Design and Operating Practice," *AIChE Loss Prevent. Series,* **8,** 20 (1974).

Laznow, J., "A Pre-Programmed Computerized System for Predicting the Atmospheric Spread of Accidental Chemical Release," Paper 2E, AIChE Winter Meeting, Orlando, FL, March 1982.

Munday, G., "Unconfined Vapor Cloud Explosions," *Chem. Eng.,* 278 (April 1976).

Palmer, K. N., *Dust Explosions and Fires,* Chapman & Hall, London, UK, 1973.

Pasquill, F., "The Estimation of the Dispersion of Airborne Material," *Met. Mag.,* **90,** 33 (1961).

Polycyn, A. J. and H. E. Hesketh, "A Review of Current Sampling and Analytical Methods for Assessing Toxic and Hazardous Organic Emissions from Stationary Sources," *J. Air Pollut. Control Assoc.,* **35,** 54 (1985).

Slade, D. H., "Meteorology and Atomic Energy," Report No. TID-24190, U.S. Atomic Energy Commission, Washington, D.C., 1968.

Small, F. H. and G. E. Snyder, "Controlling In-Plant Toxic Spills," *AIChE Loss Prevent. Ser.,* **8,** 24 (1974).

Thomas, G. L. R., "Fire Prevention/Protection," *Inst. Chem. Eng. Loss Prevent. Bull.,* **43,** 1 (1981).

Van Buijtenen, C. J. P., "Calculation of the Amount of Gas in the Explosive Region of a Vapor Cloud," *J. Haz. Mater.,* **3** (3), 201 (1980).

Vervalin, C. H., *Fire Protection Manual for Hydrocarbon Processing Plants,* Gulf Publishing Co., Houston, TX, 1983.

Wallace, M. J., "Controlling Fugitive Emissions," *Chem. Eng.,* **86,** 79 (August 27, 1979).

Wilson, E., "A Selected Annotated Bibliography and Guide to Sources of Information of Planning for and Response to Chemical Emergencies," *J. Haz. Mater.,* **4** (4) 373 (1981).

# E. Protective Systems and Related Topics

Collacott, R. A., *Vibration Monitoring and Diagnosis,* George Godwin, London, UK, 1979.

Drew, J. W., "Distillation Column Startup," *Chem. Eng.,* **90,** 221 (November 14, 1983).

Heinze, A. J., "Pressure Vessel Design for Process Engineers," *Hydrocarbon Proc.,* **58,** 181 (May, 1979).

I. Chem. E., "The Protection of Exothermic Reactors and Pressurized Storage Vessels," Institution of Chemical Engineers, Chester, UK, (1984).

Jenkins, J. M., P. E. Kelly, C. B. Cobb, "Design for Better Safety Relief," *Hydrocarbon Proc.,* **56,** 93 (August, 1977).

Kletz, T. A., "Emergency Isolation Valves for Chemical Plants," *Chem. Eng. Prog.,* **71** (9), 134 (September 1975).

Kletz, T. A., "Protect Pressure Vessels from Fire," *Hydrocarbon Processing,* **56,** 98 (Aug., 1977).

Lawley, H. G. and T. A. Kletz, "High Pressure Trip Systems for Vessel Protection," *Chem. Eng.,* **82,** 81 (May 12, 1975).

Melancon, C. L., "Improving Emergency Control and Response Systems," *AIChE Loss Prevent. Ser.*, **13**, 43 (1980).

Neale, D. F., "The Monitoring of Critical Equipment," *Chem. Eng. Prog.*, **70** (10), 53 (1974).

Niesenfeld, A. N., "Shutdown Features of In-Line Process Control," *AIChE Loss Prevent. Ser.*, **6**, 61 (1972).

Rasmussen, E. J., "Alarm and Shutdown Devices Protect Process Equipment," *Chem. Eng.*, **84**, 74 (May 12, 1975).

Sonti, R. S., "Practical Design and Operation of Vapor-Depressuring Systems," *Chem. Eng.*, **91**, 66 (January 23, 1985).

Stewart, R. M., "The Design and Operation of HIPS," *The Chemical Engineer*, 622 (October, 1974).

Verdin, A., "Process Analyzers," *The Chemical Engineer*, 683, (November, 1980).

Whelan, T. W. and S. J. Thomson, "Reduce Relief System Costs," *Hydrocarbon Proc.*, **54**, 83 (August, 1975).

Wright, T. K., "Inherent Safety by Choice of Chemistry," *AIChE Loss Prevent. Ser.*, **17**, 52 (1983).

## F.  Layout and Structures

Chemical Industries Association, "Process Plant Hazards and Control Building Design," London, UK, 1979.

Granger, J. E., "Plant Site Selections," *Chem. Eng.*, **88**, 85 (June 15, 1981).

House, F. F., "An Engineer's Guide to Process Plant Layout," *Chem. Eng.* **76**, 120 (July 28, 1969).

Kern, R., "A Series of Articles on Layout," for list see *Chem. Eng.*, **71**, 141 (August 14, 1978).

Kletz, T. A., "Plant Layout and Location: Methods for Taking Hazardous Occurrences into Account," *AIChE Loss Prevent. Ser.*, **14**, 150 (1980).

Lawrence, E. R. and E. E. Johnson, "Design for Limiting Explosion Damage," *Chem. Eng.*, **81**, 96 (January 7, 1974).

Meckelnburgh, J. (ed.), *Plant Layout*, Longmann, London, UK, 1982.

Rinder, R. M. and S. Wachtell, "Structures to Resist the Effects of Accidental Explosions," U.S. Army Technical Manual, TM5-1300, June 1969.

## G.  Selected Reports

Cremer and Warner, "An Analysis of the Canvey Report," Oyez Intelligence Reports, ISBN-0-85120-465-10, London, UK, 1980.

Her Majesty's Stationary Office (HMSO), "Report on Fire and Explosion at Chemstar Factory," London, UK, 1981.

Institution of Chemical Engineers, "Bhopal: The Company Report," *Loss Prevent. Bull.*, **63**, 1 (1985).

Rasmussen, N. C., "Nuclear Reactor Safety Study," United States Nuclear Regulatory Commission, NUREG-75/014, Washington, D.C., 1978.

Rowe, W. D., "Reactor Safety Study (Wash-1400): A Review of the Final Report," U.S. Environmental Protection Agency, Report EPA-520/3-76-009, Washington, D.C., 1976.

Simpson, W., "Canvey: An Investigation of Potential Hazards from Operations in the Canvey Island/Thurrock Area," U.K. Health and Safety Executive Report, Her Majesty's Stationary Office, London, UK, 1978.

# 10

# DEVELOPING DESIGN BASES
# FOR SYNTHETIC FUEL PLANTS

**W. ROBERT EPPERLY**

## 10.1.  INTRODUCTION

This fiftieth anniversary celebration and remembrance of Professor Frank C. Vilbrandt challenges us to consider the future of plant design. An important aspect of
this subject is the design of "first-of-a-kind" or "pioneer" commercial plants,
which has always been difficult. What constitutes an adequate database for design
of a pioneer plant? The answer involves balancing the overall economics of the
venture (including research and development costs) against incremental investment
(in both research and development and plant facilities) to reduce technical uncertainties and improve reliability (Swabb, 1979). This balance ideally should be made
for all new process developments from biotechnology and new petroleum and
chemical processes to new microelectronic devices. Synthetic fuel plants, the subject of this chapter, present particularly important challenges because of their large
size, the technologies involved, and the uncertainty of the competitive environment for the products. The delay in evolution of the synthetic fuel industry provides time to develop new and improved plant-design tools that can facilitate future
commercial development.

   This chapter describes the (1) strategic importance of synthetic fuels, (2) the
complexities of designing the large facilities involved, (3) the strategies available
for getting the database necessary to design pioneer plants, and (4) the experience
in a specific case—development of the Exxon donor solvent (EDS) coal-liquefaction process. Finally, the chapter describes directions in plant design that will be
important to future synthetic fuel technologies.

   Although this chapter is based primarily on the experience of the EDS coal-
liquefaction process development, the principles involved are broadly applicable.
In general, development of a new coal conversion process involves the following
steps: (1) definition of the chemistry of the conversion including the conditions
required; (2) definition of the process flowsheet including all of the steps required;
(3) identification of any special equipment needs or process-control problems; (4)
identification of scale-up issues associated with designing commercial-sized equipment; and (5) development of design tools that incorporate all relevant information. In the case of the EDS process, these steps were followed and are described
elsewhere (Vick and Epperly, 1982). Literature describing other developments is
included in the section on "Further Reading."

## 10.2.  SUPPLY OF HYDROCARBON LIQUIDS

Since 1970, the world oil production rate has exceeded the rate of discovery of
reserves (Figure 10.1). As a result, worldwide reserves of oil have been decreasing
for 15 yr. The production rate has decreased roughly 15% from the peak in 1979,
but the discovery rate has also dropped. Thus there is no evidence of convergence
of production and discovery rate at this point, and the statistics underline the obvious fact that oil reserves are finite (Office of Technology Assessment, 1985).

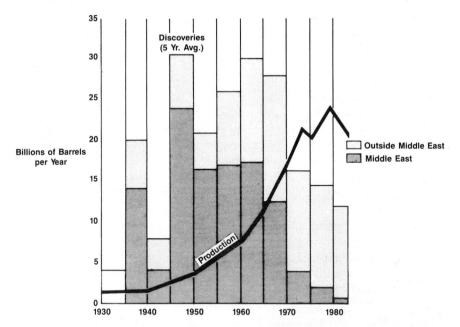

**FIGURE 10.1** Rates of discovery and production of total world oil reserves. Source: Middle East Oil and Gas, © 1984 Exxon Corporation.

In the United States today, oil and gas combined account for 80% of fossil-energy production, but only 5% of reserves. In sharp contrast, coal accounts for about 20% of fossil-energy production, but coal and oil shale combined account for 95% of reserves.

Clearly, coal and oil shale are major fossil-energy resources and will become more important as oil reserves are depleted. However, timing of the transition is difficult to predict in view of uncertainties about future energy demand and discovery rates in exploration.

## 10.3. DIMENSIONS OF A SYNTHETIC FUEL PLANT

By any standard, a full-sized commercial synthetic fuel plant would be large (National Research Council, 1982). A coal-liquefaction plant producing 50,000 barrels/day (about 6900 m³/day) of product would cost several billion dollars, would involve over 20 million construction hours and up to 10,000 workers at the peak of construction, and would take perhaps 8 yr to complete (including planning and process design). Previously, the largest projects have involved 5–10 million construction hours.

The completed plant would feed 25,000 tons/day of coal. If supplied from underground, this would require four of the largest mines ever built. Annual revenue from such a plant would approach a billion dollars.

These statistics emphasize a simple point. Synthetic fuel plants represent enormous financial commitments, and they must be safe, reliable, and environmentally acceptable. The database and the scale-up strategy supporting the plant design must be consistent with the financial stakes.

In view of the risks, pioneer plants probably will be smaller than subsequent ones (Office of Technology Assessment, 1985). However, the investments will still be very large, and reliable design bases will be required.

The task of developing a reliable plant design is complicated by the nature of the resource. Coal is a solid and, as a result, is hard to handle. It requires crushers to reduce it to the desired size range, belt conveyors, and slurry pumps (once mixed with a liquid). These steps are mechanically cumbersome compared with pumping liquid or transporting gas through a pipe, and care is needed to achieve the desired reliability (Merrow, 1985). In addition, coal dust must be collected and utilized to minimize pollution and maximize efficiency.

Coal contains mineral matter, an erosive inorganic solid. Hence, special attention is needed in design to achieve satisfactory pumps, valves, piping, and vessels. In addition, mineral matter is difficult to separate from the liquid product.

Coal also contains large amounts of sulfur, nitrogen, and oxygen, which must be removed. The sulfur and oxygen have special significance—sulfur because it increases corrosion and oxygen because it leads to the formation of phenols, which are potential water pollutants.

Hydrogen pressures of over 14 MPa (2030 psi) at roughly 450°C (842°F) are required to liquefy coal. This means that special, high-pressure equipment is required to resist hydrogen attack, erosion by the mineral matter, and corrosion.

The combination of large size, cost, and the nature of coal resource must be considered in developing a plant-design basis, including a scale-up strategy. While oil shale is different in some respects, it shares many of the same characteristics and requirements.

## 10.4.  PLANT DESIGN BASIS

The database for design includes all of the information needed for detailed heat and material balances and for a definition of process and mechanical design features and operating procedures. More specifically, it includes the effects of feed and process parameters on performance, including product yield and quality, critical mechanical features, and the effects of scale-up to large size on performance. It includes critical operating and process control factors, bases for generating all utilities, including hydrogen and plant fuel, occupational health and safety parameters, and environmental controls. Note that each of these features is a subject covered in today's chemical-engineering curriculum: stoichiometry, unit operations, kinetics and reactor design, process control, plant design, and environmental engineering. However, in each step, synthetic fuels place unusually large demands on the database currently available for chemical-engineering design.

The extremes of plant facility size are fixed. The desired size of commercial plants is known to be large, based on business planning and economics, while the bench-scale units used to obtain initial process data are as small as possible, consistent with obtaining valid data. The intermediate steps needed to obtain the required data have an important impact on cost and schedule. This is illustrated by Table 10.1, based on experience from EDS coal liquefaction.

In EDS process development, the commercialization path visualized involved three phases: a bench-scale program, an integrated development program involving a 250 T/D pilot plant, and a full-scale pioneer plant (which has not yet been built). The intermediate step costed over 300 million dollars and added 7 yr to the development time. The size of the 250 T/D pilot plant was based on an engineering analysis of the critical scale-up factors. For example, a 2-ft (0.61-m) diameter liquefaction reactor was needed to achieve liquid and gas holdups representative of commercial reactors (Vick and Epperly, 1982).

In the future, with more detailed understanding of the chemical and physical processes involved, improved scale-up and design tools should reduce both the need for large pilot plants and the scope of the operating programs for these pilot plants, when required. In addition, the pilot plants built should be more precisely designed to provide the data needed.

The sections that follow describe the results of the EDS coal-liquefaction program. This experience identified areas in which improved tools are required to reduce the need for large pilot-plant programs.

## 10.5. EXXON DONOR SOLVENT (EDS) PROCESS DESCRIPTION

The EDS coal-liquefaction process evolved from a research program started in 1966 after a thorough review of earlier work (Lowery and Rose, 1947; Sherwood, 1950; Wu and Storch, 1968).

In the EDS process (Figure 10.2), crushed coal is mixed with hydrotreated recycle solvent and recycle liquefaction bottoms and is dried by heating the mixture to about 150°C. The resulting slurry is pumped to reaction pressure, mixed

TABLE 10.1 Approximate Costs and Time Schedules for Different Phases of EDS Coal-Liquefaction Development

| Phase | Approximate Cost (millions of dollars) | Approximate Elapsed Time (yr) |
|---|---|---|
| Bench program (0.5 T/D) pilot plant | 30 | 10 |
| Integrated development (250 T/D) pilot plant | 300+ | 7 |
| Pioneer commercial plant construction (50,000 B/D) | 3000+ | 8 |

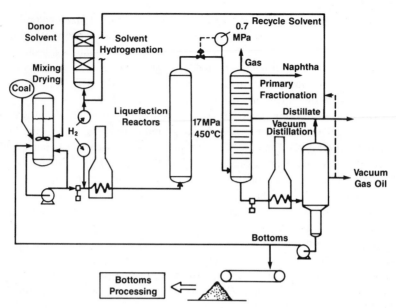

**FIGURE 10.2**   EDS (Exxon donor solvent) process.

with hydrogen treat gas, and preheated to about 425°C. Typical conditions in the liquefaction reactors are 17 MPa (2466 psi) total pressure, 14 MPa (2040 psi) hydrogen pressure, and 450°C. Liquefaction occurs as hydrogen is transferred from donor molecules in the solvent and from gaseous hydrogen to the coal. The reactor effluent is throttled by a single control valve to about 0.7 MPa (102 psi) and then sent to the distillation towers to recover the liquid products. The process solvent is recovered as a distillate stream and recycled to the solvent hydrogenation unit to replenish its donor-hydrogen content. About half of the vacuum bottoms is recycled to the process; the rest is solidified on a cooling belt and sent to bottoms processing where it is typically combusted as a fuel or gasified to produce hydrogen. With further processing, the products are suitable for use as high-octane gasoline and various distillate fuels.

## 10.6.   EDS INTEGRATED DEVELOPMENT

An integrated development program was defined to obtain the data needed for a commercial design and thus to bring the technology to a state of commercial readiness. It involved all phases of process development including bench-scale research, small pilot-plant operation, engineering design and technology studies, and a 250 T/D pilot plant.

The program was jointly funded by the U.S. Department of Energy, Exxon Company USA, Electric Power Research Institute, Japan Coal Liquefaction Development Company, Phillips Coal Company, ARCO Coal Company, Ruhrkohle A.G., and ENI (Epperly, 1981).

The description that follows focuses on the 250 T/D EDS Coal-Liquefaction Pilot Plant or ECLP (Figure 10.3), since this unit provided information on process and mechanical bases and on scale-up (Cavanaugh et al., 1984; Epperly and co-workers, 1978–1985; Epperly et al., 1983). As shown, pilot-plant performance was not fully anticipated, even after over 10 yr of research leading to this phase (Vick and Epperly, 1982).

## 10.7.  EDS DESIGN CONSIDERATIONS—MECHANICAL

Coal liquefaction raises new and significant problems compared with conventional refining. Because of the nature of coal and coal processing, some of the most important issues relate to the mechanical design of the associated equipment (e.g., pumps, furnaces, valves, materials, and piping). In many cases, the proper choices were made in designing ECLP. In other cases, important modifications that would affect the design of a commercial plant were necessary.

One area in which substantial changes were made was pumping. As mentioned earlier, coal was conveyed through the liquefaction plant as a slurry (coal mixed with process-derived solvent). Plunger pumps were used where large pressure increases were required and centrifugal pumps where modest pressure increases were needed.

**FIGURE 10.3**   250 T/D EDS coal-liquefaction pilot plant (ECLP).

Plunger pumps were used to pump the bottoms from the atmospheric distillation tower through a furnace and into the vacuum tower. The pump used for this service at ECLP initially had a packing life of 2–3 days. After redesign of the packing box (to prevent solids from getting in the packing) and selection of better packing, its life was extended to 90 days, which is considered acceptable. Longer life is anticipated for commercial plants based on all of the experience gained.

The concern with centrifugal pumps was impeller life in the erosive environment; this was found to be satisfactory when the tip speed was kept below 39.6 m/sec (130 ft/sec).

The valve that reduces pressure from 17 MPa (2466 psi) in liquefaction to less than 0.7 MPa (102 psi) in distillation is critical, and erosion was a major concern. After careful review of the alternatives available, an angle-valve design based on experiences with petroleum was used at ECLP, and this gave acceptable performance.

Block valves were also very important. The capability to isolate areas of a unit was required for safe and efficient operation. The problem here was identifying the type of valve that would work and learning how to operate and maintain it.

Materials problems were minimal at ECLP. Virtually all of the potential problems were anticipated. A section of the distillation system was subject to chloride attack, but this was controlled by specifying alloy steel, based on petroleum experience. Erosion was a more important issue. In one case, a line failed in 27 days because a long-radius elbow was not used. After installing a proper design, there was no further problem in this area.

Finally, line plugging was initially a big problem at ECLP. The process streams containing liquefaction bottoms "froze" at temperatures below about 150°C (302°F). When ECLP was shut down for short periods or was operating erratically, these lines tended to plug. In other words, the unit was "unforgiving." Modifications were required to provide more heat for these lines, to relocate instrument taps, and to flush instrument lines with solid-free solvent. A commercial plant would be less critical because of the larger size (relatively less heat loss), but these lessons should be kept clearly in focus when designing such a plant.

## 10.8. EDS DESIGN CONSIDERATIONS—PROCESS

Process issues concerned the liquefaction system, distillation, solvent hydrogenation, furnace coking, process control, and interactions of the process steps.

In liquefaction, the hydrodynamics of the reactor must be understood to predict gas and liquid holdup and, thus, conversion. Liquid holdup differed significantly from predictions, and adjustments in operating ECLP were required.

In distillation, it was necessary to provide the right boiling range for the recycle solvent, keep solids out of the solvent (since these would plug the solvent-hydrogenation unit), and remove the liquids from the coal-liquefaction bottoms. This system generally worked as expected.

In solvent hydrogenation, the goal was to maximize hydroaromatics and minimize cycloparaffins. Condensed aromatics react with hydrogen to form hydroaromatics, which can subsequently react with hydrogen to form cycloparaffins. Hydroaromatics are the "donor" compounds that provide hydrogen in the liquefaction step, thereby forming condensed aromatics. Cycloparaffins or saturates dilute the donor constituents and may promote coke formation in the slurry preheat furnace. They do not donate hydrogen in the liquefaction step.

| Condensed aromatics | Hydroaromatics (Hydrogen-donor compounds) | Naphthenes |

With furnaces, the concern was coking of the coal slurry and subsequent plugging. This turned out to be a key problem.

Finally, process control is always essential, and in this case an understanding of all the interactions among process steps was important to achieving adequate control. The ECLP experience demonstrated the importance of this issue.

## 10.9. INITIAL EDS COAL-LIQUEFACTION PILOT PLANT (ECLP) OPERATION

ECLP was successfully started in 1980 and initially achieved a coal-in factor (percentage of calendar time in which coal was fed to the unit) of about 50%. This was very close to expectations. Initially, emphasis was placed on mechanical problems to improve plant reliability. These problems included breakdown of solid conveyors and the gate-lock coal feeder, plugging of the coal-slurry heat exchangers, packing life of slurry plunger pumps, erosion in the vacuum-tower transfer line, and plugging of lines containing liquefaction bottoms. As mechanical operability improved in late 1980 and early 1981, the coal-in factor increased to 70–80%, indicating that the staff of the pilot plant was successfully overcoming mechanical problems which existed in the operation (Figure 10.4).

In the first half of 1981, it became apparent that, while the mechanical operation of the unit was satisfactory, three process issues required resolution before scheduled termination of the operation in 1982. First, coking in the preheat furnace for the coal-liquefaction slurry had to be reduced. With the frequency of coking experienced, satisfactory design and operation of a future commercial-sized furnace could not be achieved.

Second, solvent quality needed to be improved and maintained. In smaller units, the solvent-quality index (a measure of the availability of hydrogen from the solvent) had been consistently maintained at target levels, but in ECLP it was often

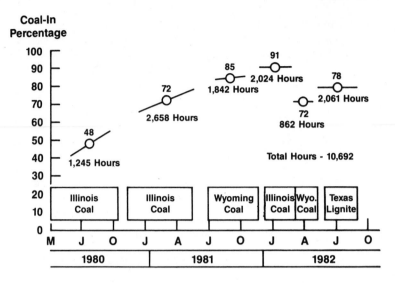

**FIGURE 10.4** ECLP operation steadily improved (Cavanaugh et al., 1984). Reprinted with permission from McGraw-Hill Company.

as low as 75% of the target levels, and was not consistently maintained. Poor solvent quality affected coking in the slurry preheat furnace (Figure 10.5) and also the level of coal conversion. However, the magnitude of the effects and reasons for unsatisfactory solvent quality in ECLP were not fully understood.

Third, the conversion of coal to liquids and gas in ECLP was significantly below the levels expected in small pilot-plants. Under typical conditions, conversions were 5–10% less than expected; this was unacceptable because of the adverse effect on economics and the question raised regarding further scale-up. It was known that poor solvent quality would reduce conversion, but not to the extent observed. It was important to understand the reasons for the difference so that commercial-sized equipment could be designed with confidence.

At first, these problems appeared to be separable, but they proved to be interrelated. There were significant process interactions which, if unmanaged, led to furnace coking, poor solvent quality, and low coal conversion.

## 10.10.  PROCESS INTERACTIONS

Intensive efforts to interpret the data available led to the definition of important process interactions. In the coking area, studies showed that furnace coking is a complex function of coil-outlet temperature, heat flux, and solvent quality. It was possible to relate the important variables to the boundary between coking and non-coking, and thus to define a region of satisfactory operation. The analysis showed that the furnace had been operating in the coking region.

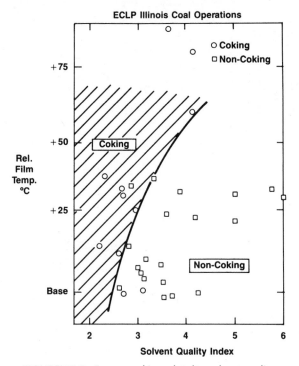

**FIGURE 10.5** Furnace coking related to solvent quality.

Solvent quality was affected by liquefaction severity (e.g., temperature) and solvent hydrogenation conditions (Figure 10.6). Liquefaction severity affected the rate of cracking (removal) of cycloparaffins from the solvent and solvent hydrogenation, the rate of formation of cycloparaffins. Thus it was possible to define liquefaction and solvent hydrogenation conditions leading to low levels of cycloparaffins and high levels of hydrogen-donor molecules.

Coal conversion in ECLP was found to be lower than expected from small pilot-plant and cold-model studies because of decreased liquid residence time resulting from higher gas holdup in the liquefaction reactor (Figure 10.7). Tracer studies that led to this conclusion provide a scale-up basis for commercial-sized reactors (Tarmy et al., 1984). It is not surprising that the primary ways identified for increasing coal conversion were longer residence time, higher temperature, and improved solvent quality. However, residence time could not be increased readily at ECLP because all four reactors were being used. Hence the options available were raising temperature and increasing solvent quality.

The analysis showed that the important independent variables were furnace heat flux and solvent hydrogenation conditions. With lower heat flux, it would be possible to operate with higher coil-outlet and liquefaction temperatures without coking. This higher temperature, coupled with proper solvent hydrogenation conditions, would lead to improved solvent quality and coal conversion.

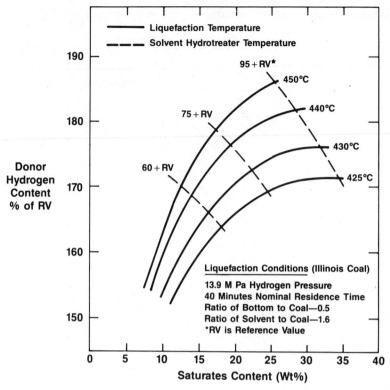

**FIGURE 10.6** Solvent quality influenced by solvent hydrogenation and liquefaction conditioning.

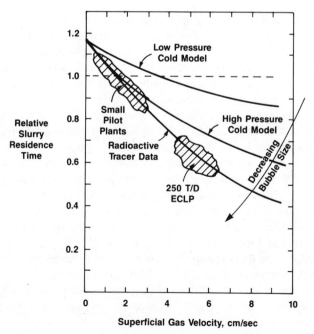

**FIGURE 10.7** Variation in slurry residence time in liquefaction reactor.

## 10.11. CHANGES EFFECTED

The slurry preheat furnace was modified for operation at a low heat flux (Figure 10.8) by adding an insulating curtain in the fire box. As a result, the higher temperature zone (where coking was occurring) was fired at a lower rate than the entrance zone. This resulted in a film-temperature reduction of about 25°C at a given bulk-fluid temperature in the higher temperature zone and permitted operation at a higher coil-outlet temperature (Figure 10.9). Thus, with the modified furnace, it was possible to increase the coil-outlet temperature without entering the coking region. The resulting increase in liquefaction temperature helped improve solvent quality and further reduce the potential for coke formation.

Solvent quality was also improved by coupling solvent hydrogenation conditions with liquefaction severity. This coupling improved process response time and minimized the concentration of cycloparaffins, as noted earlier.

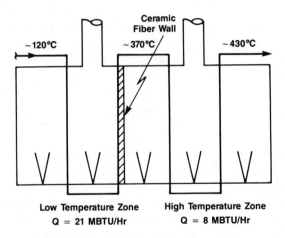

**FIGURE 10.8**  ECLP slurry preheater modifications.

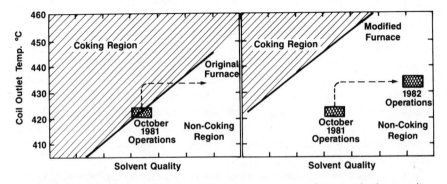

**FIGURE 10.9**  Modified furnace allowed higher temperatures and improved solvent quality.

With an Illinois No. 6 coal, it was possible to raise the average liquefaction temperature from 440 to 455°C. Combined with close control of solvent hydrogenation severity, this was an important ingredient in obtaining good solvent quality.

## 10.12. RESULTS

Satisfactory operations were achieved on an Illinois No. 6 bituminous coal, Wyoming subbituminous coal, and Texas lignite containing up to 23% mineral matter. ECLP operation was coke-free from start-up on November 28, 1981 until the end of operation on August 20, 1982. Solvent quality was maintained at a satisfactory level, and coal conversion was increased from 55 to over 60%. The reasons for differences in conversion between ECLP and smaller units are understood and can be compensated for in a commercial design. This understanding is based on the reactor tracer studies at ECLP and the model that was developed (Tarmy et al., 1984).

The data needed to design the liquefaction section of a commercial-sized plant were obtained, and the major constraints influencing the "operating window" were defined. Overall, the project was a success, and the objectives of the program were met.

## 10.13. FUTURE CHALLENGES

In the future, it should be possible to shorten and simplify the programs required to obtain design bases for pioneer commercial plants. The keys will be better engineering design and scale-up tools. The EDS development would have been facilitated by better design tools for solid-handling equipment, better scale-up tools for the process furnace and liquefaction reactor, and better understanding of the interactions among process steps. In addition to illustrating these needs, the EDS program provided important data with which to improve these tools. The knowledge base should continue to increase as the underlying chemistry and physics are better understood.

It is well-known that solid handling introduces significant risks in plant design and reduces operational reliability. The principal design problem at ECLP involved the criteria for satisfactory slurry pumping at ambient and elevated temperatures. Reliability issues related to all aspects of solid handling, including transport, size reduction, drying, dust control, and slurry handling. Additional research is needed to develop improved design tools for handling solids and liquid–solid mixtures, particularly at high temperatures with high reliability. Current programs in this area may be inadequate, and deserve additional attention (Merrow, 1985).

There is reason to be optimistic about future process-design capability in light of current trends. Much larger-scale simulations are now possible because of the increase in computing capacity and improved mathematical methods. More detailed models continue to be developed for use in these simulations. Optimization techniques with both process and mechanical features are evolving. Improved equipment-cost models, including piping and installation costs, are being developed and will enhance the reliability of these optimization techniques for predicting real-world performance. Already, sensitivity analysis techniques can be more easily used to focus on major economic and reliability issues. In addition, capabilities are improving for the dynamic simulation of large-scale systems that include utility demands, system efficiency changes through catalyst deactivation, equipment fouling, and partial-load operation.

The problems in EDS with furnace coking and liquefaction reactor performance illustrate the need for improved models that include hydrodynamics and reaction kinetics. For process furnaces, it is important to relate coke formation to flow conditions, heat transfer, and coking kinetics. For reactors, it is important to relate conversion and product selectivity to hydrodynamics and reaction kinetics. In addition, EDS clearly illustrates the need to develop overall simulations to anticipate and understand process interactions. The significant progress in modeling during the EDS program shows that these challenges can be met, but they involve additional effort.

In the next 10–15 yr, improved tools to assist the process-design engineer are likely to be developed. Large-scale database managers are evolving to facilitate the sorting and storage of physical measurements from the laboratory, design calculations, and optimization studies. When connected to models of various levels of sophistication and to reporting/graphical tools, these will give the designer an ever-increasing capability to evaluate large-scale designs. Artificial-intelligence research is leading to the development of knowledge-based managers (i.e., expert systems) in parallel with database managers (see, for example, Monarch and Carbonell, 1987). Used together, these major advances should make the design engineer more productive and should result in more economic and operable designs. These new tools will be particularly important in the development of synthetic fuels.

The challenges in process simulation should not be underestimated. As capabilities improve, it will be important to address more obscure—but important—problems, such as foaming, fouling, erosion, and corrosion. The task will not be completed until all of the physical and chemical phenomena have been simulated reliably.

Plant-design capabilities have improved dramatically since Vilbrandt's pioneering efforts 50 yr ago (Vilbrandt, 1934), and this trend should continue in view of the incentives. Improvements will become increasingly feasible as computer capabilities and the knowledge base grow. One of the challenges for chemical engineers will be to provide the integration needed to extend the fundamentals in Vilbrandt's *Chemical Engineering Plant Design* (Vilbrandt and Dreyden, 1959) to include both relevant new information and the elements of the engineering disciplines that pertain to plant design, namely chemical, mechanical, civil, and electrical engineering.

## ACKNOWLEDGMENT

The author acknowledges the contributions of J. L. Robertson and A. J. Callegari to this manuscript, particularly in Section 10.13.

## REFERENCES

Cavanaugh, T. A., W. R. Epperly, and D. T. Wade, "EDS Coal Liquefaction on Process," in *McGraw-Hill Handbook of Synfuels Technology*, Robert A. Meyer (ed.), McGraw-Hill, New York, Chap. 1-1, 1984.

Epperly, W. R., "Running Government-Industry Projects," *ChemTech*, **11**, 220 (1981).

Epperly, W. R. and co-workers, EDS Coal Liquefaction Project Technical Progress Reports prepared for the U.S. Department of Energy: Final, FE-2353-20, issued 3/15/78 in two volumes; Interim, FE-2893-16, Summary of EDS Predevelopment, issued 7/31/78; Annual, Final, FE-2893-66, Construction of Exxon Coal Liquefaction Pilot plant, issued 4/17/81; Final FE-2893-121m, Vol. 1—Laboratory Research and Development and ECLP Project Management, Procurement, Construction, and Operations, issued 10/31/83; Final, FE-2893-128, Vol. 2—Engineering Research and Development, issued 8/30/84; Annual, FE-2893-143, July 1983–June 1984; Quarterly, FE-2893-155, January 1985–March 1985.

Epperly, W. R., G. C. Lahn, R. E. Payne, and S. Zaczepinski, "Development of EDS Coal Liquefaction Process—The Final Step." 1983 Distinguished Alumni Lectures in Chemical Engineering, Virginia Polytechnic Institute & State University, Blacksburg, VA, November 1983.

Lowery, H. H. and H. J. Rose, "Pott–Broche Coal Extraction Process and Plant of Ruchrol G.m.b.h. Bottrop—Welheim, Germany," U.S. Bureau of Mines Information Circular 7420, 1947.

Merrow, E. W., "Linking R&D to Problems Experienced in Solids Processing," *Chem. Eng. Prog.*, **81** (5), 14 (1985).

National Research Council, "Safety Issues Related to Synthetic Fuels Facilities," NRC report, National Academy Press, Washington, D.C., 1982.

Monarch, I. and J. Carbonell, "CoalSORT: A Knowledge-Based Interface," *IEEE Expert*, **2**, 39 (1987).

Office of Technology Assessment, "Oil and Gas Technologies for the Arctic and Deepwater," United States Congress, Washington, D.C., May 1985.

Sherwood, P. W., "High-Pressure Hydrogenation of Carbonaceous Matter Part II. Primary Hydrogenation in the Liquid Phase," *Pet. Refiner*, **29**, 119 (1950).

Swabb, L. E., Jr., "Liquid Fuels from Coal: From R&D to an Industry," *Science*, **199**, 619 (1979).

Tarmy, B. L., M. Chang, C. A. Coulaloglou, and P. Ponzi, "The Three-Phase Hydrodynamic Characteristics of the EDS Coal Liquefaction Reactors: Their Development and Use in Reactor Scaleup," Eighth International Symposium on Chemical Reactor Engineering, Edinburgh, Scotland, September 10–13, 1984.

Vick, G. K. and W. R. Epperly, "Status of the Development of EDS Coal Liquefaction," *Science*, **217**, 311 (1982).

Vilbrandt, F. C., *Chemical Engineering Plant Design*, 1st ed., McGraw-Hill, New York, 1934.

Vilbrandt, F. C. and C. E. Dryden, *Chemical Engineering Plant Design*, McGraw-Hill, New York, 1959.

Wu, W. R. K. and H. H. Storch, "Hydrogenaton of Coal and Tar," U.S. Department of the Interior, Bureau of Mines, Bulletin 633, 1968.

## FURTHER READING

Allned, V. Dean, *Oil Shale Processing Technology*, The Center for Professional Advancement, East Brunswick, NJ, 1982.

Bisio, Attilio and Robert L. Kabel (eds.), *Scaleup of Chemical Processes*, Wiley, New York, 1985.

*Coal Conversion Systems Technical Data Book*, 4 Vols., Institute of Gas Technology, Washington, D.C., U.S. Government Printing Office, 1978–1984.

Elliott, M., (ed.), *Chemistry of Coal Utilization*, McGraw-Hill, New York, 1981.

Epperly, W. R., "Catalytic Coal Gasification," and "Coal Liquefaction," in *McGraw-Hill Encyclopedia of Science and Technology*, Sybil P. Parker (ed.), 1987.

Hendrickson, T. A., *Synthetic Fuels Data Handbook*, Cameron Engineers, Inc., Denver, Colorado, 1975.

Meyers, Robert A., (ed.), *Handbook of Synfuels Technology*, McGraw-Hill, New York, 1984.

Pelofsky, Arnold H., *Coal Conversion Technology*, ACS Symposium Series, No. 110, American Chemical Society, Washington, D.C., 1979.

Probstein, Ronald F. and R. Edwin Hicks, *Synthetic Fuels*, McGraw-Hill, New York, 1982.

Shah, Y. T., *Reaction Engineering in Direct Coal Liquefaction*, Addison Wesley, Reading, MA, 1981.

Tsonopoulos, C., J. L. Heidman, and S. C. Hwang, *Thermodynamic and Transport Properties of Coal Liquids*, An Exxon Monograph, Wiley, New York, 1986.

Whitehurst, D. D., T. O. Mitchell, and M. Farcasiu, *Coal Liquefaction*, Academic Press, New York, 1980.

Wilson, Carroll L., *Coal-Bridge to the Future*, Ballinger, Cambridge, MA, 1980.

# 11

## PROCESS AND PLANT DESIGN FOR FOOD AND BIOCHEMICAL PRODUCTION

HENRY G. SCHWARTZBERG

## 11.1. THE FOOD AND DRUG INDUSTRIES

### A. Introduction

In terms of value of product shipped, the food industry is the largest industry in the United States based on Standard Industrial Code categories (U.S. Bureau of the Census, 1984a). The drug industry, which also involves processing materials of biological origin and biochemicals, is much smaller; but it is more active in developing new technology and its profitability is much higher. Data relating to processing and production categories for the food and drug industries are listed in Table 11.1. For comparison purposes, similar data are listed for petroleum refining and for the chemical industry as a whole.

Other industries such as tobacco (value of shipments: 16.0 billion dollars), paper (value of shipments: 79.7 billion dollars), and leather (value of shipments: 9.7

**TABLE 11.1  Value of Shipments, Cost of Materials and Labor for Food Industry Production, and Process Categories for the Drug, Chemical, and Petroleum-Refining Industries (1982 Data)**[a,b]

| Category | Cost of Materials | Labor Costs | Value Added by Manufacture | Value of Shipments | Capital Expenditures |
|---|---|---|---|---|---|
| Food and kindred products | 192,117 (68.4%)[c] | 26,074 | 88,844 (31.6%)[c] | 280,961 | 6,807 (2.4%)[c] |
| Meat products | 56,693 (83.4%) | 4,990 | 11,309 (16.6%) | 68,002 | 893 (1.3%) |
| Dairy products | 30,494 (78.4%) | 2,595 | 8,384 (21.6%) | 38,878 | 713 (2.3%) |
| Preserved fruit and vegetables | 17,580 (58.9%) | 3,317 | 12,276 (41.1%) | 29,856 | 943 (3.0%) |
| Grain-mill products | 20,953 (66.8%) | 2,182 | 10,406 (33.2%) | 31,359 | 884 (2.8%) |
| Bakery products | 7,104 (40.1%) | 4,032 | 10,597 (59.9%) | 16,037 | 474 (3.0%) |
| Sugar and confections | 9,438 (60.2%) | 1,667 | 6,233 (39.8%) | 15,671 | 471 (3.0%) |
| Fats and oils | 13,932 (82.9%) | 774 | 2,871 (17.1%) | 16,803 | 394 (2.3%) |
| Beverages | 22,146 (57.1%) | 4,237 | 16,641 (42.9%) | 38,787 | 1,513 (3.9%) |
| Miscellaneous | 13,778 (57.6%) | 2,279 | 10,127 (42.4%) | 23,905 | 523 (2.2%) |
| Drugs | 7,756 (31.3%) | 3,396 | 16,951 (68.7%) | 24,707 | 1,249 (5.1%) |
| Biological products | 875 (38.7%) | 440 | 1,384 (61.3%) | 2,259 | 98 (4.3%) |
| Bulk medicines and botanicals | 1,330 (38.1%) | 462 | 2,165 (61.9%) | 3,495 | 284 (8.1%) |
| Pharmaceuticals | 5,551 (29.1%) | 2,063 | 13,509 (70.9%) | 19,060 | 868 (4.6%) |
| Chemicals | 91,966 (54.1%) | 20,675 | 78,118 (45.9%) | 170,085 | 9,080 (5.3%) |
| Petroleum refining | 178,790 (89.3%) | 3,414 | 21,405 (10.7%) | 200,195 | 6,412 (3.2%) |

[a]Data obtained or calculated from the Statistical Abstract of the United States (U.S. Bureau of the Census, 1984a), Annual Survey of Manufactures, 1980–1981 (U.S. Bureau of the Census, 1984b), and Predicasts Inc. (1984).

[b]Monetary figures are in millions of dollars.

[c]As a percentage of value of shipments.

billion dollars) also process biological materials; but specialized aspects of process and plant design for these industries are quite different than for the food and drug industries. Other industry components might be lumped with the chemical industry, for example, the paper and rubber and miscellaneous plastics industry (value of shipments: 55.3 billion dollars).

Table 11.2 provides economic data for the food and drug industries, for sources of supply for the food industry, and for retail and other outlets for food and drug products.

## B. Overall Processing and Economic Considerations

The types of products produced by food plants are specified in Table 11.1. The large ratio of the cost of materials to value of shipments indicates that, in many cases, food-processing economics are highly sensitive to product yield and production-induced yield losses, and less sensitive to equipment cost and equipment-cost minimization. Therefore, even relatively costly processes that provide significant yield or quality improvement have quickly displaced older methods of manufacture. Older processes have also been replaced by new methods that provide significant labor or energy savings.

New food products often require new and different types of plants. Products designed to provide greater convenience of preparation, and those that require more complex packaging, are entering the market with great frequency. Americans tend to eat out more frequently; consequently, processors have developed many products for the restaurant, fast-food, and institutional trades. They have also entered these businesses themselves, and have devised entree and meal-like products to compete with restaurants. Products designed to provide improved nutrition and promote better health are also entering the market at an increasing rate. Equipment and plant construction costs are particularly important for new processes and products because they greatly affect the amount of money at risk when products are introduced in markets where consumer acceptance cannot be predicted with great accuracy.

Food production is in a state of flux because of major changes occurring in food preferences. Poultry is replacing red meat. Less sucrose is being used, but more corn-syrup-based sweeteners, particularly high-fructose corn syrup, are being utilized. Though unbranded, low-cost, generic food products have been introduced by some major food retailers in recent years, branded products still dominate the retail market. The shares of the market for some branded products are 99.3% for baby foods, 97% for ready-to-eat cereals, 89.3% for canned soup, and at the low end of the range 45.4% for granulated sugar and 46% for frozen orange juice (Connors et al., 1985).

Pharmaceuticals often have a limited patent-protected life. Because of this, the need to ensure product purity, safety, integrity, and effectiveness and the need to use government registered or approved processes, expensive but highly reliable processes are frequently employed instead of less costly alternatives which might have to be modified or debugged.

**TABLE 11.2  1982 Economic Data for the Food and Drug Industries, Food Industry Supply Sources, and Outlets for Food and Drugs[a,b]**

| | |
|---|---:|
| Total number of U.S. food-manufacturing establishments | 21,316 |
| Those with 20 or more employees | 10,638 |
| Total value of shipments by food manufacturers | 280,961 |
| Shipments by top 5 companies | 35,000 |
| Shipments by top 10 companies | 59,200 |
| Shipments by top 100 companies | 173,500 |
| Total assets, food companies | 48,600 |
| Profits after taxes, food companies | 8,383 |
| Percent of sales | 3.0% |
| Percent of assets | 17.2% |
| Total number of U.S. drug-manufacturing establishments | 1,280 |
| Those with 20 or more employees | 592 |
| Total value of shipments by drug manufacturers | 24,707 |
| Total assets, drug companies | 7,309 |
| Profits after taxes, drug companies | 3,261 |
| Percent of sales | 13.2% |
| Percent of assets | 44.6% |
| U.S. agricultural products | |
| Value of animal feed used directly without sale | 16,200 |
| Sales of agricultural products | 144,762 |
| Crops | 74,623 |
| Crops not used for food or feed | 8,290 |
| Crops used for animal feed | 18,274 |
| Food Crops | 48,059 |
| Food grains | 11,548 |
| Vegetables | 8,233 |
| Fruit and nuts | 6,726 |
| Oil seeds | 13,961 |
| Other | 7,760 |
| Livestock and products derived from livestock | 70,139 |
| Poultry | 6,126 |
| Eggs | 3,437 |
| Dairy products | 18,273 |
| Cattle and calves | 29,906 |
| Hogs and pigs | 10,586 |
| Sheep and lambs | 447 |
| Other livestock | 1,620 |
| Domestically caught fish and shellfish | 2,586 |
| Imported food and live animals | 15,412 |
| Meat and meat preparations | 2,034 |
| Fish and shellfish | 3,594 |
| Vegetables and fruit | 2,920 |
| Coffee | 2,590 |
| Sugar | 1,047 |

**TABLE 11.2** *(Continued)*

| | |
|---|---|
| Cocoa | 349 |
| Tea | 132 |
| Agricultural exports including processed foods | 36,098 |
| Processed food exports | 11,079 |
| Energy consumption | |
| Food industry (919 trillion BTU or 969 trillion kJ) | 3,731 |
| Chemical industry (2630 trillion BTU or 2775 trillion kJ) | 10,678 |
| Packaging for foods | 23,200 |
| Food advertising | 23,900 |
| Retail Sales | |
| Total food and beverage expenditures by consumers | 394,800 |
| Wholesale value of food and beverages sold | 251,500 |
| Food stores | 259,400 |
| Eating and drinking places | 115,700 |
| Liquor stores | 19,700 |
| Drug stores | 38,800 |
| Drugs and pharmaceuticals (1983) | 30,080 |
| Research expenditures | |
| Food industry | 752 |
| Chemical and allied products (excluding drugs) | 5,901 |
| Drugs (1984) | 3,250 |
| Employment (thousands of workers) | |
| Food industry | 1,494 |
| Farms | 2,815 |
| Fishing | 216 |
| Food stores | 2,560 |
| Eating and drinking places | 5,159 |
| Chemical industry (including drug industry) | 866 |
| Drug industry | 166 |
| Petroleum refining | 151 |

[a]Data obtained or calculated from the Statistical Abstract of the United States (U.S. Bureau of the Census, 1984a), Annual Survey of Manufactures, 1980–81 (U.S. Bureau of the Census, 1984b), and Predicasts Inc. (1984).

[b]Monetary values are in millions of dollars.

More detailed food- and drug-product breakdowns are provided by the Annual Survey of Manufactures (U.S. Bureau of the Census, 1984b) and the *Predicast's Basebook* (Predicasts, Inc. 1984). Used in conjunction with processing flow sheets, these compilations are helpful for identifying food- and drug-processing unit operations and their frequency of use. Connors et al. (1985) analyzed the structure of the food-processing industry and economic trends in the industry from the points of view of four economists and provided much valuable economic data. Food products can be divided into the following broad categories, which are related to potential profitability.

## 11.2.  FOOD-PRODUCTION CATEGORIES

### A.  Commodities

These products have a relatively standard identity and are frequently used in other food products and processes or as animal feeds. They are sometimes bought and sold in commodity exchanges. They often command low markups and profits relative to raw-material costs. Examples include wheat flour, cleaned and polished rice, sucrose, corn-syrup solids, defatted soy meal, fluid whole milk, spray-dried skim milk, and vegetable oils. An extremely wide range of processes is used in producing these commodities. Flowsheets for some of these products are shown in Figures 11.1 to 11.4. Even though the profits made by producing food commodities are low, the plants involved are among the largest and most complex in the food industry. Detailed technical data and flowsheets for plants producing cane sugar are provided by Meade and Chen (1977) and Hugot (1972). Similar information for beet-sugar production is provided by Autorenkollektiv (1980), Schneider (1968), and McGinnis (1982). Swern (1982) presents comparable data for vegetable-oil production. Mechanistic methods for analyzing grain milling are described by Kuprits (1967).

### B.  Meat, Poultry, and Fish

Characteristic operations involved in processing these products are slaughtering, carcass disassembly, evisceration, bleeding, portion-size reduction, freezing, and extraction of enzymes and hormones from organs. A flowsheet for a slaughterhouse is shown in Figure 11.5. Though many publications deal with meat science or meat cutting (American Meat Institute, 1960; Forrest et al. 1975; Price and Schweigert, 1971), there is little coverage of slaughterhouse design. Karmas (1970b) provides some information about fresh meat processing.

Depending on how one defines food processing, the meat, poultry, and fish category might also include animal, milk, poultry, and egg production in automated, animal housing facilities and fish production by mariculture. Such operations are macroscopic analogs of systems that use microorganisms in controlled environments to produce foods, food ingredients, and pharmaceuticals.

### C.  Produce

These products are marketed in relatively unmodified form and their processing is limited to washing and cleaning, grading, inspection, and packaging. They usually have a relatively short shelf life, but it may be significantly enhanced by storage under refrigeration and/or in controlled atmospheres. A flowsheet for the processing of potatoes that are to be stored is shown in Figure 11.6. For more information on such products, see Cruess (1958), Peleg (1985), Kochesperger (1978), Ryall and Lipton (1979), and Ryall and Pentzer (1982). Engineering analysis of these operations is provided by Peleg (1985). Appropriate storage and transportation

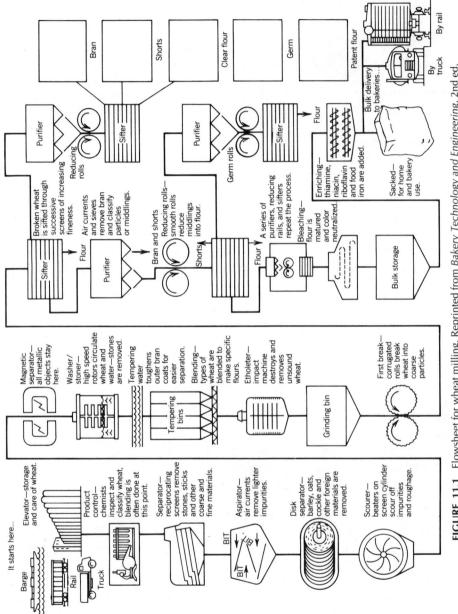

**FIGURE 11.1** Flowsheet for wheat milling. Reprinted from *Bakery Technology and Engineering*, 2nd ed. by S. A. Matz, Copyright © 1972 by AVI, all rights reserved.

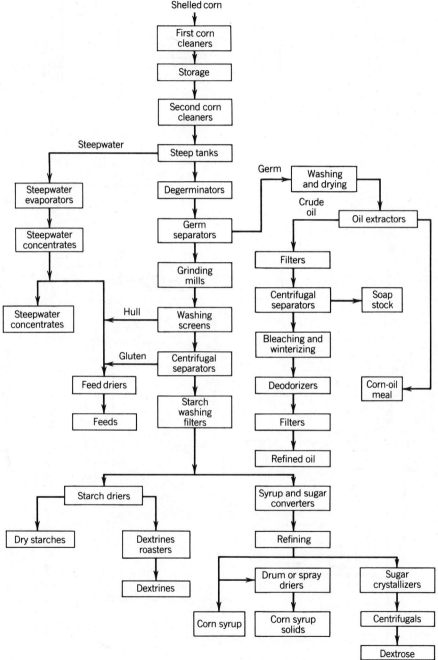

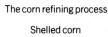

**FIGURE 11.2** Flowsheet for wet-corn milling and refining (McNicol et al., 1972; courtesy of Corn Refiners Association).

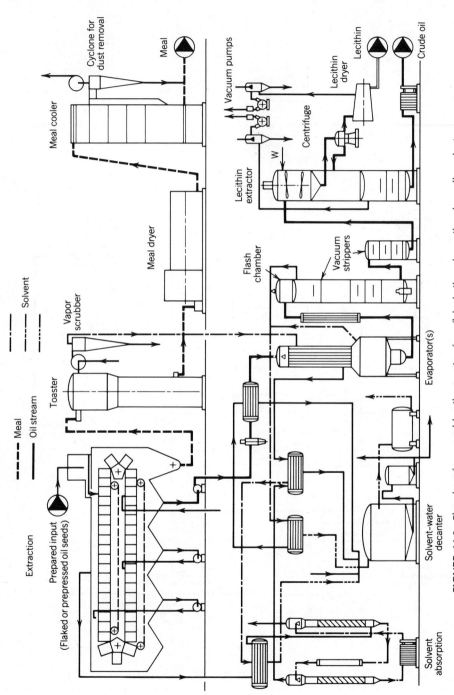

**FIGURE 11.3** Flowsheet for vegetable-oil extraction from flaked oil seeds or oilseed expeller cake (courtesy of Lurgi GmbH, 1977).

Extraction

Prepared input
(Flaked or prepressed oil seeds)

- - - Meal
Oil stream

Solvent

Toaster

Vapor scrubber

Meal dryer

Meal cooler

Cyclone for dust removal

Meal

Flash chamber

Vacuum strippers

Evaporator(s)

Solvent-water decanter

Solvent absorption

Vacuum pumps

Lecithin extractor

W

Centrifuge

Lecithin dryer

Lecithin

Crude oil

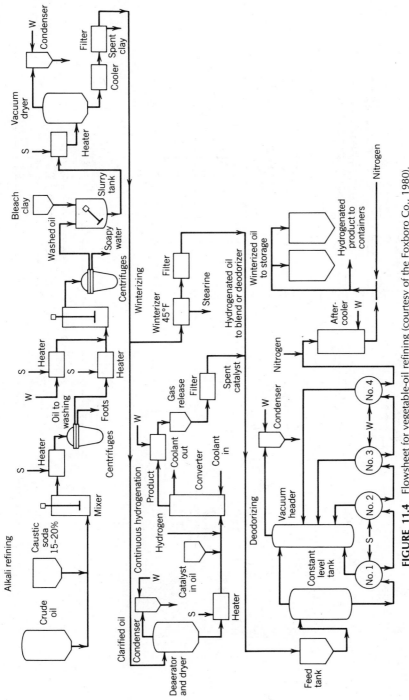

**FIGURE 11.4** Flowsheet for vegetable-oil refining (courtesy of the Foxboro Co., 1980).

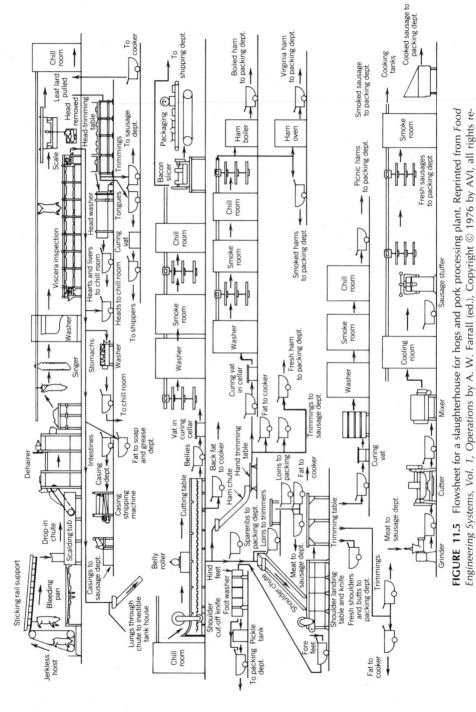

**FIGURE 11.5** Flowsheet for a slaughterhouse for hogs and pork processing plant. Reprinted from *Food Engineering Systems, Vol. 1, Operations* by A. W. Farrall (ed.), Copyright © 1976 by AVI, all rights reserved.

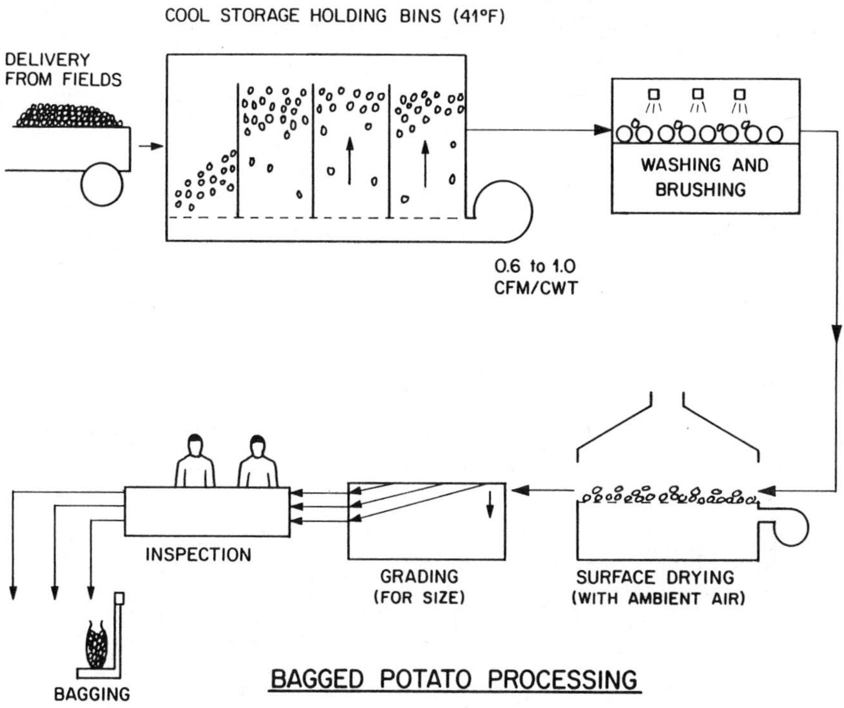

COOL STORAGE HOLDING BINS (41°F)

DELIVERY
FROM FIELDS

WASHING AND
BRUSHING

0.6 to 1.0
CFM/CWT

INSPECTION

GRADING
(FOR SIZE)

SURFACE DRYING
(WITH AMBIENT AIR)

BAGGING

**BAGGED POTATO PROCESSING**

**FIGURE 11.6**  Flowsheet for processing stored potatoes.

conditions are specified by ASHRAE (1982), the International Institute of Refrigeration (1963, 1967), Lutz and Hardenberg (1968), Patchen (1971), Redit and Hamer (1961), Ashby (1970), and the Refrigeration Research Foundation (1985).

## D.  Treated Produce

These products have been processed to enhance storage stability so they can be safely stored for periods of up to 1 or 2 yr without excessive deterioration of sensory quality for functionality. Such products are usually more processed versions of items in the previous category. They include canned, frozen, dried, and pickled fruits and vegetables, and juices derived from such fruits and vegetables. Size reduction and cutting, coring, peeling, stem removal, blending, and precooking are often carried out to increase convenience of use or to facilitate packaging, in addition to steps designed primarily to promote storage stability, such as pasteurization, sterilization, blanching, addition of salt or sugar, or infusion to osmotically inhibit microorganism growth and fermentation designed to achieve pH reduction. Even such minimal processing often changes sensory properties. Flowsheets for a typical vegetable-canning process and for an alternative freezing process are shown in Figure 11.7. Detailed informtion about the production of canned

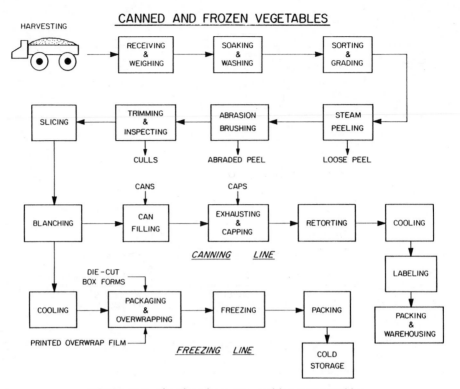

**FIGURE 11.7** Flowsheet for canning and freezing vegetables.

fruits and vegetables is provided by Lopez (1981); more general descriptions are provided by Potter (1978), Desrosier (1977), and Hall et al. (1971). Raw-material characteristics are covered succinctly in Considine and Considine (1982).

## E.  Transformed Products

These are products whose form or nature and sensory properties have been changed substantially by processing. They include cheeses, yogurt, tofu, ice cream, sausages, liquid and dried beverage and flavoring extracts, canned and bottled beverages derived from these extracts, breakfast cereals, gelatin, vegetable gums, extrusion-texturized vegetable protein, and pasta. Flowsheets for cheese, yogurt, soluble coffee, and past production are shown in Figures 11.8 to 11.11. The group also includes items such as vegetable oil, sucrose, high-fructose corn syrup, cured meats, and pickled vegetables, which also fall into other categories in this list. Detailed information about the production of cheeses and fermented milk products is provided by Kosikowski (1977). Much useful engineering design information relating to milk-based products is provided by Kessler (1981). Sausage processing

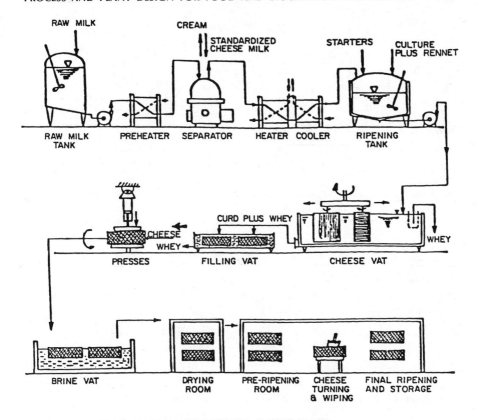

FLOW DIAGRAM OF PRESSED CHEESE TYPES

**FIGURE 11.8**  Flowsheet for manufacturing cheese. Reprinted from *Food Engineering and Dairy Technology* (1981) by H. G. Kessler, with permission from Verlag A. Kessler.

is covered by Karmas (1970b) and Pearson (1984), ice-cream production by Arbuckle (1977), and coffee extraction by Sivetz and Desrosier (1979).

## F.  Fabricated and Assembled Foods

These products use combinations of ingredients or components. They include bread, cakes and other baked items, filled candies, frozen and canned entrees and meals, packaged salads and appetizers, and sauces. A flowsheet for bread production is shown in Figure 11.12. The borderline between this group and the trans-formed-product category is not always clear, but the transformed products usually are based on one or two main ingredients and the assembled products involve many ingredients or components and a more complex sequence of processing steps. For more detailed information about baking see Matz (1972), for candy manufacture see Alikonis (1978), for extrusion-texturized protein see Altschul (1974), and for chocolate production see Cook (1972) and Zoumes and Finnegan (1979).

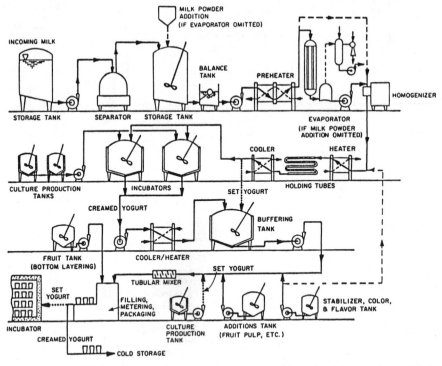

**FIGURE 11.9** Flowsheet for manufacturing yogurt. Reprinted from *Food Engineering and Dairy Technology* (1981) by H. G. Kessler, with permission from Verlag A. Kessler.

## G.  Fermented and Distilled Beverages

Flowsheets for beer and wine production are shown in Figures 11.13 and 11.14. For more information about the manufacture of fermented and distilled beverages, consult Amerine et al. (1980), Vine (1981), Hoyrup (1978), and Peppler and Perlman (1979).

## 11.3.  PHARMACEUTICAL PRODUCTS

The pharmaceutical industry produces biologically active synthetic drugs, synthesized food ingredients, micronutrients and flavoring agents, botanical and animal tissue extracts and chemically or physically modified forms of such extracts, fractionated and treated blood, antibiotics, and metabolites produced by microorganisms or derivatives of these products. Flowsheets for the production of citric acid and some antibiotics are shown in Figures 11.15 to 11.17. Fermentation processes for making pharmaceuticals are described in some detail in Peppler and Perlman (1979) and Prescott and Dunn (1982) and journals such as *Biotechnology and Bioengineering*, the *Journal of Bioengineering*, *Applied Chemistry and Biotech-*

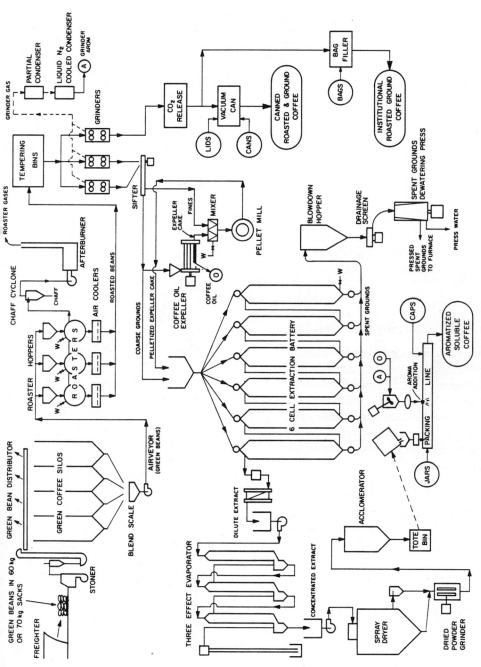

**FIGURE 11.10** Flowsheet for manufacturing roasted and ground coffee and soluble coffee.

**399**

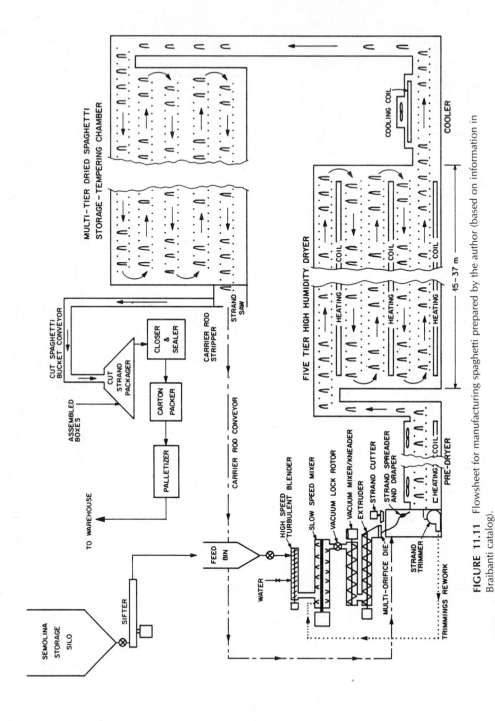

**FIGURE 11.11** Flowsheet for manufacturing spaghetti prepared by the author (based on information in Braibanti catalog).

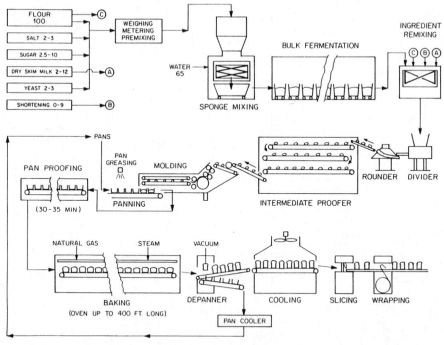

**FIGURE 11.12**    Flowsheet for the manufacture of bread. Reprinted from *Bakery Technology and Engineering*, 2nd ed. by S. A. Matz, Copyright © 1972 by AVI, all rights reserved.

*nology*, the *Journal of Fermentation*, and review series such as *Applied Microbiology, Applied Biochemistry and Microbiology, Progress in Industrial Microbiology*, and *Advances in Biochemical Engineering*.

## 11.4.  PROCESS AND PLANT DESIGN

What problems are peculiar to designing processes and plants for producing food and pharmaceutical products?

Process and plant designs specify:

- The equipment to be used.
- Performance requirements for that equipment.
- Interconnections and raw-material flows in terms of flow sheets and plant layouts.
- The placement of equipment, storage spaces, shop facilities, office spaces, delivery and shipping facilities, access ways on floor and site plans, and elevation drawings.

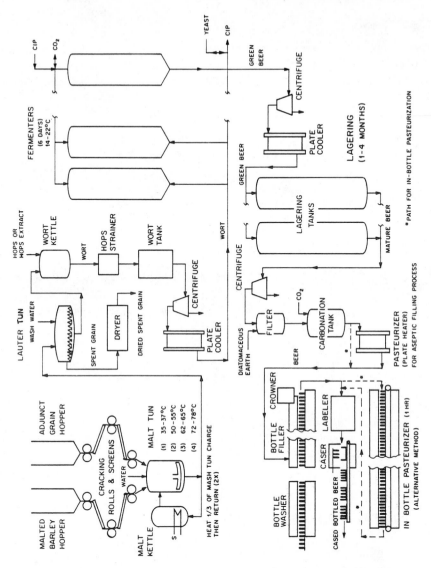

**FIGURE 11.13** Flowsheet for the manufacture of beer.

Modern wine making

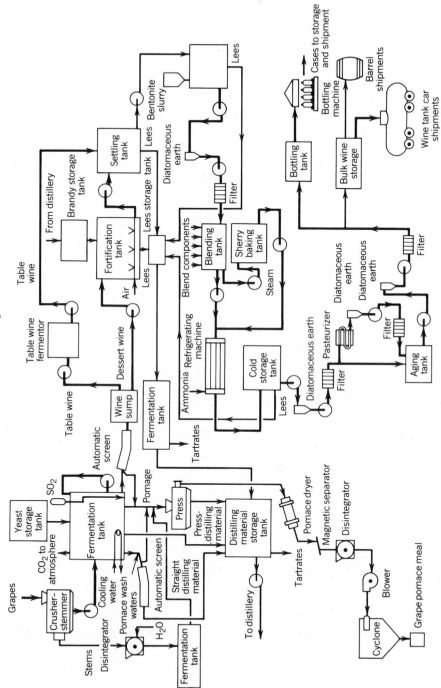

**FIGURE 11.14** Flowsheet for manufacturing wine (Hull, Kite, and Auerbach, 1951). Reprinted with permission from the American Chemical Society.

403

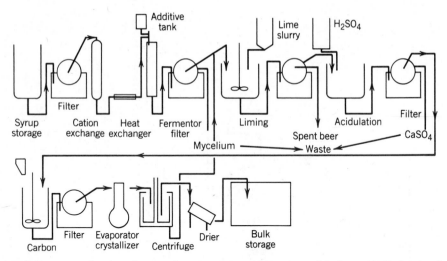

**FIGURE 11.15** Flowsheet for manufacturing citric acid (Peppler and Perlman, 1979). Reprinted with permission from Academic Press.

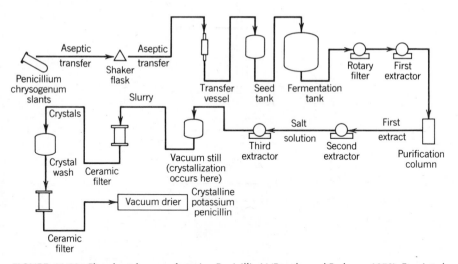

**FIGURE 11.16** Flowsheet for manufacturing Penicillin V (Peppler and Perlman, 1979). Reprinted with permission from Academic Press.

- Required instrumentation and controls, and processing monitoring and control interconnections.
- Utility and waste-treatment requirements, connections, and facilities.
- The rationale for site selection.
- The basis for selecting and sizing critical pieces of equipment.

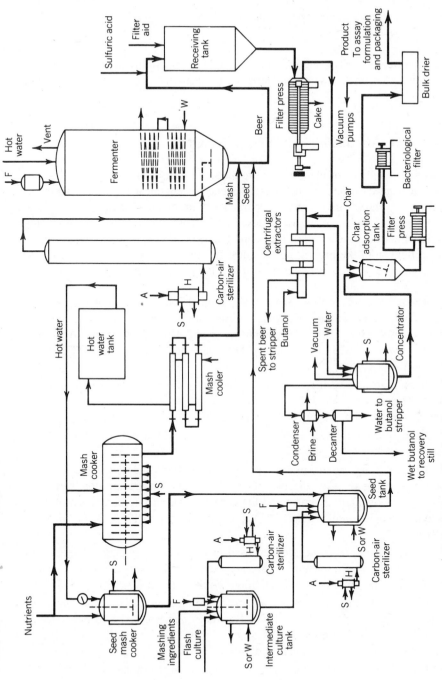

**FIGURE 11.17** Flowsheet for manufacturing Bacitracin (Inskeep et al., 1951). A, raw air from compressors; S, steam; W, cooling water; F, antifoam agent; H, heat exchanger. Reprinted with permission from the American Chemical Society.

- Ways in which the design was optimized and the engineering basis for such optimization.

They also often provide economic analyses of plant profitability in terms of various product demand, price, and raw-material cost scenarios.


## 11.5. SPECIAL DESIGN CONSIDERATIONS

Many of the elements of plant design are the same for plants processing food and biochemicals as they are for other plants designed by chemical engineers. But there are also many significant differences, particularly in the areas of equipment selection and sizing, and in working-space design. These differences stem from the ways in which the processing of foods and biochemicals, particularly pharmaceuticals, differs from the processing of industrial chemicals. For foods, processing and design differences occur because of the following considerations.

1. The storage life of foods is relatively limited and strongly affected by temperature, pH, water activity, maturity, prior history, and initial microbial contamination levels (Karel et al., 1975; Fennema et al., 1973; Stewart and Amerine, 1982; Duckworth, 1975; Desrosier and Desrosier, 1977; Desrosier, 1977; Hawthorn and Rolfe, 1968).

2. Very high and verifiable levels of product safety and sterility have to be provided (Karel et al., 1975; Desrosier and Desrosier, 1977; Tannenbaum, 1970; Lopez, 1981; Stumbo, 1973).

3. Foods are highly susceptible to microbial attack and insect and rodent infestation (Imholte, 1984; Troller, 1983).

4. Fermentations are used in producing various foods and biochemicals. Successful processing requires the use of conditions that ensure the dominance of desired strains of microorganism and the correct mode of microorganism growth or activity (Peppler and Perlman, 1979; Prescott and Dunn, 1982; Bailey and Ollis, 1986).

5. Enzyme-catalyzed processes are used or occur in many cases. These, like microbial growth and fermentation, are very sensitive to temperature, pH, water activity, and other environmental conditions (Acker, 1962; Karel et al., 1975).

6. Many foods are still living organisms or are biochemically active long after harvest or slaughter (Purr, 1966, 1970; Hawthorn and Rolfe, 1968; Fennema et al., 1973). Sucrose depletion versus time for sweet corn (see Figure 11.18) is typical of such activity (Appleman and Arthur, 1919).

7. Living foods generate heat and consume and transform their own substance at rates that are strongly temperature-dependent. Rates of heat generation as functions of temperature for various fruits and vegetables (ASHRAE, 1982) and for chickens (Longhouse et al., 1953) are shown in Figures 11.19 and 11.20.

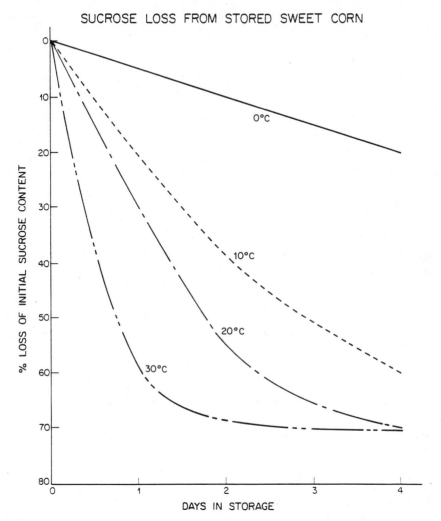

**FIGURE 11.18**  Sucrose depletion in sweet corn vs. time at different storage temperatures (Appleman and Arthur, 1919). Reprinted from *J. Agr. Res.*

8. In other cases, foods (e.g., ripening cheeses) contain active living microorganisms which induce chemical transformations for long periods of time (Kosikowski, 1977).

9. Crop-based food raw materials may only be available in usable form on a seasonal basis. Therefore plant design may involve the modeling of crop availability and growing processes (France and Thornley, 1984).

10. Food raw materials are highly variable and their variability is enhanced by the aging of raw materials and uncontrollable variations in climatic conditions.

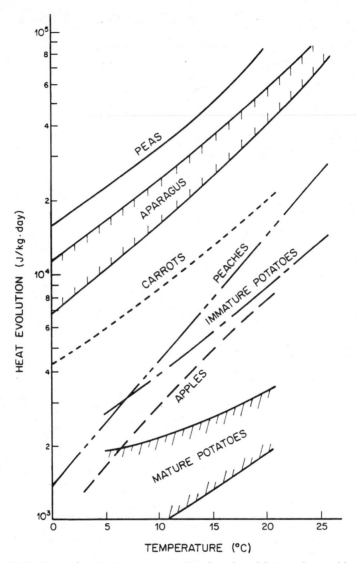

**FIGURE 11.19** Heats of respiration vs. temperature for selected fruits and vegetables (adapted from data published by ASHRAE).

Therefore, the use of statistical methods is frequently required when processing foods (Kramer and Twigg, 1970, 1973; Gould, 1977; Bender et al., 1982; Bozyk and Rudzki, 1972).

11. The biological and cellular nature and the structural complexity of foods cause special heat and mass-transfer and component-separation problems (Schubert and Heideker, 1985).

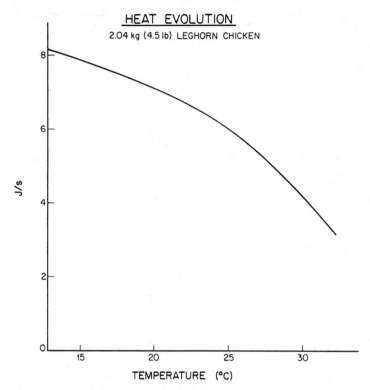

**FIGURE 11.20**   Heat production vs temperature for a 4.5-lb chicken (Longhouse et al., 1953). Reprinted with permission from *Agricultural Engineering*.

12. Foods are frequently solid. Heat- and mass-transfer problems in solids have to be treated in ways that are different from those used for liquid and gas streams (Schwartzberg, 1982, 1983; Bruin and Luyben, 1984). As shown in Figure 11.21, the kinetics of microbe and enzyme inactivation during thermally-induced sterilization and blanching and heat-transfer in the solids being sterilized or blanched are strongly linked (Lenz and Lund, 1977a–c; Stumbo, 1973; Thijssen et al., 1978; Thijssen and Kochen, 1980).

13. Food processing generates wastes with high BOD loads (Farrall, 1979; Sell, 1981; Jones, 1974a, b; Jones et al., 1979; EPA, 1978).

14. Foods are often chemically complex systems whose components tend to react with one another. Certain types of reactions (e.g., oxidative and hydrolytic rancidification and enzymatic and nonenzymatic browning) tend to occur with a high degree of frequency (Berk, 1976; Fennema, 1985). Reactions induced by processing of foods are discussed in Hoyem and Kvale (1977), Priestley (1979), and Richardson and Finley (1985).

15. The engineering properties of foods and biological material are less well-known and more variable than those of pure chemicals and simple mixtures of

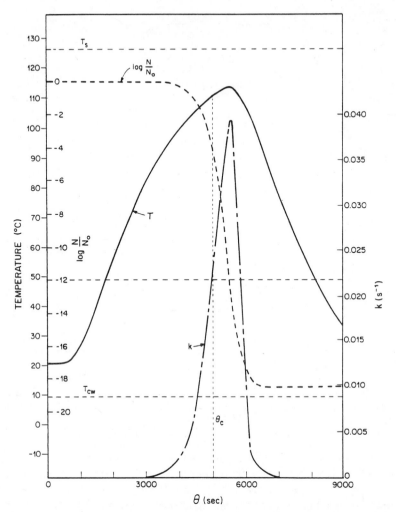

**FIGURE 11.21** Temperature, death-rate constant, $k$ ($s^{-1}$), and log ($N/N_0$) (log of fractional survival) of C. *Botulinum* at the center of a can of solid food subjected to thermal sterilization. $\theta$ is the elapsed processing time and $\theta_c$ is the time when cooling starts. $T_s$ = steam temperature in retort and $T_{cw}$ = temperature of cooling water.

chemicals. Table 11.3 lists data sources for some properties of foods. Other data can frequently be obtained by searching the *Food Science and Technology Abstracts* (International Food Information Service, 1986).

16. Vaguely defined sensory attributes often have to be preserved, generated, or modified. Raw-material variation, minor processing changes, and trace contaminants leached from processing equipment and packages can often cause significant changes in these attributes. Frequently, we do not have mechanistic bases for linking these attributes for processing conditions and equipment design. Much current

**TABLE 11.3   Sources of Data Relating to the Physical Properties of Foods and Related Materials**

| Product | Property or Method | Source |
|---|---|---|
| Grain and other crops | Density, hydrodynamic resistance, particle sizes, friction coefficients, dielectric coefficients, thermal properties | ASAE (1985) |
| Cattle and poultry | Weight, dimensions, and growth rates | ASAE (1985) |
| Cold-stored foods | Recommended storage temperatures, heats of respiration versus temperature, bulk and absolute densities, standard container sizes | ASHRAE (1982) |
| Cans and jars | Standard sizes, grades of coatings | Judge (1985) |
| Foods | Thermal properties, heat capacities, and thermal conductivities | Lentz (1961), Qashou et al. (1972), Tschubik and Maslow (1973) |
| | Mechanical and rheological properties | Mohsenin (1980); Jason and Jowitt (1968); Peleg and Bagley (1983) |
| | Flowability, mixing characteristics, internal friction, cohesiveness | Peleg (1977); Peleg et al. (1982, 1985); Peleg and Hollenbach (1984) |
| | Textural and rheological properties | Peleg (1980, 1983); Peleg et al. (1982); Bourne (1966, 1975, 1982); de Man et al. (1976); Rha (1975); Sherman (1970, 1977); Szczesniak (1966); Rao and Anastheswaran (1982) |
| | Water sorption isotherms Moisture diffusivity | COST 80 and 90 projects (Jowitt 1983); Iglesias and Chirife (1982) |
| | Properties of water in food | Duckworth (1975); Rockland and Stewart (1981); Simatos and Multon (1985); Wolf et al. (1985) |
| | Solute diffusion | Schwartzberg and Chao (1982) |
| | Dielectric properties | Mudgett (1982); Ohlsson and Bengtsson (1975) |
| | Methods of chemical analysis | Williams (1984) |
| | Physical, physicochemical and biological testing methods | Gruenwedel and Whitaker (1984a, b; 1985) |
| | Proximate composition and nutrient content | Watt and Merrill (1975); Scherz and Kloos (1981) |

411

**TABLE 11.3**  (*Continued*)

| Product | Property or Method | Source |
| --- | --- | --- |
| | Kinetic data for thermal inactivation of microorganisms and destruction of nutrients | Karel et al. (1975); Thompson (1982) |
| | Color characteristics | Frances and Clydesdale (1975); Hunter (1975) |
| | Sensory testing methods | Amerine et al. (1965); Birch et al. (1977); Stone and Sidel (1985); Moskowitz (1983) |
| Frozen foods | Enthalpies and ice content versus moisture content and temperature, heat capacities, bound water, freezing points | Riedel (1949, 1951, 1956, 1965a, b, 1959, 1961, 1968, 1976); Jason and Long (1955); Fleming (1969); Whiteman (1957) |
| | Thermal property correlations | Schwartzberg (1977); Succar and Hayakawa (1983) |
| Meats | Thermal properties | Morley (1972) |
| Milk fat | Drop size, density, composition | Mulder and Walstra (1974) |
| Additives | Chemical properties | Furia (1972) |
| Moist foods | Expression properties | Schwartzberg (1983) |
| Aroma | Relative volatility, retention | Bomben et al. (1973); Menting et al. (1970a, b) |

food engineering and food-science research activity at universities is designed to provide such linkages. For example, workers in the Food Engineering Department at the University of Massachusetts have studied (1) how particle size, shape, and porosity and related processing conditions affect light reflectance from foods (Schwartzberg, 1977); (2) how the kinetics of protein cross-linking reactions affect the texture of extruded soy flour (Huang and Schwartzberg, 1984; Bouvier et al., 1985); (3) how processing affects the melting properties of cheese (Lazaridis and Rosenau, 1979; Lazaridis et al., 1981; Mahoney et al., 1981; Rosenau et al., 1978) and the puffing of extrudates (Clayton and Huang, 1984); (4) the mechanics of texture perception (Calzada and Peleg, 1978; Peleg, 1980); (5) varietal, seasonal, and processing-induced changes in fish texture (Johnson et al., 1980a, b, 1981); and (6) how evaporation conditions affect aroma retention (Chardon and Quemarais, 1985).

17. Pilot-plant production of new food and pharmaceutical products is almost invariably needed before such products are introduced. In the case of foods, prototype products have to be consumer-tested so as to assure market acceptability before plants for large-scale production are built. In the case of pharmaceuticals,

prototype production is needed to provide material for clinical testing for efficacy, safety, and dosage-level determination. Such pilot-plant runs should be used to generate data for plant design and to insure reliable scale-up.

18. Mechanical working is sometimes used to induce desired textural changes. Examples include kneading and sponge mixing during the making of bread, the calendering of pastry dough, shearing during extrusion texturization, and the pulling of taffy.

19. Packaging in small containers is often used or required, and strong package–product interactions exist (Griffin and Sacharow, 1985; Schmitt, 1985). Many different types of containers are used: multi- and single-seam, tin-plated and lacquer-coated steel cans capable of withstanding positive and negative pressure differentials, aerosol containers that can stand higher pressures, thin-walled aluminum cans that must be internally pressurized to retain their shape when externally stressed, cans with scored tabs and other easy opening devices, pouches and/or semirigid containers made of clear plastic and of laminated films combining several different types of plastics and aluminum foil, paperboard boxes and containers, paper bags, plastic and glass bottles, shrink and cling-type transparent plastic wrappings, blister packages made of cardboard and transparent plastic film, foamed plastic cartons, corrugated cardboard cartons, and wooden crates.

Packaging prevents processed foods from becoming contaminated; helps maintain environments that promote storage stability; prevents light and radiation-induced damage when needed; retards heat transfer; sometimes allows selective exchange of gases and moisture; cushions products and prevents or minimizes impact-induced damage; helps prevent or minimize tampering; permits attractive display of products; provides display surfaces for advertising and informative labeling; permits compact, convenient stacking and shelving; sometimes facilitates convenient in-package cooking; and can sometimes be used as serving dishes or trays.

Packaging involves filling to specified product weights; providing correct proportions or dosing of ingredients and product components; inserting promotional literature and premiums; placing product pieces so as to attractively display a product in certain cases; purging head spaces and replacing air in head spaces with nitrogen, carbon dioxide, or steam; forming containers in some cases; sterilizing containers when aseptic packaging is used; capping, closing, and sealing containers; inspecting to insure that correct weights are obtained and that container, closure, and seal integrity is maintained; overwrapping; labeling; loading cartons; and palletizing. Packaging also may include in-package pasteurization and sterilization, and subsequent cooling of products in jars and cans. Great care must be taken to ensure sterility, maintain integrity of closure and correctness of fill, eliminate air from head spaces, and prevent subsequent moisture and oxygen transfer. Providing sterility is so important that the thermal processing of low-acid foods is regulated. Segregation often causes problems in the packaging of powdered foods (Barbosa-Canovas et al., 1985).

Packaging is frequently carried out in plant areas that are separated from processing areas; but in-container sterilization by thermal treatment of products in cans, jars, and retortable pouches is frequently carried out in processing areas.

Aseptic packaging is starting to be widely used, and Toledo (1984) provides a good review of the state of the art. Aseptic packaging requires the use of separate packaging areas and of enclosures to maintain sterility.

20. Despite continued public resistance, it appears that food irradiation will be used to an increasing extent. For updates on this process, see Diehl (1983) and Josephson and Peterson (1982).

21. Food-processing techniques and formulations are sometimes constrained by standards of identity and by good manufacturing practice regulations and codes. Judge (1985) provides a convenient summary of such regulations. The USDA requires submittal of equipment plans and inspection during the processing of meat and poultry. Dairy plants are inspected for compliance with respect to milk receipts and pasteurization. The FDA regulates ingredients, indirect additives, and contamination. The EPA regulates pollution and OSHA occupational safety. In some cases, accepted types of equipment are specified (U.S. Department of Agriculture, 1983).

22. Food plants should be located near sources of raw materials when such materials are highly perishable and near markets to be served when products have a short shelf life. Products made from stable raw materials and those that have a long shelf life can be manufactured in plants that are optimally positioned with respect to minimizing combined shipping costs for both raw materials and products.

Food processing is an art to a certain extent. Engineers are frequently uncertain as to whether portions of that art are really justified or necessary. It is sometimes difficult for engineers to translate the necessary portions of that art into quantifiable heat- and mass-transfer and chemical-reaction processes on which rational designs can be based.

## 11.6. PHARMACEUTICAL PROCESSING CONSIDERATIONS

Pharmaceuticals must be also safe and sterile. Furthermore, they must be clinically effective and that effectiveness must be verified and quantified so that proper dosage levels can be provided. Pharmaceutical production systems must adequately deal with various types of potential challenges to product safety, for example, pyrogenicity, contamination and infection of biological raw materials, such as blood and animal organs, allergens, and other sources of biological incompatibility. When fermentations are used, fermentors must be sterilized and maintained in a sterile state, excess foaming must be prevented, metabolic heat must be removed, and sterile nutrients and air or oxygen must be supplied in doses or at rates that do not excessively limit microorganism growth or the production of desired metabolites. Fermentation conditions must be shifted from those that favor microorganism growth to those that favor metabolite production at the proper time, and microorganism strain integrity must be preserved (Bailey and Ollis 1986; Wang et al., 1979; Aiba et al., 1973). Relatively dilute cultures are used to facilitate gas transport and to prevent nutrient depletion and concentration buildup of inhibitory products; but this means that the product must be subsequently concentrated to a great extent. Problems in scale-up of fermentations are considered in a special issue of *Biotechnology and Bioengineering* (Gaden, 1966). Some of the scale-up problems encountered when using molecular biology techniques are discussed by Galliher (1985).

Good background knowledge in biochemistry and cellular biology is almost mandatory for engineers planning to work in biochemical processing. Texts such as those by Lehninger (1982), Stryer (1981), and Albert et al. (1983) can be used to help acquire this knowledge.

Pharmaceuticals produced by fermentation have to be separated from residual nutrients, other metabolites, and cells or cell debris, materials that are nonvolatile. Therefore, the production of such pharmaceuticals frequently involves the use of filtration, centrifugation, liquid–liquid extraction, adsorption, ion exchange, chromatography, selective precipitation, and crystallization and derivatization to enhance separability (Belter, 1979; Jakoby, 1971; Stryker and Waldman, 1978). Affinity chromatography is beginning to be used to provide high specificity of separation (Lowe and Dean, 1974; Jakoby and Wilcheck, 1974; Nishikawa, 1979). Separations of optical isomers are sometimes required. The unwanted isomer is then usually racematized and recycled to the isomer separation step to enhance product recovery. Pharmaceuticals often must be separated from water or solvents sooner or later. Therefore, their production involves the use of drying under relatively gentle conditions to prevent product degradation (Masters, 1979).

In general, biochemical reactions, including those that occur in foods, are much slower and have much greater activation energies than reactions involving industrial chemicals. They take place only over narrow ranges of temperature, pH, and water activity and are very sensitive to environmental conditions, which must be closely controlled to provide favorable reaction rates and selectivity. Productivity and selectivity are also enhanced by using selected strains of microorganisms or mutants, ensuring dominance of that strain and preventing unwanted mutations. Fermentation kinetics are often complex and difficult to elucidate.

## 11.7.  PRODUCT INTEGRITY

When foods and pharmaceuticals are processed, it is often necessary to provide easily identifiable batches or lots of finished products, verify the adequacy of processing and product safety, and ensure that raw materials are free from contamination. Thus batch rather than continuous processing is used to a great extent in food and drug production. For pharmaceutical products, great care must be exercised to achieve high levels of product purity and to preserve product integrity, that is, to prevent different products from being substituted for or mixed with one another. To provide high levels of product safety and to prevent unremediable contamination of large amounts of product and the endangerment of large user populations, production batch sizes for sensitive products (e.g., blood plasma fractions) may have to be quite small. Very stringent environmental control (e.g., the use of "clean rooms") is often required during pharmaceutical processing. It may be necessary to use and not deviate from procedures registered with or certified by the Food and Drug Administration. Similar restrictions may apply to procedures used for manufacturing purchased reagents and ingredients and to methods for producing certain equipment and equipment parts used in drug manufacture.

In the case of foods, it is sometimes necessary to verify that raw materials will function as desired during processing, use raw material blends that will ensure constant sensory or functional characteristics and/or provide processing conditions which can be readily adjusted to accommodate raw-material variation. Close environmental control is often needed during food processing, but the level of control is usually less stringent than in pharmaceutical processing.

## 11.8. SANITARY CONSIDERATIONS

Processed foods are derived from natural raw materials which almost invariably contain at least small amounts of spoilage and pathogenic microorganisms and which are contaminated with dirt and insect parts to some extent. Therefore, processing to a large extent involves eliminating or greatly reducing such contamination, or inactivating contaminants. Since microbial populations, if unchecked, grow with great rapidity, processing environments that prevent microbial populations and pests from growing and entering food-process streams must be provided.

Equipment and work spaces in both food and pharmaceutical processing plants must be readily cleanable, free from stagnant zones and pockets, and not susceptible to soil buildup and fouling. Therefore, highly polished, completely drainable, stainless-steel tanks with rounded corners and highly polished stainless-steel dairy tubing must be used in many food plants. The tubing is pitched to ensure complete drainage and has fittings designed to eliminate product-wetted dead spaces. Since food-growing and animal-rearing environments are far from clean, delivery areas and certain adjacent areas in food plants will not be as clean as those devoted to later stages of processing. It is sometimes difficult to decide the exact interface between areas in which a food material is still an agricultural raw material which requires low levels of sanitation, and areas where it is a foodstuff requiring legally-specified or code-regulated levels of sanitation. Processing areas for foodstuffs, that are subsequently sterilized or subjected to processing conditions which incidentally produce sterility, are often not as sanitary as is desired in light of current standards, particularly in older plants. However, air-flow and human traffic-flow patterns that eliminate possibilities of contaminant transfer from "dirty" areas to clean ones should be maintained; and very high levels of sanitation must be provided for foodstuffs that provide good substrates for the growth of microorganisms and when processing temperatures and conditions favor such growth (Imholte, 1984; Troller, 1983).

Extremely rigorous sanitary precautions must be used to maintain sterility in fermentations and other drug-manufacturing operations (Wang et al., 1979; Bailey and Ollis, 1986; Aiba et al., 1973; Blakebrough, 1968; Webb, 1964; Steel, 1958).

## 11.9.  FOOD-PROCESSING UNIT OPERATIONS

Food processing involves many conventional unit operations, but it also involves several that differ greatly from those usually encountered in the production of industrial chemicals. These include: freezing, thawing, and other temperature-induced phase transitions or phase-transition analogs; freeze drying; freeze concentration; curd and gel formation; development of structured gels; cleaning and washing (the operation that occurs with the greatest frequency in food-processing plants); leavening, puffing, and foaming; slaughtering; carcass disassembly; component excision; slicing and dicing; peeling and trimming; grading; cell disruption and maceration; pasteurization and sterilization; blanching; baking; cooking (for purposes of tenderization or textural modification); roasting (for purposes of flavor generation); radiation sterilization; mechanical expression; structure-based component separation; filling and packaging; canning and bottling; coating and encapsulation; sausage and flexible casing stuffing; controlled atmosphere storage; fumigation and smoking; churning; artificially induced ripening; fermentation; pureeing; emulsification and homogenization; biological waste treatment; and controlled feeding of confined animals, poultry, and fish.

Some of these operations are covered in texts dealing with food processing and food-processing unit operations. Most food-engineering texts, such as Charm (1978), Heldman and Singh (1981), Karel et al. (1975), Jackson and Lamb (1981), Loncin and Merson (1979), Blakebrough (1968), Kessler (1981), and Lopez (1981), provide reasonably detailed coverage of thermal processing. Charm (1978), Kessler (1981), and Heldman and Singh (1981) provide similar coverage for freezing and thawing. Geankoplis (1983) discusses sterilization, freezing, and thawing and gives many examples of food processing in a chemical engineering unit-operations text. Kessler (1981), Loncin and Merson (1979), and Mulder and Walstra (1974) consider emulsification and homogenization. Qualitative descriptions of a wide variety of processing operations are included in Farrall (1976), Leniger and Beverloo (1975), Brennan et al. (1976), and Woollen (1969). Schwartzberg (1982) covers the engineering of some less commonly treated food-processing operations.

Some food-processing operations are used in parts of the chemical industry, for example, emulsification in cosmetics and latex paint manufacture, flocculation in waste treatment, and mechanical expression in clay, peat, and waste-treatment sludge dewatering. Various chemical-engineering unit operations are used in characteristically different ways in food processing. These include: extrusion; evaporation; reverse osmosis and ultrafiltration; adsorption and ion exchange; solid–liquid extraction; various types of drying; crystallization, hydraulic, and pneumatic classification; centrifugation; fluming; heat transfer; vapor (aroma) stripping and recovery; grinding; filtration; calendering (sheeting); agglomeration; and inspection. Some examples of the types of differences encountered are presented below.

1.  Sugar crystallization is usually a batch process in which large crystals are grown from injected seed crystals; most crystallizations in the chemical industry

are carried out continuously with seed crystals being produced by secondary nucleation.

2. Product temperatures and hold-up times must be limited to prevent fouling and product degradation in food evaporations; hold-up time and temperature limits are much less stringent in evaporations carried out in the chemical industry. Spray drying is frequently used in the food industry because of similar hold-up time and temperature limitations.

3. Food aromas exhibit very high relative volatilities. Therefore, food evaporators act as stills. Small amounts of vapor are withdrawn from fruit-juice evaporators to permit aroma recovery; but the presence of noncondensibles in the withdrawn vapor reduces the efficiency of aroma recovery. Aroma losses can be completely avoided by using freeze concentration instead of evaporation.

4. Hydraulic and pneumatic classification can be used to provide food-component enrichment if the classification step is preceded by suitable grinding. In food processing, fluming is often used as a classification, cleaning, or cooling step as well as a conveying step. Centrifugation is used to produce controlled levels of emulsion separation and to standardize the dispersed phase content of emulsions such as milk and cream.

5. In food processing, grinding is used to rupture cells, produce textural changes, and disengage constituents as well as to reduce particle size.

6. In food processing, calendering and extrusion is used to generate desirable textures as well as to form shaped products.

7. When heat treatment impairs product quality, membrane filters, which do not pass particles greater than 0.2 $\mu$m in diameter, can be used to provide sterile, bacteria-free food, and pharmaceutical products. While membrane filtration is also used in manufacturing ultrapure electronic components, it is rarely used elsewhere in the chemical industry.

It is interesting to note that a reasonably large number of chemical-engineering unit operations were initially developed for use in food processing. These include distillation, drying, grinding, filtration, solid–liquid extraction, fermentation, multiple-effect evaporation and perhaps evaporation itself, adsorption, crystallization, and mechanical expression. Stewart and Amerine (1982) provide chronological charts showing the antiquity of various food-processing operations.

## 11.10. UNIT OPERATIONS FOR BIOCHEMICAL PRODUCTION

As previously noted, biochemical and drug-related production and processing involve fermentations, liquid–liquid extraction, chromatographic separations, adsorption and ion exchange, centrifugation, crystallization, filtration, solid–liquid extraction, freeze and vacuum drying, mixing, tableting, coating and encapsulation, cleaning and washing, radiation and fumigation-based sterilization (for disposable medical devices), sterilization, bottling, and packaging. Fermentations are covered in the standard biochemical engineering texts, of which the most up-to-date are Wang et al. (1979), Aiba et al. (1974), and Bailey and Ollis (1986). The annotated bibliographies in Bailey and Ollis provide particularly good references for other work in the field. Belter (1979), Blakebrough (1968), and Webb (1964) represent the main sources of information on processing following fermentation.

## 11.11.  OVERALL DESIGN CONSIDERATIONS

Food- and drug-plant designs must provide necessary levels of sanitation, means of preventing product and material contamination, and means of preventing or limiting product, raw material, and intermediate deterioration due to naturally occurring processes, in addition, of course, to accomplishing the desired processing.

### A.  Prevention of Contamination

Prevention of contamination involves the provision or use of: filtered air; air locks; piping layouts that ensure complete drainage and absolutely prevent cross-stream contamination (particularly contamination of finished products by unsterilized or unpasteurized raw materials and cleaning solutions); solid-material and human traffic-flow layouts that also prevent such contamination; suitably high curbs when pipes, conveyors, or equipment pass through floors and where gangways pass over processing areas; bactericides in cooling water; culinary (i.e., contaminant-free) steam whenever direct contact between a product and steam is used; impermeable covers for insulation; dust covers over conveyors and clear plastic covers for electric lights; methods for washing bottles and containers; suitable barriers against pest entry (e.g., metal-lined overhangs on loading-dock platforms, screens and filters on air intakes); windowless construction; solid instead of hollow walls, or completely tight enclosure of hollow spaces in walls; air-circulation systems and external roof and wall insulation that prevent the formation of condensates which can drip into products or favor mold growth; ultraviolet irradiation of tank-head spaces; electric light traps for flying insects; impactors for killing insect eggs, larvae, pupae, and adults in grain; carbon dioxide and nitrogen fumigation of dry-food storage bins; screening systems to remove insects and insect parts; magnetic traps and iron screens for sieving equipment (so that screen fragments can be picked up by magnetic traps); metal detectors for rejecting packaged product that contains unwanted metal; and methods for storing and keeping track of segregated batches of raw materials and finished goods until necessary quality assurance tests have been carried out. Many of these precautions and those cited in the section dealing with sanitation are based on the recommendations of Imholte (1984) and Troller (1983).

## B.  Sanitation

Sanitation, which helps prevent contamination, should be facilitated by providing or using: impermeable coated or tiled floors and walls; at least one floor drain per every 400 ft$^2$ of wet processing area and special traps for such drains; pitched floors that ensure good drainage; polished vessels and equipment that do not contain dead spaces and that can be drained and automatically cleaned in place; sanitary (dairy) piping; clean-in-place (CIP) systems; plate heat exchangers and other types of equipment which can be readily disassembled for cleaning if necessary; clearances (usually 36- to 60-in. clear borders and clearances of 12 in. or more beneath equipment) for cleaning under and around equipment; grouting to eliminate crevices at the base of equipment support posts and building columns; tubular pedestals instead of support posts constructed from beams; and methods for removing solid particles that fall off conveyors.

## C.  Deterioration

To minimize product and raw-material deterioration, provisions should be made for: refrigerated and controlled-environment storage areas; space and facilities for product inspection (including federal inspection) and for carrying out quality-assurance tests; surge vessels for processed material between different operations (particularly operations subject to breakdown); equipment for precooling material stored in such vessels; means of cooling, turning over, or rapidly discharging the contents of bins and silos when excessive temperature rises occur; and standby refrigeration and utility arrangements which are adequate to prevent product and raw material deterioration in case of power interruptions or unusual climatic conditions.

Imholte (1984), Troller (1983), and Jowitt (1980) provide good guidelines for sanitary design practice. Hyde (1984) and Clem (1984) give up-to-date coverage of clean-in-place (CIP) system practice and design.

## D.  Seasonal Production

Food plants must be able to accommodate peak seasonal flows of products without excessive delay; and in some cases, these plants have to be highly flexible so as to handle different types of fruit and vegetables. Flexible production lines are often used in plants that produce frozen fruits and vegetables and those that produce strained fruits and vegetables for baby food. In such plants, the processing lines are rapidly switched around to accommodate the product that has to be produced at the moment.

Crops such as sugar beets are processed over a period of roughly 100 days; but some types of fruits and vegetables may only be available in suitably fresh form for a month or two. Processing seasons can sometimes be extended by using both late- and early-ripening forms of a given fruit. Thus the orange-packing season in Florida is roughly 6 months long because early- and late-ripening varieties are used; but there are two seasonal peaks in fruit availability, and the best-quality fruit is available mainly near the time of peak availability.

Since many foods are highly perishable, feed material cannot be allowed to accumulate, and must be promptly processed. Often, field agents for a processing plant arrange purchasing agreements with farmers, which, in return for a slight increase in price and guaranteed level of purchasing, allows the plant management to set planting and harvesting dates and schedules so that a smooth flow of raw materials is provided. Modeling of crop and animal-growth processes (France and Thornley, 1984) can be of great help in scheduling production and adequately determining the size of plants.

Sometimes, the period over which food raw materials can be used can be greatly extended by limited amounts of processing. Drying stabilizes grain so that it can be milled later and used for baking, brewing, and pasta production on a year-round basis. Dried beans can similarly be used for baked-bean production. Hardy vegetables (e.g., potatoes) can be stored for long times at relatively low temperatures and, after tempering to reduce their sugar level, can be used to produce products such as potato chips, which have a limited storage life.

## E.  Dusty Conditions

Adequate provision against dust explosions must be provided when dry and dusty foods such as grains are handled or stored. This involves: using spark proof tools and conveyors; providing suitable electrical grounding at frequent intervals; preventing dust entrainment; avoiding frictional generation of heat (e.g., by lubricating bearings with sufficient frequency); monitoring bins so as to detect temperature rises that could lead to spontaneous combustion; and providing for the turnover of the contents of bins so as to prevent temperature buildup. Unfortunately, since high moisture contents lead to mold growth in grains, the use of high humidity environments to prevent static electrical charge buildup may not be feasible, but excessively dry environments should be avoided.

## 11.12.  DESIGN OF UNIT OPERATIONS

Aside from the aforementioned general plant design considerations, the main differences between the design of food and drug plants and plants for producing industrial chemicals stem from variations in unit operations employed or the particular ways these operations are used in food and drug plants. Therefore, the rest of this chapter concentrates on methods for selecting, sizing, and specifying operating conditions for the equipment used in selected unit operations encountered in food and drug processing.

To design and size the equipment needed for a given operation, we frequently have to answer the following questions.

## A.  Quality or Completeness Indices

What measures or indices will tell us how well the process is done or indicate its degree of completeness? In chemical processes and in many food processes, these indices include component yields and purities, residual levels of undesired impurities, fractional recoveries of heat, and the extent of completion or selectivity of a reaction. In food processing, residue-based indices are frequently used to specify residual fraction levels (based on initial contamination levels) for pathogenic and spoilage microorganisms. For potent pathogens, these residual fractions are extremely small, for example, $1 \times 10^{-12}$ for $C.$ $Botulinum$ in thermally-processed, low-acid canned foods (Teixeira et al., 1969; Karel et al., 1975; Lenz and Lund, 1977a; Stumbo, 1973). Because microbial death during thermal sterilization is a first-order reaction whose rate depends on the existing level of contamination, residual fractions for microorganisms are independent of the initial contamination. Processing is also designed to ensure very low residual fractions for potential pathogenic contaminants in drug products, for example, $1 \times 10^{-6}$ for hepatitis virus in human-blood fractions used for therapeutic purposes.

Other indices in food processing involve functional or sensory responses and attributes which are harder to objectively specify or quantitatively measure (Potter, 1978; Moskowitz, 1983; Amerine et al., 1965; Stone and Sidel, 1985; Rha, 1975; Bourne, 1975; Sherman, 1977; Szczesniak, 1966). Examples include firmness of texture and other textural attributes, and gel strength and the maintenance of taste identity or taste-preference ratings. In many cases, rapid, reliable empirical tests have been developed for measuring these attributes. Examples of such empirical tests include: grade viscosity for pectin; Bloom gel strength for gelatin; Bostwick consistency for tomato concentrates (Marsh et al., 1980); Farinograph, Amylograph, Extensometer, and Alveograph readings and patterns for doughs (Matz, 1972); Marschall cup test readings and curd tension for cheese (Kosikowski, 1977); and reflectance color for roasted coffee (Sivetz and Desrosier, 1979). However, other tests, such as taste tests, taste-profile analysis, texturometer measurements, and shear-strength reading, are time-consuming, subjective, and/or far from reliable (Bourne, 1975; Peleg, 1980).

## B.  Throughput and Quality

How are the indices of product quality or process completeness related to equipment size, hold-up time, and throughput rate?

## C.  Secondary Effects

How are secondary indices of quality affected by the completeness of processing and the processing conditions used? For example, a great many combinations of heating and cooling conditions produce acceptable levels of sterility during thermal processing, but these combinations are not equally acceptable because they produce different amounts of nutrient breakdown and different losses of sensory quality. Nutrient breakdown processes usually are first order, but their activation energy is much smaller than for thermally induced microbial death. Therefore, sterilization processes can frequently be optimized to maximize nutrient retention. Teixeira et al. (1969a,b; 1975a,b), Thijssen et al. (1978, 1980) and Lenz and Lund (1977a–c) provide methods for carrying out such optimization. Food-drying processes cause many other changes in addition to water removal. Some of these changes are reviewed by Bimbenet and Guilbot (1966). The handling of complex, multiproperty changes often has to be dealt with by sensory evaluation (Amerine et al., 1965; Moskowitz, 1983; Stone and Sidel, 1985) and appropriate statistical methods (Gould, 1977; Kramer and Twigg, 1970, 1973; Bender et al., 1982; Bozyk and Rudzki, 1972).

## D.  Equipment Size

How is the size of equipment related to the amount of material which must be processed per unit time; and how is the performance of the equipment related to increases or decreases in processing rate? Usually the equipment volume or transfer-surface area is linearly proportional to the throughput rate.

## E.  Material Balances

What are the material balances involved; and how are they related to the particular type of equipment selected and its size, and to production rates? In food processing, losses of useful raw materials and degradation into materials of low value are particularly important economic considerations. For example, recoverable yields of fish flesh when producing fish fillets are frequently as low as 30%. This not only represents a great economic loss, but also greatly increases waste-treatment loads and problems. Useful by-products can frequently be recovered from food-processing wastes (Green and Kramer, 1979).

## F.  Energy Needs

How much energy is required to carry out the process and how is this amount related to throughput rates and the quality-of-performance indices? How can energy requirements be reduced or minimized without seriously impairing process performance or product quality? A great deal of work designed to determine how much cold-storage temperatures can be safely raised has been carried out in Europe in recent years. Casper (1977) lists much information which can be used to estimate energy needs for various types of food-processing plants.

## G.  Utility Requirements

What utilities are required? Farrall (1979) and Slade (1967) provide coverage of utility requirements for food plants.

## H. Design Details

What detailed design characteristics must we specify for equipment and its internal parts; and how do such design details affect satisfactory performance of the equipment? Examples from food processing include: filler-machine head spacing and port sizes; knife shapes and cutting frequency for beet-cosset slicing; sizes of openings in conveyor screens and for screens designed to retain particles in solid–liquid extractors and other wet-processing vessels; flume cross-sections; air duct spacings for air cooling of respiring produce stored in bins; defroster specifications for refrigeration coils; blower characteristics for dryers; and length-to-diameter ratios for diffusion batteries and ion-exchange columns. Frequently such details are the province of design specialists, but reasonable estimates for preliminary design purposes can be made on the basis of published rules of thumb (ASHRAE, 1982; Patchen, 1971; Peleg, 1985; Slade 1967).

## I. Control

How is the process going to be controlled and what hardware and hardware interconnections are required to achieve the desired control? McFarlane (1983), Green (1984), Lloyd (1984), McLellan (1985), Spanbauer (1984), Tarnawski et al. (1984), Yano (1985), and Dinnage (1985) discuss control in food-processing plants, but improved instrumentation is needed to monitor many important food-processing variables.

## J. Costs

How are the preceding considerations related to processing and equipment costs; and how can the design be manipulated to minimize overall cost, while still satisfying necessary processing constraints? Costs depend very much on sources of supply. Food ingredient sources are listed by Judge (1983), Chilton (1985), and Thomas (1986b).

## 11.13. MODELING

To answer some of the questions listed above, we must understand the underlying mechanisms controlling the process. We can usually proceed effectively if we can model these mechanisms in mathematical terms which permit us to clearly relate the outcome of the process to operating conditions and raw-material properties. Practical modeling almost invariably involves focusing on those factors that are most important and neglecting some that are less significant. In food and drug processing, we often find that the number of variables we can afford to neglect is smaller than in industrial chemical processing.

Sometimes models can be broken down into partly independent components and results can be combined by iteratively solving sets of simultaneous algebraic and differential equations. An example of this approach is the simultaneous and progressive solving of the heat-, mass-, and momentum-transfer coefficient and rate equations, mass and energy balances, phase-equilibrium relationships, drop size and shape correlations, and drop-trajectory calculations involved in designing a spray dryer (Kerkhoff and Schoeber, 1974; Masters, 1979). In other cases, interactions are a one-way process. In thermal processing, for example, heat transfer affects the microbial death rate, but only rarely does the microbial death rate affect heat transfer. Great difficulty may be experienced when we do not know how much relative value to assign to the different attributes of a product, particularly sensory attributes, and when we are dealing with a process that can only be manipulated to improve one attribute at the expense of another. For example, in pressing juice from cranberries, conditions that favor high juice yields tend to reduce the desired extraction of red color (Helfferich, 1985).

## 11.14.  UNIT-OPERATIONS CATEGORIES

Food-processing unit operations can be broken down into the following sometimes overlapping categories, which can be used to provide organized bases for process design.

### A.  Complex Mechanical Operations

These require relatively complex mechanical manipulation and solid handling (e.g., coring, pitting, and carcass disassembly). Chemical engineers are not well equipped to deal with these operations and may have to seek the assistance of mechanical, food, and agricultural engineers or specialists who have relevant design experience. Special equipment manufacturers and vendors often provide capacity-rating and space-requirement information which is adequate for preliminary design purposes. The names of suitable vendors and manufacturers can usually be obtained from sources such as the Thomas register (Thomas, 1986a) and the Food Engineering Master (Chilton, 1985). Many companies provide catalogs which are useful for identifying wide ranges of needed equipment. Examples are the Food Machinery Corporation (FMC Corp, 1984) and Dixie Canner (1986), in the case of canning-related processes.

### B.  Modular Operations

These require the use of a great many units in parallel. Modularity also applies for operations carried out by humans, such as vegetable, meat, and fish trimming, cutting, and inspections. In some cases, modular units (e.g., hydroclones, baghouse elements, filter frames, cheese-press hoops, cartridge filters, and reverse osmosis and ultrafiltration modules) can be readily manifolded to make up an assembly that can be treated as a single piece of equipment. In other cases, small to moderately large numbers of free-standing units are used in parallel, for example, centrifugal (Urschel) slicers, packaging machines, grinders, pulpers, screening centrifuges, juice reamers and presses for citrus fruits (which are pressed individually or as fruit halves), finishers, homogenizers, scraped-surface freezers for producing ice-cream slushes, kettles for cooking treated hides (for gelatin), jelly, jam, and soups, fermenters for beer and wine, lagering tanks for beer, sedimentation tanks for wine, and steeping tanks for corn.

In still other cases, units of different size can be used to accommodate moderate differences in throughput rate, but parallel units are needed once larger throughput rates are used. For example, a suitably-sized, scroll-type, decanting centrifuge might be able to carry out a typical clarification or sedimentation operation at flow rates between 7,000 and 90,000 L/hr. Above 90,000 L/hr, several centrifuges would have to be used in parallel. Similar considerations apply for continuous and large-batch presses for expressing juice, expellers, rotary-vacuum filters, filter presses, vibrating screens, drum dryers, sterilization retorts, batch pasteurizers, continuous cookers, batch freeze dryers, and electrooptical sorters. Small-scale tests can usually be carried out (frequently at vendors' laboratories) to size modular units and to determine the numbers of units needed and the effectiveness of processing. However, because of raw-material variability, such tests may not always be reliable for foods and biological materials. If possible, tests should be carried out using the full range of raw materials likely to be encountered in practice; or adjustments in calculated equipment capacity should be made, based on known types of raw-material variation which will occur.

## C. Large-Scale Equipment

One to four units of large-scale equipment usually accommodates the throughput needed for even very large processing plants. Changes in design throughput capacity are usually accommodated by varying the size (usually the diameter or width of the equipment). Changes in throughput in existing plants can usually be accommodated within reasonable limits by changing stream-feed rates or processing temperatures. This generally involves accepting some loss in quality or yield for overcapacity operation, or some loss in productivity and some increase in labor cost per unit weight of product for under-capacity operation. Similar trade-offs may have to be made to accommodate variations in raw materials, for example, progressive deterioration in sugar-beet firmness and progressive increases in the sugar-to-acid ratio of citrus juice as the processing season proceeds.

Large-scale equipment used in the food and drug industries includes: solid-liquid extractors, which range up to 10,000 T/day in capacity (Schwartzberg, 1980); multiple-effect evaporators with capacities ranging from 5,500 to 55,000 kg (12,000 to 120,000 lb) of water evaporated per hour for orange juice and milk evaporators (Hansen, 1985) and up to 136,000 kg (300,000 lb) per hour for sugar evaporators (Schwartzberg, 1977b); spray dryers; fermentors for producing food acids and antibiotics; hydrostatic cookers; cooking, texturizing, and dough-shaping extruders; disk centrifuges, including automatic desludging types; liquid–liquid extractors; rotary dryers; moving-belt and other conveyor dryers; continuous-freeze dryers; fluidized-bed freezers; spiral-conveyor freezers; flaked-ice makers; continuous proofers; continuous dough makers; band and moving-hearth baking ovens; stills; automated cheese-making vats (22,700 kg or 50,000 lb of feed per batch, 2 hr hold-up per batch); draining, matting conveyors for cheese curds (22,700–34,000 kg or 50,000–75,000 lb of feed per hour capacity); continuous pasteurizers for bottled beer; and mashing kettles for beer and toaster-desolventizers for extracted oilseeds.

## D.   Heat- and Mass-Transfer Processes in Solids

In these operations, required hold-up times are regulated by mass- or heat-transfer processes in solids. They include solid–liquid extraction, infusion of curing agents and salt, osmotic dehydration, and the drying of solids (particularly the last stages of drying).

Selective diffusive transfer of water can permit the retention of volatile food aromas and flavors during the drying of foods; and drying conditions that maximize selectivity frequently have to be used. Consequently, the mechanisms involved have been studied by King (1984), Thijssen (1979), Thijssen and Rulkens (1968), Menting et al. (1970a, b), Flink (1975), Flink and Karel (1970a, b; 1972), Bomben et al. (1973), Voilley et al. (1977), Voilley and Loncin (1978), and Fritsch et al. (1971).

Sensible heat transfer in solids and analogous mass-transfer processes can readily be analyzed using methods provided by Carslaw and Jaeger (1959); alternatively, Crank (1975) provides techniques for mass-transfer analysis.

Heat-transfer processes in solids are often linked with kinetically regulated processes. These include reaction-induced solute production during solid–liquid extraction, baking, heating to induce tenderization and flavor development, sterilization, pasteurization, and enzyme inactivation. One can analyze these processes in terms of dimensionless parameters $At/L^2$, $(T_s - T_0)/Z$, and $[A/L^2k] \cdot [\ln (N_0/N_f)]$. Here, $t$ is the heating time; $N_0$ and $N_f$ are, respectively, the initial and final reactant concentration; $L$ is the smallest dimension of the solid; $k$ is the reaction rate constant at the process heat-source temperature $T_s$; $T_0$ is the initial solid temperature; $Z$ is the temperature drop that will cause the reaction-rate constant to decrease by a factor of 10; and $A$ is the thermal diffusivity. If the third dimensionless parameter is large, reaction-rate limitations will require appreciable holdup at near-maximum temperatures; but if it is small, heat transfer is rate-limiting and satisfactory completion of the desired reaction will occur shortly after some desired end-point temperature is reached at the center of the solid. Computational methods for dealing with complicating factors for such processes have been presented by Hiddink et al. (1976) and Manson and Cullen (1974). Procedures for optimizing them with respect to quality retention have been described by Lenz and Lund (1977a–c), Teixiera et al. (1969a, b; 1975a, b), and Thijssen et al. (1978, 1980). Hoyem and Kvale (1977), Priestly (1979), and Richardson and Finley (1985) consider changes produced in foods by thermal processing.

Multistage, diffusion-regulated, mass-transfer processes in solids, such as solid–liquid extraction, cannot be correctly analyzed by the same methods used for gas–liquid and liquid–liquid multistage processes. Appropriate solutions of partial differential equations governing diffusion are readily available (Crank 1975), but they must be used in conjunction with Duhamel's theorem (superposition) in order to account for boundary-condition changes that occur when stage-to-stage transfer takes place. Schwartzberg (1983, 1987) and Schwartzberg et al. (1983) developed superposition methods for various frequently used multistage, solid–liquid extraction (leaching) systems. Axial dispersion and flow instability and maldistribution also should be taken into account when designing bed-based, solid–liquid extraction systems. Sivetz and Desrosier (1979) describe coffee extraction. Oil-seed extraction is discussed in Swern (1982).

Combined heat and mass transfer occur in a wide variety of processes used to dry foods. Spray drying is discussed by Masters (1979), Crowe (1980), and Kerkhof and Schoeber (1974); freeze drying is described by King (1971), Mellor (1978), Kessler (1962), Spicer (1974), and Goldblith et al. (1975). Other aspects of food drying are discussed by Keey (1972), Kneule (1975), Krischer and Kroll (1958, 1963), Bruin and Luyben (1980), and the Society of Chemical Industry (1958). Combinations of in-solid heat and mass transfer (and accompanying reactions in some cases) can cause puffing and expansion.

Microwave heating is being used to an increasing extent in food processing. The characteristics of such heating are covered by Okress (1968), Bengtsson (1976), Copson (1975), Schiffmann (1976), and Mudgett and Schwartzberg (1982).

## E. Reactions and Growth Processes

These include: enzymatic reactions (e.g., starch conversions to simple sugars and oligosaccharides, sugar breakdown in respiring fruits and vegetables, and rennet-induced casein coagulation); leavening and proofing; malting, yeast production and fermentations, including microbial treatment of wastes; conversions of lactose to lactic acid during pickling (Prescott and Dunn, 1982); and cheese making and other sugar-to-acid conversions in foods (Peppler and Perlman, 1979; Kosikowski, 1977). They also include complex biosyntheses, steroid modification, and amino-acid, vitamin, antibiotic, and enzyme production (Peppler and Perlman, 1979; Wang et al., 1979; Bailey and Ollis, 1986; Aiba et al., 1973; Blakebrough, 1968; Webb, 1964; Steel, 1958). Food storage and food processing also involve many naturally occurring enzymatic processes (Acker, 1962; Berk, 1976; Fennema, 1985). All these processes are regulated by reaction or growth kinetics at near isothermal conditions. Necessary sizes for equipment for such processes can be determined fairly readily from desired production rates, levels of conversion, and reaction rates. Such rates can usually be conveniently calculated from known rate constants and substrate, enzyme, and/or growing organism concentrations. Factors that should be considered when designing equipment include: tradeoffs between maximization of productivity and yield; the production of inhibitors and their effect on reaction rates; the provision of needed nutrients and sterile air (or the avoidance of air contact when anaerobic reactions are desired); the elimination of contamination by undesired microorganisms or the suppression of their growth; the possibility of mutations; and the effect of environmental conditions on growth rates and metabolic paths. Tissue culture is an attractive new processing area, but methods for scaling it up are still in the development stage (Kruse and Patterson, 1973; Bailey and Ollis, 1986; Shuler, 1981; Staba, 1980; Sahai and Knuth, 1985). Other reactions, such as hydrogenation (Bern and Schoon, 1975; Bern et al., 1976; Swern 1982), are more chemical than biochemical in nature and can be analyzed in terms of the reaction and mass-transfer rates involved.

## F.  Penetrations

Hold-up times for these operations are determined by desired deposit depths or by penetration depths for desired temperature or concentration changes near the surface of a solid. These include the deposition of melt-based coatings and steam and lye peeling. The hold-up times needed are proportional to the square of the desired penetration or deposition depth. These times are also inversely proportional to solute or thermal diffusivities in the solid and to the desired dimensionless concentration or temperature change squared. By carefully controlling heat penetration, modern steam peelers, such as that shown in Figure 11.22, have greatly reduced peeling losses and achieved rapid peeling.

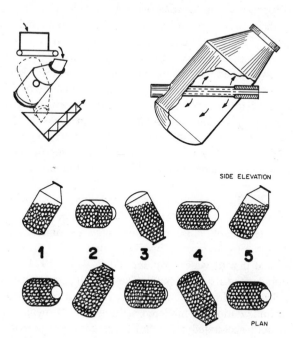

FIGURE 11.22 Operating cycle for a steam peeler. Numbers 1 through 5 indicate successive positions of material in rotating retort. (Courtesy of Odenberg Engineering).

## G. Operations Involving Stable Dispersions

The creation of stable liquid–liquid dispersions almost invariably involves the use or presence of a surfactant to coat dispersed droplets and prevent coalescence. Power requirements and throughput rates for equipment designed to create stable dispersions can sometimes be determined in terms of critical Weber numbers and the dimensionless parameter $K_c t_s Q / (d_p^3)$, where $t_s$ is the surfactant redistribution time, $K_c$ is similar to a Smoluchowski coalescence rate constant (corrected for surface-repulsion effects), $Q$ is the dispersed phase-volume fraction, and $d_p$ is the desired dispersed droplet size (Schwartzberg, 1985). For normal Smoluchowski coalescence, $K_c$ is proportional to $RT$ and inversely proportional to viscosity; but in highly turbulent, dispersion-creating environments, $K_c$ is probably a function of the power dissipation per unit volume, instead of $RT$. Other operations, such as churning, which involve the breaking of emulsions or phase inversions for emulsions, may be analyzed in terms of coalescence kinetics and the creation of conditions that favor particle-to-particle adhesion. Relevant aspects of colloid chemistry are discussed by Friberg (1976), Dickinson and Stainsby (1982), and Mulder and Walstra (1974).

## H. Sedimentation, Elutriation, and Filtration

These include processes for separating solids and liquids, and solids and gases and liquid–liquid dispersions, and for fractionating solids according to density or size. In many cases, these processes are similar to those encountered when processing industrial chemicals. Pneumatic elutriation is frequently used to separate dry-food solids from stones, and, in somewhat different equipment, from twigs, chaff, and insect parts. It can be used to separate food components (Schubert and Heideker, 1985) and flat from round fish during the unloading of fishing boat holds (Whitney and Correa, 1979; Carey et al., 1981). Flotation is used to separate moist food solids from stones and sweet peas from denser, starchy peas. Hydraulic elutriation is used to separate parts that exhibit high drag from more compact parts with low drag. An example is the separation of green pepper cores from the denser, outer portions of green pepper. Separations of sand, grit, and stones can be carried out during fluming.

Many biological solids, including most microorganisms (but not viruses), are highly hydrated and have densities close to that of water. The density of fats and edible oils is also reasonably close to that of water. In some cases (e.g., milk) fat globules are encapsulated by protein lipid films that reduce the difference between globule density and that of water (Mulder and Walstra, 1974). Cells, cell parts, and these fat globules are also very small. These small sizes and density differences reduce gravity-driven sedimentation rates, leading to a frequent need for centrifugally-driven sedimentation. If hindered settling (and Brownian motion in some cases) is adequately accounted for, centrifuges can be sized or scaled up with reasonable accuracy based on settling-rate data obtained from batch centrifugations in clinical centrifuges. Equipment-sizing procedures and methods for calculating collection efficiency for hydroclones are available (Svarovsky, 1981), but effects produced by different levels of solid loading are difficult to predict. Classification by centrifugal sedimentation in fluids with density gradients may become an important method of separating subcellular components.

Pneumatic or hydraulic classification can be used to achieve some degree of food-ingredient fractionation (e.g., protein enrichment of wheat flour or gossypol removal from cottonseed meal), but the success of such processes depends strongly on suitable prior grinding (Schubert and Heideker, 1985).

Many biological solids are slimy, highly compressible, and susceptible to breakup, particularly when exposed to conditions that cause protopectin breakdown. Therefore, filter aids are frequently used during filtration; and calculations of achievable filtration rates have to account for compression-induced permeability changes and other changes which can greatly influence filtration resistance (e.g., proximity to isoelectric points). Methods for calculating equipment sizes for some juice- and oil-expression operations have been developed by Schwartzberg (1983) and Schwartzberg et al. (1985). These require prior measurement of presscake and media-filtration resistances, wall-friction coefficients, presscake-compressive stress, and the pressures at which cake extrudes through the media as functions of the insoluble solid-based bulk-specific volume of presscakes. Test equipment for obtaining these measurements has been developed. Ultrafiltration is being used to an increasing extent to replace "fining" for the clarification of fruit juices (Milnes et al., 1985; Nielsen et al., 1984).

## I.  Deep-Bed Processes

These include many grain-drying systems, various solid–liquid extraction processes, ion exchange, and chromatographic and other sorption-based processes. Deep-bed grain dryers are susceptible to moisture redeposition ahead of the advancing desorption front. This leads to conditions that favor mold growth. Therefore, air-flow rates and temperatures that limit susceptibility to redeposition and the duration of redeposition have to be used, even though this results in greater energy expenditure. A specialized text dealing with grain drying has been written by Brooker et al. (1975).

When sorption isotherms are concave downward with respect to the solution-concentration axis and fine sorbent particles are used, sorption processes can often be designed on the basis of saturated-zone displacement and the shape of the sorption wave characterizing the breakthrough. These can be determined from tests in small beds. If the sorption isotherms favor sorption-wave sharpness and selectivity, they will be unfavorable in terms of sharpness and ease of desorption. Hence, to ensure clean desorption, it may be worthwhile to displace a desired sorbed species by a more strongly sorbed species. Textbook coverage of sorption and ion exchange is provided by Bowen (1975), Treybal (1980), and McCabe et al. (1985).

Processes involving the flow of liquid through deep beds of small particulate solids are greatly affected by axial dispersion and displacement stability, which is usually favorable during upflow sorption or infusion and conditionally unstable during downflow extraction or desorption. Displacement instability and axial dispersion during downflow will be very marked if the displacement velocity $V$ exceeds $V_c = g(d\rho/d\mu)/K$, where $\rho$ and $\mu$ are, respectively, the solution density and the viscosity near the desorption or extraction front, $g$ is the acceleration of gravity, and $K$ is the bed permeability (Dumore, 1964). Similar considerations apply during the washing of deep beds. Therefore sorption and extraction beds and wash columns should be designed to maintain $V$ smaller than $V_c$. Since $K$ is proportional to particle size squared, displacement stability, flow maldistribution, and axial dispersion are greatly affected by particle size as well as bed compressibility and direction of flow. Upflow extraction is more unstable than downflow; but operating conditions that aggravate instability during downflow can sometimes minimize the effects of instability during upflow. Therefore, sorption processes, particularly those involving compressible solids, are difficult to scale up reliably. Special designs, such as in-series arrays of shallow beds (called pancake beds), have to be used when beds are excessively compressible.

Schwartzberg (1983, 1986), Schwartzberg et al. (1983), and Flores and Schwartzberg (1985) have recently developed design methods for single and multistage fixed-bed infusions and extractions and sorptions involving linear regions of isotherms. These methods appear to adequately account for effects of axial dispersion when such dispersion is not excessive. Further work is required to extend these methods to cases where sorption isotherms are not linear, axial dispersion is very great, or flow maldistribution cannot be accounted for in terms of axial dispersion. Recent progress in this area has been made by Lee (1985).

## J. Cleaning, Washing, and Rinsing

Good designs and flow layouts for clean-in-place (CIP) installations are available from various design houses and from consulting firms that specialize in this area (Clem, 1984). Food-processing equipment is usually off-stream for cleaning and washing from 2 to 25% of available working time. Reductions in capacity due to such losses of on-stream time should be accounted for in designing food and drug plants. Washing and rinsing can frequently be described in terms of combined displacement and kinetic models (Loncin, 1977; Loncin and Merson, 1979; Thor and Loncin, 1978; Bourne and Jennings, 1963a, b; 1965) and cleaning times usually are linearly proportional to the surface area to be cleaned divided by cleaning and rinsing fluid flow rates. This provides a theoretical basis for predicting cleaning time and water requirements. These requirements also depend on types of soil, soil adhesion, and accidental fouling. Several recent, widely attended conferences (e.g., Lund et al., 1985; Hallstrom et al., 1981) attest to a high level of interest in this area. The cleaning of bottles and containers is dealt with by Jacobson and Lindqvist (1964), Hoffman (1970), and Leipner (1975). Clean-in-place (CIP) methods and design procedures are covered by Clem (1984), Hyde (1984), and Imholte (1984).

## K.    Solidification and Melting Processes

These include freezing and thawing, the hardening of butter and fats, crystallization, gelling, curd formation and consolidation, and the hardening of glassy sugar melts.

Freezing and thawing times can usually be predicted within 20% by using Plank's equation (Plank, 1960). Thermal properties during freezing and thawing have been correlated using equations developed by Schwartzberg (1977) and Schwartzberg et al. (1977). Figures 11.23 and 11.24 show some results of these correlations. The equations can be used to predict refrigeration and heat-transfer loads for freezing and thawing processes; but, if freezing-point depression data are available, heating and cooling loads can also be calculated by relatively simple

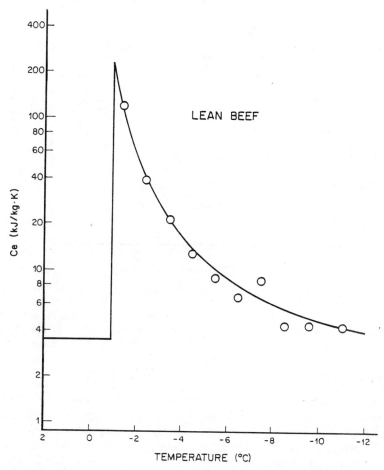

**FIGURE 11.23**  Effective heat capacity, Ce, versus temperature during freezing.

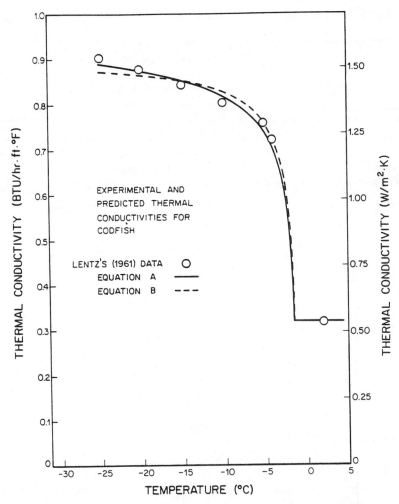

**FIGURE 11.24** Thermal conductivity, $k$, versus temperature, $T$, during freezing. Equation A: $k = k_f + (k_0 - k_f)(T_0 - T_i)/(T_0 - T)$. Equation B: $k = k'_f + B(T_i - T) + (k_0 - k'_f)(T_0 - T_i)/(T_0 - T)$, where $T_0$ is the freezing point of pure water, $T_i$ is the initial freezing point of the food, $k_f$ is the thermal conductivity in the fully frozen state, $k_0$ is the thermal conductivity in the thawed state, $B$ is the temperature coefficient of conductivity in the fully frozen state, and $k'_f$ is the value of $k_f$ extrapolated to $T_0$.

enthalpy balances. The thermal properties equations have been used to derive shortcut equations (Schwartzberg et al., 1977; Schwartzberg, 1981; Reinick and Schwartzberg, 1985a, b; Hayakawa et al., 1983), which predict freezing and thawing times more accurately than Plank's equation. More rigorous, but time-consuming, predictive methods were developed by Comini et al. (1974), Hayakawa et al. (1983, 1987), Nonino and Hayakawa (1986), Cleland and Earle (1976b, 1984a), Reinick (1985), and Schwartzberg et al. (1977). Cleland (1985) and Cleland and

Earle (1977, 1984b) reviewed the relative merits of various rigorous and shortcut freezing time-computation methods. The determination of surface heat-transfer coefficients during freezing and thawing is dealt with by Cleland and Earle (1976a) and Reinick and Schwartzberg (1985c). Metastable effects caused by freezing are dealt with by Maltini (1975).

Since glassy sugars are really highly viscous supercooled liquids, the hardening of glassy sugar melts does not involve a phase transition. Therefore, hardening-time requirements can be determined by standard methods for analyzing unsteady-state heat transfer in solids (Carslaw and Jaeger, 1959).

Fat solidification, while similar to freezing, involves more phase transitions as different types of triglycerides solidify, and is often subject to delayed phase transitions and metastability. The melting of cheeses is a gradual process unlike the melting of pure materials. It is greatly influenced by calcium ion levels and the presence of "emulsifying" salts which sequester calcium (Lazaridis and Rosenau, 1979; Lazaridis et al., 1981; Mahoney et al., 1981; Rosenau et al., 1978).

Much work is being done on the kinetics and theory of gelling and curd formation (Carlson, 1985; Dickinson and Stainsby, 1982; Brown and Enstrom, 1984). Curd consolidation by pressing can, to some extent, be analyzed by methods developed to deal with mechanical expression, but mechanistic analysis of curd shrinkage due to syneresis and whey exudation have not yet been developed. Nevertheless, empirical methods (e.g., controlling cooking temperatures, starter-culture addition levels, and the size of curd cubes) for controlling whey exudation have been developed (Kessler, 1981). These permit reliable production of soft (68–73% water), semihard (55–68% water), and hard cheeses (less than 55% water). New methods, such as direct acidification (Rosenau, 1984) and the use of ultrafiltration, which improve the retention of whey solids in cheeses, have been developed and work well for some cheeses, but sensory problems must be overcome in other types (Kosikowski, 1977).

Sugar crystallization is the second largest use of crystallization in all types of industry. It is almost always carried out by growth from injected seeds in batch-type crystallizing evaporators. Tolerable saturation levels and rates of growth are well known. Designs for crystallizing evaporators are fairly diverse, but widely available (Meade and Chen, 1977; McGinnis, 1982; Hugot, 1972), and energy requirements for such evaporators can be readily predicted. Bennett (1969) discusses some of the difficulties encountered when attempting to use continuous crystallization for sugar production. Ion-exchange treatment of evaporator mother liquors increases yields of recoverable sucrose and has been increasingly used in recent years.

## L.  Coating and Encapsulation

Foods are frequently coated or encapsulated. Some flow characteristics important in coating are covered by Groenveld (1970a–e, 1971) and Tallmadge (1969, 1970). Other aspects of coating are discussed by Linke (1969) and Middleman (1977).

## M.  Shaping and Texturizing

Foods are shaped by a wide variety of techniques—molding, sheeting (calendering), extrusion, drawing, rolling, and prilling. The methods developed for polymer processing can be used to analyze sheeting (Middleman, 1977) and extrusion (Tadmor and Klein, 1970; Schenkel, 1966). Harper (1985) reviews current work dealing with food extrusion. Rates for molding processes are regulated by rates of removal of heat or by the kinetics of gelling. Textures can be developed by extrusion, leavening, foaming, and the formation of structured gels by using combinations of polyvalent cross-linking ions and freezing and thawing. Possible kinetic bases for scaling up extrusion texturization have been presented by Huang and Schwartzberg (1984), Linko et al. (1985), and Hagan et al. (1985). Other recent work on food extrusion has been done by Diosady (1985), Janssen (1985), Mange and Gelus (1985), Bouvier et al. (1985), Rossi and Peri (1985), and van Zuilichem et al. (1985).

## N.  Concentration

Liquid foods are concentrated to facilitate subsequent drying or crystallization, provide concentrates that are stable enough to be stored, generate sugar glasses, produce cheeses without losses of whey, and provide thick sauces. Methods used to concentrate liquid foods have been reviewed by Schwartzberg (1977b). The principle means of concentration used is multiple-effect evaporation. Thermal recompression (the use of steam-jet compressors to recycle steam) is widely utilized to improve the thermal efficiency of such evaporation. In recent years, mechanical vapor recompression has been increasingly employed, or the number of effects used in a typical evaporation system has been increased from about three to roughly five to seven.

Though thousands of short, submerged-tube calandria evaporators with internal recirculation are still in use in sugar refineries, most new food-evaporation systems use single-pass falling-film evaporators to minimize product holdup and thermal degradation (Hansen, 1985). Film-based Reynolds numbers of greater than 500–700 must be used to prevent incomplete tube-wall wetting in falling-film evaporators. Therefore, 6.0- to 15.0-m long tubes are used to provide adequate evaporation in a single pass. Very viscous, non-Newtonian products (e.g., tomato paste, banana puree) require the use of evaporators employing forced recirculation. Mechanically-assisted heat transfer (e.g., scraped surfaces) is used to achieve high concentrations with such products. Flavor and aroma loss during evaporation has to be dealt with when liquid foods are concentrated by evaporation (Thijssen, 1974a). Evaporator vapor and condensate streams (Bomben et al., 1973; Chardon and Quemarais, 1985) can be used to recover these aromas in a concentrated form which can be added back to the concentrate or dried product.

Freeze concentration circumvents the aroma-loss problem and can theoretically provide high thermal efficiency (Thijssen, 1974b; van Pelt, 1975). Recent improvements have increased actual thermal efficiencies and throughput rates for the process (van Pelt and Swinkels, 1985), but concentrations higher than 40–50% have not yet been achieved at commercially attractive rates.

After a long induction period, reverse osmosis and ultrafiltration are being used to an increasing extent in food processing. A good review of food-related uses for these processes is provided in the December 1980 issue of *Desalination* (Balaban, 1980). More recent developments in the field are covered by Pepper (1984) and Hansen (1985).

## 11.15.  CONCLUSIONS

The preceding discussions of unit operations leave a great deal of ground uncovered. Nevertheless, they should provide some feeling for the unique problems likely to be encountered when designing operations and processes for plants that produce foods and drugs. Many of these problems have not yet been adequately resolved and the design of plants for producing these materials will have to be based on information generated in pilot plants.

## REFERENCES

Acker, L., "Enzymic Reactions in Foods of Low Moisture Content," *Adv. Food Res.,* **11,** 263 (1962).

Aiba, S., A. E. Humphrey, and N. F. Millis, *Biochemical Engineering*, 2nd ed., Academic Press, New York, 1973.

Albert, B., D. Bray, J. Lewis, M. Raff, and J. D. Watson, *Molecular Biology of the Cell*, Garland Publishing Co., New York, 1983.

Alikonis, J. J., *Candy Technology*, AVI, Westport, CT, 1978.

Altschul, A. M., *New Protein Foods, Vol. 1A, Technology*, Academic Press, New York, 1974.

American Meat Institute, *The Science of Meat and Meat Products*, W. H. Freeman, San Francisco, CA, 1960.

Amerine, M. A., H. W. Berg, R. E. Kunkee, C. S. Ough, V. L. Singleton, and A. D. Webb, *Technology of Wine Making*, 4th ed., AVI, Westport, CT, 1980.

Amerine, M. A., R. M. Pangborn, and E. B. Roessler, *Principles of Sensory Evaluation*, Academic Press, New York, 1965.

Appleman, C. O. and J. M. Arthur, "Carbohydrate Metabolism in Green Sweet Corn," *J. Agr. Res.*, **17,** 137 (1919).

Arbuckle, W. S., *Ice Cream*, 3rd ed., AVI, Westport, CT, 1977.

ASAE, *Agricultural Engineers Yearbook, 1985*, American Society of Agricultural Engineers, St. Joseph, MI, 1985 (issued annually).

Ashby, B. H., *Protecting Perishable Foodstuffs During Transportation by Motor Truck, U. S. D. A. Agricultural Handbook No. 105*, United States Department of Agriculture, Supt. of Documents, U. S. Govt. Printing Office, 1970.

ASHRAE, *Handbook—Applications 1982*, American Society of Heating Refrigeration and Air Conditioning Engineers, New York, 1971, Chaps. 21, 22, and 24–47, 1982 (issued every four years).

Autorenkollektiv, *Die Zuckerherstellung*, VEB Fachbuchverlag, Leipzig, 1980.

Bailey, J. E. and D. F. Ollis, *Biochemical Engineering Fundamentals*, 2nd ed., McGraw-Hill, New York, 1986.

Balaban, M. (ed.), "Proceedings of the Symposium on Membrane Technology in the 80's, Ystad, Sweden. Sept. 29–Oct. 1, 1980," *Desalination*, **35**, Nos. 1/2/3 (1980).

Barbosa-Canovas, G. V., J. Malave, and M. Peleg, "Segregation in Food Powders," *Biotechnol. Prog.*, **1**, 140, (1985).

Belter, P. A., "General Procedures for the Isolation of Fermentation Products," in *Microbial Technology, Vol. 2 Fermentation Technology*, H. J. Peppler and D. Perlman (eds.), Academic Press, New York, 1979.

Bender, F. E., L. W. Douglass, and A. Kramer, *Statistical Methods for Food and Agriculture*, AVI, Westport, CT, 1982.

Bengtsson, N. E., "Dielectric Heating as a Unit Operation in Food Processing. Heating Fundamentals and Applications of Radio Frequency and Microwaves," *Confructa*, **21**, 7 (1976).

Bennett, R. C., "Continuous Sugar Crystallization: A Chemical Engineer's Viewpoint," *Chem. Eng. Prog. Symp. Ser.*, No. 95, 34 (1969).

Berk, Z., *Braverman's Introduction to the Biochemistry of Foods*, Elsevier, Amsterdam, 1976.

Bern, L., J. O. Lidefelt, and N. N. Schoon, "Mass-Transfer and Scale Up in Fat Hydrogenation," *J. Am. Oil Chem. Soc.*, **53**, 463 (1976).

Bern, L. and N. N. Schoon, "Kinetics of Hydrogenation of Rapeseed Oil: I. Influence of Transport Steps in Kinetic Study," *J. Am. Oil Chem. Soc.*, **52**, 182 (1975).

Bimbenet, J. J. and A. Guilbot, "Biochemical and Physicochemical Changes During Drying," (in French) *Chim. Ind., Genie Chim.*, **96**, 925 (1966).

Birch, G. G., J. G. Brennan, and K. J. Parker, *Sensory Properties of Foods*, Applied Science Publishers, London, 1977.

Blakebrough, N., *Biochemical and Biological Engineering Science*, Academic Press, New York, 1968.

Bomben, J. L., S. Bruin, H. A. C. Thijssen, and R. L. Merson, "Aroma Recovery and Retention in Concentration and Drying of Foods," *Adv. Food Res.*, **20**, 1–111 (1973).

Bourne, M. and W. G. Jennings, "Kinetic Studies of Detergency, I. Analysis of Cleaning Curves," *J. Am. Oil Chem. Soc.*, **40**, 517 (1963a).

Bourne, M. and W. G. Jennings, "Kinetic Studies of Detergency. II. Effect of Age Temperature and Cleaning Time on Rates of Soil Removal," *J. Am. Oil Chem. Soc.*, **40**, 523 (1963b).

Bourne, M. and W. G. Jennings, "Kinetic Studies of Detergency. III. Dependence of the Dupré Mechanism on Surface Tension," *J. Am. Oil Chem. Soc.*, **42**, 546 (1965).

Bourne, M. C. "A Classification of Objective Methods for Measuring Texture and Consistency of Foods," *J. Food Sci.*, **33**, 223 (1966).

Bourne, M. C., "Is Rheology Enough for Food Texture Measurement," *J. Texture Stud.*, **6,** 297 (1975).

Bourne, M. C., *Food Texture and Viscosity*, Academic Press, New York, 1982.

Bouvier, J. M., G. Fayard, R. Mahoney, and H. G. Schwartzberg, "Protein Reactions During Extrusion," AIChE Annual Meeting, Chicago, IL (November 1985).

Bowen, J. H., in *Chemical Engineering*, Vol. 3, J. M. Coulson and J. F. Richardson (series eds.), J. F. Richardson and D. G. Peacock, (vol. eds.), Pergamon, Oxford, 1975, Chap. 7.

Bozyk, Z. and W. Rudzki, *Quality Control of Food by Methods of Mathematical Statistics* (in German), Fachbuchverlag, Leipzig, 1972.

Brennan, J. G., J. R. Butters, N. D. Cowell, and A. E. V. Lilly, *Food Engineering Operations*, 2nd ed., Applied Science Publishing Co. London, 1976.

Brooker, D. B., F. W. Bakker-Arkema, and C. W. Hall, *Drying Cereal Grains*, AVI, Westport, CT, 1975.

Brown, R. J. and C. A. Ernstrom, "Use of Computers in the Cheese Industry: Modelling of Cheese Coagulation," in *Engineering and Food, Vol. 2, Processing Applications*, B. M. McKenna (ed.), Elsevier Applied Science Publishers, Barking, Essex, England, 1984, pp. 781–789.

Bruin, S. and K. C. A. M. Luyben, "Drying of Food Materials: A Review of Recent Developments," in *Advances in Drying*, Vol. 1, A. S. Mujumdar (ed.), Hemisphere Publishing Co., Washington, D. C., 1984, pp. 155–215.

Calzada, J. F. and M. Peleg, "Mechanical Interpretation of Compressive Stress–Strain Relationships of Solid Foods," *J. Food Sci.*, **43,** 1087 (1978).

Carey, R. L., L. F. Whitney, and L. R. Correa, "Fish Orientation and Feed Rate Effects on Air Separation by Shape," *J. Food Sci.*, **48,** 1582 (1981).

Carlson, A., "Kinetics of Gel Forming in Enzyme Coagulated Milk," *Biotechnol. Prog.*, **1,** 47 (1985).

Carslaw, H. S. and J. C. Jaeger, *Conduction of Heat in Solids*, 2nd ed., Clarendon Press, Oxford, 1959.

Casper, M. E. (ed.), *Energy-Saving Techniques for the Food Industry*, Noyes Data Corporation, Ridge Park, NJ, 1977.

Chardon, S. and B. Quemarais, "Stripping and Recovery of Model Aroma Compounds During Batch Evaporation," Report, Food Engineering Department, University of Massachusetts, Amherst, MA, 1985.

Charm, S. E., *Fundamentals of Food Engineering*, 3rd ed., AVI, Westport, CT, 1978.

Chilton Publishing Co., *Food Engineering Master 86*, Chilton Publishing Co., Radnor, PA, 1985.

Clayton, J. T. and C. T. Huang, "Porosity of Extruded Foods," in *Engineering and Food Vol. 2, Processing Applications*, B. M. McKenna (ed.), Elsevier Applied Science Publishers, Barking, Essex, England, 1984, pp. 611–620.

Cleland, A. C., "A Review of Methods for Predicting the Duration of Freezing Processes," Fourth International Congress on Engineering and Food, Edmonton, Alberta, Canada (July 1985).

Cleland, A. C. and R. L. Earle, "A New Method for Predicting Surface Heat Transfer Coefficients in Freezing," *Bull. Int. Inst. Refrig.*, **56**, Suppl. 1976-1, 361 (1976a).

Cleland, A. C. and R. L. Earle, "A Comparison of Freezing Calculations Including Methods to Take into Account Initial Superheat," *Bull. Int. Inst. Refrig.*, **56**, Suppl. 1976-1, 369 (1976b).

Cleland, A. C. and R. L. Earle, "A Comparison of Analytical and Numerical Methods of Predicting Freezing Times for Foods," *J. Food Sci.*, **42**, 1390 (1977).

Cleland, A. C., and R. L. Earle, "Assessment of Freezing Time Methods," *J. Food Sci.*, **49**, 1034 (1984).

Cleland, A. C. and R. L. Earle, "Freezing Time Predictions for Different Final Product Temperatures," *J. Food Sci.*, **49**, 1230 (1984b).

Clem, L. W., "CIP Concepts in Processing Equipment Design," AIChE National Meeting, Philadelphia, PA (August 1984).

Comini, G., S. Del Giudice, R. W. Lewis, and O. C. Zienkiewicz, "Finite Element Solution of Nonlinear Heat Conduction Problems With Special Reference to Phase Change," *Int. J. Num. Method Eng.*, **8**, 613 (1974).

Connors, J. M., R. T. Rogers, B. W. Marion, and W. F. Mueller, *The Food Manufacturing Industries*, Lexington Books, D. C. Heath and Company, Lexington, MA, 1985.

Considine, D. M. and G. D. Considine, *Foods and Food Production Encyclopedia*, Van Nostrand Reinhold, New York, 1982.

Cook, L. R., *Chocolate Production and Use*, Magazines for Industry, Inc., New York, 1972.

Copson, D. A., *Microwave Heating*, 2nd ed., AVI, Westport, CT, 1975.

Crank, J., *The Mathematics of Diffusion*, 2nd ed. Oxford University Press (Clarendon) London, 1975.

Crowe, C. T., "Modeling Spray-Air Contact in Spray Drying Systems," in *Advances in Drying*, Vol. 1, A. S. Mujumdur (ed.), Hemisphere Publishing Co., Washington, D. C., 1980, Chap. 3.

Cruess, W. V., *Commercial Fruit and Vegetable Production*, McGraw-Hill, New York, 1958.

de Man, J. M., P. W. Voisey, V. F. Rasper, and D. W. Stanley (eds.), *Rheology and Texture in Food Quality*, AVI, Westport, CT, 1976.

Desrosier, N. W. (ed.), *Elements of Food Technology*, AVI, Westport, CT, 1977.

Desrosier N. W. and J. N. Desrosier, *The Technology of Food Preservation* 4th ed., AVI, Westport, CT, 1977.

Dickinson, E. and G. Stainsby, *Colloids in Foods*, Applied Science Publishers, London, 1982.

Diehl, J. F., "Food Irradiation," in *Developments in Food Preservation—2*, S. Thorne (ed.), Applied Science Publishers, London, 1983, pp. 25–59.

Dinnage, D., "Microprocessor Control in the Food Industries," Fourth International Congress on Engineering and Food, Edmonton, Alberta, Canada (July 1985).

Diosady, L. L., "Extrusion of Wheat Starch," Fourth International Congress on Engineering and Food, Edmonton, Alberta, Canada (July 1985).

Dixie Canner, *Equipment and Supply Catalog #82*, Dixie Canner Equipment Co., Athens, GA, 1986, prices updated periodically.

Duckworth, R. B. (ed.), *Water Relations in Foods*, Academic Press, New York, 1975.

Dumoré, J. M., "Stability Considerations in Downward Miscible Displacement," *Soc. Pet Eng. J.*, 357–362 (1964).

Environmental Protection Agency (EPA), *Proceedings, Ninth National Symposium on Food Processing Wastes*, Denver, CO, March 29–31, 1978, EPA—600/2-78-188, Industrial Environmental Res. Lab., U. S. Environmental Protection Agency, Cincinnati, OH. Available from NTIS, Springfield, VA, 1978 (eight prior proceedings are available).

Farrall, A. W. (ed.), *Food Engineering Systems, Vol. 1, Operations*, AVI, Westport, CT, 1976.

Farrall, A. W. (ed.), *Food Engineering Systems, Vol. 2, Utilities*, AVI, Westport, CT, 1979.

Fennema, O. (ed.), *Food Chemistry*, 2nd ed., Marcel Dekker, New York, 1985.

Fennema, O., W. D. Powrie, and E. H. Marth, *Low Temperature Preservation of Foods and Living Matter*, Marcel Dekker, New York, 1973.

Fleming, A. K., "Calorimetric Properties of Lamb and Other Meats," *J. Food Tech.*, **4,** 199 (1969).

Flink, J., "The Retention of Volatile Components During Freeze Drying: A Structurally Based Mechanism," in *Freeze Drying and Advanced Food Technology*, S. A. Goldblith, L. Rey, and W. W. Rothmayr (eds.), Academic Press, London, 1975, pp. 143–160.

Flink, J. and M. Karel, "Effect of Process Variables on Retention of Volatiles in Freeze Drying," *J. Food Sci.*, **35,** 444 (1970a).

Flink, J. and M. Karel, "Retention of Organic Volatiles in Freeze Dried Solutions of Carbohydrates," *J. Agric. Food Chem.*, **18,** 295 (1970b).

Flink, J. and M. Karel, "Mechanisms of Retention of Organic Volatiles in Freeze-Dried Systems," *J. Food Technol.*, **7,** 199 (1972).

Flores, S. and H. G. Schwartzberg, "Modelling of Countercurrent, Cross-Flow, Solid-Liquid Extractors and Experimental Verification," 4th International Congress on Engineering and Food, Edmonton, Canada, July 1985.

FMC Corp., *Catalog 184*, Food Processing Machinery Division, FMC Corp., Madera, CA, 1984.

Forrest, J. C., E. D. Aborte, H. P. Hedrick, M. D. Judge, and R. A. Merkel, *Principles of Meat Science*, W. H. Freeman and Sons, San Francisco, CA, 1975.

Foxboro Company, "Application Engineering Data, Vegetable Oil Processing," AED 209–600, 1980.

France, J. and J. H. M. Thornley, *Mathematical Models in Agriculture*, Butterworths, London, 1984.

Frances, F. J. and F. M. Clydesdale, *Food Colorimetry: Theory and Applications*, AVI, Westport, CT, 1975.

Friberg, S., *Food Emulsions*, Marcel Dekker, New York, 1976.

Fritsch, R., W. Mohr, and R. Heiss, "Conservation of Flavor of Foodstuffs in Various Drying Processes," *Chem. Ing. Tech*, **43**, 445–453 (1971).

Furia, T. E. (ed.), *Handbook of Food Additives*, 2nd ed., CRC Press, Cleveland, OH, 1972.

Gaden, E. L. (ed.), "Symposium, Scale-Up Problems in Fermentation Processes," *Biotech. Bioeng.*, **8** (February 1966).

Galliher, P. M., "Process Development and Scale Up of an rDNA Pharmaceutical Product," Fourth International Congress on Engineering and Food, Edmonton, Alberta, Canada, July 7–10, 1985.

Geankoplis, C. J., *Transport Processes and Unit Operations*, 2nd ed., Allyn and Bacon, Inc., Boston, MA, 1983.

Goldblith, S. A., L. Rey, and W. W. Rothmayr, *Freeze Drying and Advanced Food Technology*, Academic Press, New York, 1975.

Gould, W. A., *Food Quality Evaluation*, AVI, Westport, CT, 1977.

Green, D. A., "Use of Microprocessors for Control and Instrumentation in the Food Industry," in *Engineering and Food, Vol. 2, Processing Applications*, B. M. McKenna (ed.), Elsevier Applied Science Publishers, Barking, Essex, England, 1984, pp. 757–772.

Green, J. H. and A. Kramer (eds.), *Food Processing Waste Management*, AVI, Westport, CT, 1979.

Griffin, R. C. and S. Sacharow, *Principles of Package Development*, 2nd ed., AVI, Westport, CT, 1985.

Groenveld, P., "High Capillary Number Withdrawal from Viscous Newtonian Liquids by Flat Plates," *Chem. Eng. Sci.*, **25**, 33 (1970a).

Groenveld, P., "Low Capillary Number Withdrawal," *Chem. Eng. Sci.*, **25**, 1259 (1970b).

Groenveld, P., "Laminar Withdrawal with Appreciable Inertial Forces," *Chem. Eng. Sci.*, **25**, 1267 (1970c).

Groenveld, P., "The Shape of the Air–Liquid Interface During the Formation of Liquid Viscous Films by Withdrawal," *Chem. Eng. Sci.*, **25**, 1571 (1970d).

Groenveld, P., "Withdrawal of Power Law Liquid Films," *Chem. Eng. Sci.*, **25**, 1579 (1970e).

Groenveld, P., "Drainage and Withdrawal of Liquid Films," *AIChE J.*, **17**, 489 (1971).

Gruenwedel, D. W. and J. R. Whitaker, *Food Analysis: Principles and Techniques, Vol. 1, Physical Characterization*, Marcel Dekker, New York, 1984a.

Gruenwedel, D. W. and J. R. Whitaker, *Food Analysis: Principles and Techniques, Vol. 2, Physicochemical Techniques*, Marcel Dekker, New York, 1984b.

Gruenwedel, D. W. and J. R. Whitaker, *Food Analysis: Principles and Techniques, Vol. 3, Biological Techniques*, Marcel Dekker, New York, 1985.

Hagan, R. C., S. R. Dahl, and R. Villota, "Effects of Molecular Weight Distribution on Soy Protein Products by Twin Screw Extrusion," Fourth International Congress on Engineering and Food, Edmonton, Alberta, Canada, July 7–10, 1985.

Hall, C. W., A. W. Farrall, and A. L. Rippen, *Encyclopedia of Food Engineering*, AVI, Westport, CT, 1971.

Hallstrom B., D. B. Lund, and C. Tragardh, *Fundamentals and Applications of Surface Phenomena Associated with Fouling and Cleaning in Food Processing*, Proceedings International Workshop Tylosand, Sweden, April 6–9, 1981, Department of Food Science, University of Wisconsin, 1981.

Hansen, R. (ed.), *Evaporation, Membrane Filtration and Spray Drying in Milk Powder and Cheese Production*, North European Dairy J. Vanlose, Denmark, 1985.

Harper, J. M., "Processing Characteristics of Food Extruders," Fourth International Congress on Engineering and Food, Edmonton, Alberta, Canada, July 7–10, 1985.

Hawthorn, J. and E. J. Rolfe, *Low Temperature Biology of Foodstuffs*, Pergamon, London, 1968.

Hayakawa, K., C. Nonino, and J. Succar, "Two Dimensional Heat Conduction in Food Undergoing Freezing. Predicting Freezing Time of Rectangular or Finitely Cylindrical Food," *J. Food Sci.*, **48**, 1841 (1983).

Hayakawa, K., K. R. Scott, and J. Succar, "Theoretical and Semi-Theoretical Methods for Estimating Freezing or Thawing Time," *ASHRAE Trans.*, in press (1987).

Heldman, D. R. and R. P. Singh, *Food Process Engineering*, 2nd ed., AVI, Westport, CT, 1981.

Helfferich, J., "Effect of Temperature on Juice Solids and Color Yield During Cranberry Juice Expression," personal communication, Ocean Spray Cranberries Inc., Middleboro, MA, 1985.

Hiddink, J., J. Schenk, and S. Bruin, "Natural Convection Heating of Liquids in Closed Containers," *Appl. Sci. Res.*, **32**, 217 (1976).

Hoffman, S., "Cleaning and Disinfection of Beer Containers," *Brauwelt*, **110**, 1339 (1970).

Hoyem, T. and O. Kvale (eds.), *Physical, Chemical and Biological Changes in Food Caused by Thermal Processing*, Applied Science Publishers, London, 1977.

Hoyrup, H. E., "Beer," in *Kirk–Othmer Encyclopedia of Chemical Technology*, Vol. 3, Wiley-Interscience, New York, 1978, pp. 693–735.

Huang, B. W. and H. G. Schwartzberg, "Amine Group Disappearance During Extrusion Texturization," AIChE Summer National Meeting, Philadelphia, PA (August 19–22, 1984).

Hugot, E., *Handbook of Sugar Cane Engineering*, 2nd ed., Elsevier, Amsterdam, 1972.

Hull, W. G., W. E. Kite and R. C. Auerbach, "Modern Winemaking, *Ind. Eng. Chem.*, **42**, 2182 (1951).

Hunter, R. S., *The Measurement of Appearance*, Wiley, New York, 1975.

Hyde, J. M., "New Developments in CIP Practice," AIChE National Meeting, Philadelphia, PA (August 1984).

Iglesias, H. A. and J. Chirife, *Handbook of Food Isotherms: Water-Sorption Parameters for Food and Food Components*, Academic Press, New York, 1982.

Imholte, T. J., *Engineering for Food Safety and Sanitation*, Technical Institute of Food Safety, Crystal, MN, 1984.

Inskeep, C. C., R. E. Bennet, J. F. Dudley, and M. W. Shepard, "Bacitracin, Product of Biochemical Engineering," *Ind. Eng. Chem.*, **43**, 1488 (1951).

International Food Information Service, *Food Science and Technology Abstracts*, International Food Information Service, Commonwealth Information Bureau, Farnham Royal, Bucks, England, 1986, issued monthly.

International Institute of Refrigeration, *Recommended Conditions for Land Transport of Perishable Foodstuffs*, 2nd ed., International Institute of Refrigeration, Paris, 1963.

International Institute of Refrigeration, *Recommended Conditions for Cold Storage of Perishable Produce*, 2nd ed., International Institute of Refrigeration, Paris, 1967.

Jackson, A. T. and J. Lamb, *Calculations in Food and Chemical Engineering*, The Macmillan Press Ltd., London, 1981.

Jacobsson, B. and B. Lindqvist, "Systematic Experiments on the Cleaning of Bottles," *Brauwelt*, **104,** 181 (1964).

Jakoby, W. B. (ed.), "Enzyme Purification and Related Techniques," *Methods in Enzymology* **22**, Academic Press, New York, 1971.

Jakoby, W. B. and M. Wilcheck (eds.), "Affinity Techniques," *Methods in Enzymology* **34**, Academic Press, New York, 1974.

Janssen, L. P. B. M., "Model for Cooking Extrusion," Fourth International Congress on Engineering and Food, Edmonton, Alberta, Canada (July 1985).

Jason, A. C. and R. Jowitt, "Physical Properties of Food in Relationship to Engineering Design," in *3rd European Symposium of the European Federation of Chemical Engineering, Bristol, England, April 1968, DECHEMA Monograph 63*, DECHEMA, Frankfurt, West Germany, 1968.

Jason, A. C. and R. A. K. Long, "The Specific Heat and Thermal Conductivity of Fish Muscle Eutectics," *Proc. IX Int. Cong. of Refrig.*, **2,** 160 (1955).

Johnson, E. A., M. Peleg, R. A. Segars, and J. G. Kapsalis, "A General Phenomenological Rheological Model for Fish Flesh," *J. Texture Stud.*, **12,** 413 (1981).

Johnson, E. A., R. A. Segars, J. G. Kapsalis, M. D. Normand, and M. Peleg, "Evaluation of the Compressive Deformability of Fresh and Cooked Fish Flesh," *J. Food Sci.*, **45,** 1318 (1980a).

Johnson, E. A., R. A. Segars, J. G. Kapsalis, M. D. Normand, and M. Peleg, "Textural Variability in Fish Fillets," *Proceedings of 5th Annual Tropical and Subtropical Fisheries Technological Conference of the Americas*, 1980b, pp. 205–215.

Jones, H. R., *Pollution Control in Meat, Poultry and Seafood Processing*, Noyes Data Corp., Park Ridge, NJ, 1974a.

Jones, H. R., *Pollution Control in the Dairy Industry*, Noyes Data Corp., Park Ridge, NJ, 1974b.

Jones, J. L., M. C. T. Kuo, P. E. Kyle, S. B. Redding, K. T. Semrau, and L. P. Somogyi, *Overview of the Environmental Control Measures and Problems in the Food Processing Industries*, SRI International, Menlo Park, CA, Contract No. R8046212-01, Industrial Environmental Research Lab, Office of Research and Development, U. S. Environmental Protection Agency, Cincinnati, OH, 1979.

Josephson, E. S. and M. S. Peterson, *Preservation of Food by Ionizing Radiation*, Vol. I, CRC Press, Boca Raton, FL, 1982.

Jowitt, R. (ed.) *Hygienic Design and Operation of Food Plants* (for the Society of Chemical Industry, London), AVI, Westport, CT, 1980.

Jowitt, R., *Physical Properties of Foods*, Applied Science Publishers, London, 1983.

Judge, E. E., *The Almanac of the Food, Freezing, Preserving Industries*, E. E. Judge and Sons, Inc., Westminster, MD, 1985 (issued annually).

Judge, J. J., *The Directory of the Canning, Freezing, Preserving Industries 1982–83, 9th Biennial Edition*, James J. Judge, Inc., Westminster, MD, 1983.

Karel, M., O. R. Fennema, and D. B. Lund, *Principles of Food Science, Part II, Physical Principles of Food Preservation*, Marcel Dekker, Inc., New York, 1975.

Karmas, E., *Fresh Meat Processing*, Noyes Data Corp., Park Ridge, NJ, 1970.

Keey, R. B., *Drying, Principles and Practice*, Pergamon Press, Oxford, 1972.

Kerkhoff, P. J. A. M. and W. J. A. H. Schoeber, "Theoretical Modelling of the Drying Behavior of Drops in Spray Dryers," in *Advances in Preconcentration and Dehydration of Foods*, A. Spicer (ed.), Wiley, New York, 1974, pp. 349–397.

Kessler, H. G., *Freeze Drying*, Leybold, Cologne, 1962.

Kessler, H. G., *Food Engineering and Dairy Technology*, Verlag A. Kessler, Freising, Federal Republic of Germany, 1981.

King, C. J., *Freeze Drying of Foods*, CRC Press, Cleveland, OH, 1971.

King, C. J., "Transport Processes Affecting Food Quality in Spray Drying," in *Engineering and Food, Vol. 2, Processing Applications*, B. M. McKenna (ed.), Elsevier Applied Science Publishers, Barking, Essex, England, 1984, pp. 559–574.

Kneule, F., *Das Trocknen* (Drying), Verlag Sauerlander, Aarau, Switzerland, 1975.

Kochesperger, R. H., *Food Warehousing and Transportation*, Chain Store Publishing Co., New York, 1978.

Komen + Kuin Incorporated, "Odenberg K + K, The Multi Flash Steam Peeling Concept," Komen + Kuin Incorporated, Sacramento, CA, undated.

Kosikowski, F. W., *Cheese and Fermented Food Products*, F. W. Kosikowski and Associates, Brocktondale, NY, 1977.

Kramer, A. and B. A. Twigg, *Quality Control for the Food Industry, Vol. 1—Fundamentals*, AVI, Westport, CT, 1970.

Kramer, A. and B. A. Twigg, *Quality Control for the Food Industry, Vol. 2—Applications, AVI*, Westport, CT, 1973.

Krischer, O. and K. Kroll, *Trochnungstechnik, II. Trockner und Trocknungsverfahren*, Springer Verlag, Berlin, 1958.

Krischer, O. and K. Kroll, *Trocknungstechnik, I. Die Wissenschaftlichen Grundlagen der Trocnungstechnik*, 2nd ed., Springer Verlag, Berlin, 1963.

Kruse, P. F., Jr. and M. K. Patterson, Jr., *Tissue Culture: Methods and Applications*, Academic Press, New York, 1973.

Kuprits, Y. N. (ed.), *Technology of Grain Processing and Provender Milling*, Izdatel'stvo Kotas, Moscow, English Translation by Israel Program for Scientific Translation, Jerusalem, 1967.

Lazaridis, H. N. and J. R. Rosenau, "Effects of Emulsifying Salts and Carrageenan on The Rheology of Cheese-Like Products Produced by Direct Acidification," *J. Food Sci.*, **45**, 595 (1979).

Lazaridis, H. N., J. R. Rosenau, and R. R. Mahoney, "Enzymatic Control of Meltability in a Direct Acidified Cheese Product," *J. Food Sci.*, **46,** 332 (1981).

Lee, Y. C., "Axial Dispersion During Countercurrent Extraction—Analytical Solution for Dispersion in Both Phases," Internal Report, Food Engineering Department, University of Massachusetts, 1985.

Lehniger, A. L., *Principles of Biochemistry*, Worth Publishers, Inc., New York, 1982.

Leipner, W., "Soaking Time in Bottle-Washing Machines (in German), *Brauwelt*, **115** (7) 173 (1975).

Leniger, H. A. and W. A. Beverloo, *Food Process Engineering*, D. Reidel Publishing Co., Boston, MA, 1975.

Lentz, C. P., "Thermal Conductivity of Meats, Fats, Gelatin Gels and Ice," *Food Technol.*, **15**, 243 (1961).

Lenz, M. K. and D. B. Lund, "The Fourier-Lethality Number Method: Experimental Verification of a Model for Calculating Temperature Profiles and Lethality in Conduction-Heating Canned Foods," *J. Food Sci.*, **42**, 989 (1977a).

Lenz, M. K. and D. B. Lund, "The Fourier-Lethality Number Method: Experimental Verification of a Model for Calculating Average Quality Factor Retention in Conduction Heating Canned Foods," *J. Food Sci.*, **42**, 997 (1977b).

Lenz, M. K. and D. B. Lund, "The Fourier-Lethality Number Method: Confidence Intervals for Calculated Lethality and Mass-Average Quality Retention in Conduction Heating Canned Foods," *J. Food Sci.*, **42**, 1002 (1977c).

Linke, L., "Flow Characteristics of Coating Processes," *Lebensm. Ind.*, **16**, 452 (1969).

Linko, Y. Y., S. Hakulin, and P. Linko, "An Extruder as a Continuous Biochemical Reactor," Fourth International Congress on Engineering and Food, Edmonton, Alberta, Canada (July 1985).

Lloyd, A. K., "Plant Control Systems," in *Engineering and Food, Vol. 2, Processing Applications*, B. M. McKenna (ed.), Elsevier Applied Science Publishers, Barking, Essex, England, 1984, pp. 809–817.

Loncin, M., "Modelling in Cleaning, Disinfection and Rinsing," in *Mathematical Modelling in Food Processing, Proceedings of the European Federation of Chemical Engineers Mini-Symposium*, Lund Institute of Technology, Sweden, 1977, pp. 301–335.

Loncin, M. and R. L. Merson, *Food Engineering*, Academic Press, New York, 1979.

Longhouse, A. D., H. Ota, and J. G. Taylor, "Heat and Moisture Design Data for Poultry Housing," *Agric. Eng.*, **41**, 756 (1953).

Lopez, A., *A Complete Course in Canning*, Books 1 and 2, 11th ed., The Canning Trade, Baltimore, MD, 1981.

Lowe, C. R. and P. P. G. Dean, *Affinity Chromatography*, Wiley-Interscience, New York, 1974.

Lund, D., E. Plett, and C. Sandu (eds.), *Fouling and Cleaning in Food Processing*, Department of Food Science, University of Wisconsin, Madison, WI, 1985.

Lurgi GmbH, "Continuous Solvent Extraction of Oil Seeds and Lecithin Production," Lurgi Express Information T1136/11.77, Lurgi Umwelt und Chemotechnik GmbH, Frankfurt (Main) West Germany, 1977.

Lutz, J. M. and R. E. Hardenburg, *The Commercial Storage of Fruits, Vegetables and Florist and Nursery Stocks: Agriculture Handbook No. 66*, U. S. Department of Agriculture, U. S. Government Printing Office, Washington, D. C., 1968.

Mahoney, R. R., H. N. Lazaridis, and J. R. Rosenau, "Protein Size and Meltability in Enzyme-Treated Direct-Acidified Cheese Products," *J. Food Sci.*, **47,** 670 (1981).

Maltini, E., "Thermal Phenomena and Structural Behavior of Fruit Juices in the Pre-Freezing Stage of the Freeze Drying Process," in *Freeze Drying and Advanced Food Technology*, S. A. Goldblith, L. Rey, and W. W. Rothmayer (eds.), Academic, London, 1975, pp. 121–139.

Mange, C. and M. Gelus, "Modeling the Residence Time Distribution in Twin Screw Cooking Extruder," Fourth International Congress on Engineering and Food, Edmonton, Alberta, Canada (July 1985).

Manson, J. E. and J. F. Cullen, "Thermal Process Simulation for Aseptic Processing of Foods Containing Discrete Particulate Matter," *J. Food Sci.*, **39,** 1084 (1974).

Marsh, G. L., J. E. Beuhlert, and S. J. Leonard, "The Effect of Composition upon Bostwick Consistency of Tomato Concentrate," *J. Food Sci.*, **45,** 703 (1980).

Masters, K., *The Spray Drying Handbook*, 3rd ed., Halsted Press, Wiley, New York, 1979.

Matz, S. A., *Bakery Technology and Engineering*, 2nd ed. AVI, Westport, CT, 1972.

McCabe, W. L., J. C. Smith, and P. Harriot, *Unit Operations of Chemical Engineering*, 4th ed., McGraw-Hill, New York, 1985, Chaps. 24 and 26.

McFarlane, I., *Automatic Control of Food Manufacturing Processes*, Applied Science Publishers, London, 1983.

McGinnis, D. S., "Model for Predicting the Performance of a Countercurrent Cyclone System for Separating Food Components," Fourth International Congress on Engineering and Food, Edmonton, Alberta, Canada (July 1985).

McGinnis, R. A., *Beet Sugar Technology*, 3rd ed., Beet Sugar Development Foundation, Fort Collins, CO, 1982.

McLellan, M. R., "New Developments in Computers as Related to Process Control," Fourth International Congress on Engineering and Food, Edmonton, Alberta, Canada, July 7–10, 1985.

McNicol, D. G., "Corn Starch Utilization Study," Corn Refiners Association, September 1972.

Meade, G. P. and J. C. P. Chen, *Cane Sugar Handbook*, 10th ed., Wiley-Interscience, New York, 1977.

Mellor, J. D., *Fundamentals of Freeze Drying*, Academic Press, New York, 1978.

Menting, L. C., B. Hoogstad, and H. A. C. Thijssen, "Diffusion Coefficient of Water and Organic Volatiles in Carbohydrate-Water Systems," *J. Food Technol.*, **5,** 111 (1970a).

Menting, L. C., B. Hoogstad, and H. A. C. Thijssen, "Aroma Retention During The Drying of Liquid Foods," *J. Food Technol.*, **5,** 127 (1970b).

Middleman, S., *Fundamentals of Polymer Processing*, McGraw-Hill, New York, 1977, Chap. 7.

Milnes, B. A., B. R. Breslau, and J. L. Short, "Wine and Fruit Juice Processing via Ultrafiltration," Fourth International Congress on Engineering and Food, Edmonton, Alberta, Canada (July 1985).

Mohsenin, N. N., *Physical Properties of Plant and Animal Material*, Vol. 1, Gordon and Breach, New York, 1980.

Morley, M. J., *Thermal Properties of Meat*, Meat Research Institute, Bristol, England, 1972.

Moskowitz, H. R., *Product Testing and Sensory Evaluation of Foods*, Food and Nutrition Press, Inc., Westport, CT, 1983.

Mudgett, R. E., "Electrical Properties of Foods in Microwave Processing," *Food Technol.*, **36**, 109 (1982).

Mudgett, R. E. and H. G. Schwartzberg, "Microwave Food Processing: Pasteurization of Liquids and Semi-Liquids," in *Food Process Engineering, AIChE Symposium Series 218*, Vol. 78, H. G. Schwartzberg, D. B. Lund and J. L. Bomben (eds.), 1982.

Mulder, H. and P. Walstra, *The Milk Fat Globule*, Commonwealth Agricultural Bureaux, Farnham Royal, Bucks, England, 1974.

Nielsen, C. E., D. K. Christensen, and T. Nordbaek, "Fining of Apple Juice by Ultrafiltration," AIChE National Meeting, Philadelphia, PA (August 1984).

Nishikawa, A. H., "Chromatography, Affinity," in *Kirk–Othmer Encyclopedia of Chemical Technology*, 3rd Ed., Vol. 6, Wiley-Interscience, New York, 1979, pp. 35–54.

Nonino, C. and K. Hayakawa, "Thawing Time of Frozen Food of a Rectangular or Finitely Cylindrical Shape," *J. Food Sci.*, **51**, 116 (1986).

Ohlsson, T. and N. Bengtsson, Dielectric Food Data for Microwave Sterilization Processing, *J. Microwave Power*, **10**, 93 (1975).

Okress, E. C. (ed.), *Microwave Power Engineering Vol. 2 Applications*, Academic Press, New York, 1968.

Patchen, G. O., *Storage For Apples and Pears, Marketing Research Report No. 924*, Agricultural Research Service, U. S. Dept. of Agriculture, Superintendent of Documents, U. S. Govt. Printing Office, Washington, D. C., 1971.

Pearson, A. M., *Processed Meats*, 2nd ed., AVI, Westport, CT, 1984.

Peleg, K., *Produce Handling, Packaging and Distribution*, AVI, Westport, CT, 1985.

Peleg, M., "Flowability of Food Powders and Methods For Its Evaluation—A Review," *J. Food Proc. Eng.*, **1**, 303 (1977).

Peleg, M., "Theoretical Analysis of the Relationship Between Mechanical Hardness and Its Sensory Evaluation," *J. Food Sci.*, **45**, 1156 (1980).

Peleg, M., "The Semantic of Rheology and Texture," *Food Technol.*, **37**, 54, 1983.

Peleg, M. and E. B. Bagley, *Physical Properties of Foods*, AVI, Westport, CT, 1983.

Peleg, M. and A. M. Hollenbach, "Flow Conditioners and Anticaking Agents," *Food Technol.*, **38**, 93 (1984).

Peleg, M., R. Moreyra, and E. Scoville, "Rheological Characteristics of Food Powders," in *Food Process Engineering*, H. G. Schwartzberg, D. B. Lund, and J. L. Bomben (eds.), *AIChE Symposium Series*, **78**, No. 218 (1982).

Peleg, M. and M. D. Normand, "A Computer Assisted Analysis of Some Theoretical Effects in Mastication and Deformation Testing in Foods," *J. Food Sci.*, **47**, 1572 (1982).

Peleg, M. and M. D. Normand, "Characterization of the Ruggedness of Instant Coffee Particles by Natural Fractals," *J. Food Sci.*, **50**, 829 (1985).

Pepper, D., "Improvements in Reverse Osmosis for the Concentration of Dilute Solutions," in *Engineering and Food, Vol. 2, Processing Applications*, B. M. McKenna (ed.), Elsevier Applied Science Publishers, Barking, Essex, England, 1984, pp. 621–627.

Peppler, H. J. and D. Perlman, *Microbial Technology*, Vol. 1—*Microbial Processing*, and Vol. 2—*Fermentation Technology*, Academic Press, New York, 1979.

Plank, R., *Handbook of Refrigeration* (in German), Springer-Verlag, Berlin, 1960 (Original Edition, 1913).

Potter, N. N., *Food Science*, 3rd ed., AVI, Westport, CT, 1978.

Predicasts Inc., *Predicast's Basebook*, Predicasts Inc., Cleveland, OH, 1984.

Prescott S. C. and C. G. Dunn, *Industrial Microbiology*, 4th ed., G. Reed (ed.), AVI, Westport, CT, 1982.

Price, J. E. and B. S. Schweigert, *The Science of Meat and Meat Production*, W. H. Freeman and Co., San Francisco, CA, 1971.

Priestly, R. J. (ed.), *Effects of Heating on Foodstuffs*, Applied Science Publishers Ltd., London, 1979.

Purr, A., "Chemical Changes in Foods with Low Water Content, I. The Enzymatic Degradation of Fat at Low Water Vapor Pressures," *Fette, Seiffen, Anstrichem.*, **68,** 145 (1966).

Purr, A., "Chemical Changes in Foods with Low Water Content. II. Experiments with Lipoxygenase as well as Mixtures of Enzymatically Active Principles and their Impact on Autoxidative Deterioration of Fats in Dry Foods as a Function of Equilibrium Water Content," *Fette, Seifen, Anstrichem.*, **72,** 725 (1970).

Qashou, S., R. I. Vachon, and Y. S. Touloukian, "Thermal Conductivity of Foods," *ASHRAE Trans.*, **78,** 165 (1972).

Rao, M. A. and R. C. Anantheswaran, "Rheology of Fluids in Food Processing," *Food Technol.*, **36,** 116 (1982).

Redit, W. H. and A. A. Hamer, *Protection of Rail Shipments of Fruit and Vegetables, U. S. Department of Agriculture Handbook No. 195*, U. S. Department of Agriculture, U. S. Government Printing Office, Washington, D. C., 1961.

Refrigeration Research Foundation, *The Commodity Storage Manual*, The Refrigeration Research Foundation, Washington, D. C., 1985 (periodically updated).

Reinick, A., "Mathematical Modelling of Freezing and Thawing," Ph.D. Thesis, University of Massachusetts, Amherst, MA, 1985.

Reinick, A. and H. G. Schwartzberg, "Coefficients for Air to Solid Heat Transfer for Uniformly Spaced Arrays of Rectangular Foods," 4th International Congress on Engineering and Food, Edmonton, Canada (July 1985a).

Reinick, A. and H. G. Schwartzberg, "Non-Uniform Freezing and Thawing of Slab-Like Foods," AIChE Annual Meeting, Chicago, IL (November 1985b).

Reinick, A. and H. G. Schwartzberg, "Freezing and Thawing Temperatures Versus Time Evaluation for Rectangular Foods," IFT National Meeting, Atlanta, GA (June 1985c).

Rha, C. K. (ed.), *Theory, Determination and Control of Physical Properties of Foods*, Reidel, Dordrecht, Holland, 1975.

Richardson, T. R. and J. W. Finley, *Chemical Changes in Food During Processing*, AVI, Westport, CT, 1985.

Riedel, L., "Refractive Index and Freezing Temperature of Fruit as a Function of Temperature," *Zeit f. Lebensmittel-Unter und Forsch.*, **89**, No. 2, Springer-Verlag, Berlin, pp. 289–299 (1949).

Riedel, L., "The Refrigerating Effect Required to Freeze Fruits and Vegetables," *Ref. Eng.*, **59**, 289–299 (1951).

Riedel, L., "Calorimetric Investigation of the Freezing of Fish Meat," *Kaltetechnik*, **8**, 374 (1956).

Riedel, L., "Calorimetric Investigation of the Meat Freezing Process," *Kaltetechnik*, **9**, 38 (1957a).

Riedel, L., "Calorimetric Investigation of the Freezing of Egg White and Yolk," *Kaltetechnik*, **9**, 342 (1957b).

Riedel, L., "Calorimetric Investigation of the Freezing of White Bread and Wheat Products," *Kaltetechnik*, **11**, 41 (1959).

Riedel, L., "On the Problem of Bound Water in Meat," *Kaltetechnik*, **13**, 122 (1961).

Riedel, L., "Calorimetric Investigation of Yeast," *Kaltetechnik-Klimatisierung*, **29**, 291 (1968).

Riedel, L., "Calorimetric Investigation of Milk and Milk Products," *Chem. Mikrobiol. Technol. Lebensmittel*, **4**, 177 (1976).

Rockland, L. B. and G. F. Stewart, *Water Activity Influence on Food Quality*, Academic Press, New York, 1981.

Rosenau, J., "Non Rennet/Non Fermentative Methods of Process Cheese Manufacture," in *Engineering and Food, Vol. 2, Processing Applications*, B. M. McKenna (ed.), Elsevier Applied Science Publishers, Barking, Essex, England, 1984, pp. 857–862.

Rosenau, J. R., J. F. Calzada, and M. Peleg, "Some Rheological Properties of a Cheese-Like Product Prepared by Direct Acidification," *J. Food Sci.*, **43**, 948 (1978).

Rossi, M. and C. Peri, "Extrusion Cooking Effects on Structural and Functional Characteristics of Sunflower Protein," Fourth International Congress on Engineering and Food, Edmonton, Alberta, Canada (July 1985).

Ryall, A. L. and W. J. Lipton, *Handling, Transportation and Storage of Fruits and Vegetables*, Vol. 1—*Vegetables and Melons*, 2nd ed., AVI, Westport, CT, 1979.

Ryall A. L. and W. T. Pentzer, *Handling, Transportation and Storage of Fruits and Vegetables*, Vol. 2—*Fruits and Tree Nuts*, 2nd ed., AVI, Westport, CT, 1982.

Sahai, O. and M. Knuth, "Commercializing Plant Tissue Culture Processes: Economics, Problems and Prospects," *Biotechnol. Prog.*, **1**, 1 (1985).

Schenkel, G., *Plastics Extrusion Technology and Theory*, American Elsevier, New York, 1966.

Scherz, H. and G. Kloos, *Souci/Fachman/Kraut Food Composition and Nutrition Tables 1981/82*, 2nd ed., Wissenschaftliche Verlagsgesellschaft gmbH, Stuttgart, 1981.

Schiffmann, R. F., "An Update on the Applications of Microwave Power in the Food Industry," *J. Microwave Power*, **11**, 221 (1976).

Schmitt, P., *Packaging Design 2*, AVI, Westport, CT, 1985.

Schneider, F., *Technologie des Zuckers*, Verlag M. and H. Schaper, Hannover, 1968.

Schubert, H. and H. T. Heideker, "Protein Displacement of Vegetable Food by Selective Comminution and Air Classification," Fourth International Congress on Engineering and Food, Edmonton, Alberta, Canada (July 1985).

Schwartzberg, H. G., "Lightness and Darkness: The Reflectance of Dried Porous Foods," in *Water Removal Process: Drying and Concentration of Foods and Other Materials, AIChE Symposium Series*, **73**, No. 163, C. Judson King and J. Peter Clark (eds.), 1977a.

Schwartzberg, H. G., "Energy Requirements for Liquid Food Concentration," *Food Technol.*, **31**, 77 (March 1977b).

Schwartzberg, H. G., "Effective Heat Capacities for the Freezing and Thawing of Food," *Freezing Frozen Storage and Freeze Drying*, International Institute of Refrigeration, 1977c, p. 303.

Schwartzberg, H. G., "Continuous Counter-Current Extraction in the Food Industry," *Chem. Eng. Prog.*, **67**, 67 (April 1980).

Schwartzberg, H. G., "Mathematical Analysis of Freezing and Thawing," AIChE National Meeting, Detroit, MI (August 1981).

Schwartzberg, H. G., *A Compilation of Readings and Problems in Food Engineering*, Department of Food Science, University of Wisconsin, 1982.

Schwartzberg, H. G., "Progress and Problems in Solid–Liquid Extraction," *Lat. Am. J. Heat Mass Transfer*, **7**, 317–344 (1983a).

Schwartzberg, H. G., "Expression Related Properties," in *Engineering Properties of Foods*, M. Peleg and E. Bagley (ed.), AVI, Westport, CT, 1983b, pp. 423–471.

Schwartzberg, H. G., "Unit Operation Course Lecture Notes—Emulsification and Homogenization," University of Massachusetts, Amherst, MA, 1985.

Schwartzberg, H. G., "Leaching—Organic Materials," in *Handbook of Separation Process Technology*, R. Rousseau (ed.), Wiley, 1987, Chap. 10, pp. 540–577.

Schwartzberg, H. G. and R. Y. Chao, "Solute Diffusivities in Leaching Processes," *Food Technol.*, **36**, 73 (February 1982).

Schwartzberg, H. G., S. Flores, and S. Zaman, "Analysis of Multi-Stage Belt Extraction," AIChE National Meeting, Denver, CO, August 1983.

Schwartzberg, H. G., B. W. Huang, V. Abularach, and S. Zaman, "Force Requirements for Water and Juice Expression From Cellular Plant Material," *Latin Am. Rev. Chem. Eng. Appl. Chem.*, **15**, 141–176 (1985).

Schwartzberg, H. G., J. R. Rosenau, and J. R. Haight, "The Prediction of Freezing and Thawing Temperature vs. Time Behavior Through the Use of Effective Heat Capacity Equations," in *Freezing, Frozen Storage, and Freeze Drying*, International Institute of Refrigeration, 1977, p. 311.

Sell, N. J., *Industrial Pollution Control, Issues and Techniques*, Van Nostrand Reinhold, New York, 1981, Chap. 13.

Sherman, P., *Industrial Rheology*, Academic Press, New York, 1970.

Sherman, P. (ed.), *Food Texture and Rheology*, Academic Press, New York, 1977.

Shuler, M. L., "Production of Secondary Metabolites from Plant Tissue Culture—Problems and Prospects," *Ann. NY Acad. Sci.*, **369**, 65 (1981).

Simatos, D. and J. L. Multon, *Properties of Water in Foods*, NATO ASI Series, Series E, Applied Science No. 90, Martinus Nijhoff Publishers, Dordrecht, The Netherlands, 1985.

Sivetz, M. and N. Desrosier, *Coffee Technology*, AVI, Westport, CT, 1979.

Slade, F. H., *Food Processing Plant*, Vols. 1 and 2, CRC Press, Cleveland, OH, 1967.

Society of Chemical Industry, *Fundamental Aspects of the Dehydration of Foodstuffs*, Society of Chemical Industry, London, 1958.

Spanbauer, J., "Advanced Sensor Based Computer Control of Sugar Beet Pulp Rotary Drum Dryers," in *Engineering and Food, Vol. 2, Processing Applications*, B. M. McKenna (ed.), Elsevier Applied Science Publishers, Barking, Essex, England, 1984, pp. 819–828.

Spicer, A. (ed.), *Advances in Preconcentration and Dehydration of Foods*, Halsted Press, Wiley, New York, 1974.

Staba, E. J. (ed.), *Plant Tissue Culture as a Source of Biochemicals*, CRC Press, Boca Raton, FL, 1980.

Steel, R. (ed.), *Biochemical Engineering (Unit Processes in Fermentation)*, Heywood and Company, Ltd., London, 1958.

Stewart, G. F. and M. A. Amerine, *Introduction to Food Science and Technology*, 2nd ed., Academic Press, New York, 1982.

Stone, H. and J. L. Sidel, *Sensory Evaluation Practices*, Academic Press, Orlando, FL, 1985.

Stryer, L., *Biochemistry*, 2nd ed., W. H. Freeman and Co., San Francisco, CA, 1981.

Stryker, M. and A. A. Waldman, "Blood Fractionation," in *Kirk–Othmer Encyclodpeia of Chemical Technology*, 3rd ed., Vol. 4, Wiley-Interscience, New York, 1978, pp. 25–61.

Stumbo, C. R., *Thermobacteriology in Food Processing*, 2nd ed., Academic Press, New York, 1973.

Succar, J. and K. Hayakawa, "Empirical Formulae for Predicting Thermal Physical Properties of Food at Freezing or Defrosting Temperatures," *Lebensm.-Wiss. Tech.*, **16**, 326–331 (1983).

Svarovsky, L. (ed.), *Solid–Liquid Separation*, 2nd ed., Butterworths, London, 1981, pp. 162–188.

Swern, D., *Bailey's Industrial Oil and Fat Products*, 4th ed. Vol. 2, Wiley-Interscience, New York, 1982.

Szczesniak, A. S., "Texture Measurement," *Food Technol.*, **20**, 1292 (1966).

Szczesniak, A. S., "Rheological Problems in the Food Industry," *J. Texture Stud.*, **8**, 119 (1977).

Tadmor, Z. and I. Klein, *Engineering Principles of Plasticating Extrusion*, Van Nostrand Reinhold, New York, 1970.

Tallmadge, J., "The State of the Surface of Metals and its Importance in Food Engineering," *Ind. Aliment. Agric.*, **89**, 817 (1972).

Tallmadge, J. A. "A Variable-Coefficient Plate Withdrawal Theory For Power Law Fluids," *Chem. Eng. Sci.*, **24**, 471 (1969).

Tallmadge, J. A. "Withdrawal of Flat Plates from Power Law Fluids," *AIChE J.*, **16**, 925 (1970).

Tannenbaum, S. R., *Nutritional and Safety Aspects of Food Processing*, Marcel Dekker, New York, 1979.

Tarnawski, V., P. Jelen, and P. Barnoud, "Instrumentation of Ultrafiltration Plant for Laboratory Experiments with Cottage Cheese Whey," in *Engineering and Food, Vol. 2, Processing Applications*, B. M. McKenna (ed.), Elsevier Applied Science Publishers, Barking, Essex, England, 1984, pp. 801–808.

Teixeira, A. A., J. R. Dixon, J. W. Zahradnik, and G. E. Zinsmeister, "Computer Determination of Spore Survival in Thermally-Processed Conduction-Heated Foods," *Food Technol.*, **23**, 352 (1969a).

Teixeira, A. A., J. R. Dixon, J. W. Zahradnik, and G. E. Zinsmeister, "Computer Optimization of Nutrient Retention in Thermal Processing of Conduction-Heated Foods," *Food Technol.*, **23**, 845 (1969b).

Teixeira, A. A., G. E. Zinsmeister, and J. W. Zahradnik, "Computer Simulation of Variable Retort Control and Container Geometry as a Possible Means of Improving Thiamine Retention in Thermally Processed Foods," *J. Food Sci.*, **40**, 656 (1975a).

Teixeira, A. A., C. R. Stumbo, and J. W. Zahradnik, "Experimental Evaluation of Mathematical and Computer Models for Thermal Process Evaluation," *J. Food Sci.*, **40**, 653 (1975b).

Thijssen, H. A. C., "Fundamentals of Concentration Processes," in *Advances in Preconcentration and Dehydration of Foods*, A. Spicer (ed.), Wiley, New York, 1974a, pp. 13–44.

Thijssen, H. A. C., "Freeze Concentration," in *Advances in Preconcentration and Dehydration of Foods*, A. Spicer (ed.), Wiley, New York, 1974b, pp. 115–149.

Thijssen, H. A. C., "Optimization of Process Conditions During Drying with Respect to Quality Factors," *Lebensm. Wiss. Technol.*, **12**, 308 (1979).

Thijssen, H. A. C., P. J. A. M. Kerkhof, and A. A. A. Leifkens, "Short-Cut Method for the Calculation of Sterilization Conditions Yielding Optimum Quality for Conduction Type Heating of Packed Foods," *J. Food Sci.*, **43**, 1096 (1978).

Thijssen, H. A. C. and L. H. P. J. M. Kochen, "Short-Cut Method for the Calculation of Sterilizing Conditions for Packed Foods Yielding Optimum Quality Retention for Conduction Type Heating at Variable Temperature of Heating and Cooling Medium," in *Food Process Engineering, Vol. 1, Food Processing Systems*, P. Linko, Y. Malkki, J. Olkku, and J. Larinkari (eds.), Elsevier Applied Science Publishers, Barking, Essex, England, 1980.

Thijssen, H. A. C. and W. H. Rulkens, "Retention of Aromas in Drying Liquid Foods," *De Ingenieur*, **80**, 45 (1968).

Thomas, *Thomas Register of American Manufacturers and Thomas Register Catalog File*, 75th ed., Thomas Publishing Co., New York, 1986a.

Thomas, *Thomas Grocery Register*, 86th ed., Thomas Publishing Co., New York, 1986b.

Thompson, D. R., "The Challenge in Predicting Nutrient Retention Changes During Food Processing," *Food Technol.*, **36**, 97 (February 1982).

Thor, W. and M. Loncin, "Optimization of Rinsing Processes," Proceedings of the 5th International Congress on Food Science Technology, Kyoto, Japan, 1978.

Toledo, R. T., "Aseptic Packaging in Non-Metal Containers: A 1983 Update," AIChE National Meeting, Philadelphia, PA (August 1984).

Treybal, R. E., *Mass-Transfer Operations*, 3rd ed., McGraw Hill, New York, 1980, Chap. 11.

Troller, J. A., *Sanitation in Food Processing*, Academic Press, New York, 1983.

Tschubik, I.A. and A. M. Maslow, *Warmephysikalische Konstanten von Lebensmitteln und Halbfabrikaten*, VEB Fachbuchverlag, Leipzig, 1973.

U. S. Bureau of the Census, *Statistical Abstract of the United States, 1985*, 105th ed., Department of Commerce, Superintendent of Documents, U. S. Government Printing Office, Washington, D. C., 1984a.

U. S. Bureau of the Census, *Annual Survey of Manufactures*, U. S. Department of Commerce, Superintendent of Documents, U. S. Government Printing Office, Washington, D. C., 1984b.

U. S. Department of Agriculture, "Accepted Meat and Poultry Equipment, Bulletin MPI-2, U. S. Government Printing Office, Washington, D. C., May 1983.

van Pelt, W. H. J., "Freeze Concentration of Vegetable Juices," in *Freeze Drying and Advanced Food Technology*, S. A. Goldblith, L. Rey, and W. W. Rothmayr (eds.), Academic Press, London, 1975, Chap. 33.

van Pelt, W. H. J. and W. J. Swinkels, "Recent Developments in Freeze Concentration," Fourth International Congress on Engineering and Food, Edmonton, Alberta, Canada, July 7–10, 1985.

van Zuilichem, D. J., W. J. Tempel, W. Stolp, K. van't Riet, "The Processing of Confectionery Doughs in Twin Screw Cooking Extruders," Fourth International Congress on Engineering and Food, Edmonton, Alberta, Canada (July 1985).

Vine, R., *Commercial Winemaking, Processing and Controls*, AVI, Westport, CT, 1981.

Voilley, A. and M. Loncin, "Aroma Retention During the Drying of Foodstuffs," *Proceedings of the 5th International Congress on Food Science Technology*, Kyoto, Japan, 1978.

Voilley, A., D. Simatos, and M. Loncin, "Retention of Volatile Trace Components in Freeze-Drying Model Solutions," *Lebensm.-Wiss. Technol.*, **10**, 45, 1977.

Wang, D. I. C., C. L. Cooney, A. L. DeMain, P. Dunnill, A. E. Humphrey, and M. D. Lilly, *Fermentation and Enzyme Technology*, Wiley, New York, 1979.

Watt, B. K. and A. L. Merrill, *Composition of Foods, Agricultural Handbook 8*, U. S. Dept. of Agriculture, Superintendent of Documents, U. S. Government Printing Office, Washington, D. C., 1975.

Webb, F. C., *Biochemical Engineering*, Van Nostrand, New York, 1964.

Whiteman, T. M., *Freezing Points of Fruits, Vegetables and Florist Stock, U. S. D. A. Marketing Reseearch Report No. 196*, U. S. Department of Agriculture, Superintendent of Documents, U. S. Government Printing Office, Washington, D. C., 1957.

Whitney, L. F. and L. R. Correa, "Fluid Separation of Fish by Shape," ASAE Paper No. 79-5042, ASAE CSAE Joint Summer Meeting, Winnepeg, Canada, 1979.

Williams, S. (ed.), *Official Methods of Analysis*, 14th ed. 1984, Association of Official Analytical Chemists, Arlington, VA, 1984 (reissued periodically).

Wolf, W., W. E. Spiess, and G. Jung, *Sorption Isotherms and Water Activity of Food Material*, Science and Technology Publishers, Hornchurch, Essex, England, 1985.

Woollen, A. (ed.), *Food Industries Manual*, 25th ed., Leonard Hill, London, 1969.

Yano, T., "Engineering Control of Food Quality," Fourth International Congress on Engineering and Food, Edmonton, Alberta, Canada (July 1985).

Zoumes, B. L. and E. J. Finnegan, "Chocolate and Cocoa," *Kirk–Othmer Encyclopedia of Chemical Technology*, 3rd ed. Vol. 6, Wiley-Interscience, New York, 1979, pp. 1–19.

# 12

## CHEMICAL ENGINEERING IN ELECTRONIC MATERIALS PROCESSING: SEMICONDUCTOR REACTION AND REACTOR ENGINEERING

T. W. FRASER RUSSELL

457

## 12.1. INTRODUCTION

In Bell Telephone Laboratories on December 23, 1947, a transistor device was connected into a circuit and amplified a signal by a factor of just over 40. The first transistor was made of germanium and metal-to-semiconductor contacts were used to produce the device. The inventors, J. Bardeen, W. Brattain, and W. Shockley, were awarded the Nobel Prize for this invention in 1956. It was well deserved. The transistor can lay claim to being the most significant invention of this century. In its various modified forms, it can be used as a very rapid switch or an amplifier and is the key component in all microelectronic circuits. One individual transistor in a typical microcircuit today occupies something on the order of $10^{-10}$ of the volume of the miniature vacuum tube used to perform the same function in circuits of the 1930s and 1940s. This beginning of the "Revolution in Miniature" (Braun and MacDonald, 1982) had its roots in the following:

- The invention of wireless telegraphy and the subsequent development of radio and television.
- The invention of the vacuum tube and the development of electronic circuits employing them.
- Fundamental research in solid-state physics.

Limited manufacture of transistors was achieved by the Western Electric Company by the fall of 1951. Their use in telephones and hearing aids was initiated in late 1952. The technology was sufficiently well developed by the mid-1950s to allow the International Business Machine Corporation to market a commercial computer containing some 2200 transistors instead of vacuum tubes. In 1953, there were about 60 types of transistors being marketed; by 1957, this number had grown to about 600.

In the late 1950s, the first integrated circuit, or chip, was made by J. Kilby for Texas Instruments and by R. Noyce for Fairchild Semiconductor. An integrated circuit or a chip has all its components integrated within the semiconductor wafer itself, thus allowing the components and the circuit to be manufactured simultaneously. Even the very early integrated circuits increased the number of components per unit volume by a factor of 1000 over what could be done in the mid-1940s. The process of transferring a circuit pattern to a chip so that the individual microelectronic components can be defined and built up by deposition of various layers is shown in a simple form in Figure 12.1 (Larrabee, 1985).

The essential processes can be summarized as follows (Elliot, 1982):

1. **IMAGING** to replicate the integrated circuit patterns on the various wafer surfaces.
2. **DEPOSITION AND GROWTH** processes to obtain the layers or regions of semiconductor, conductor, and/or insulating materials.
3. **MASKING AND ETCHING** operations in which selective removal or addition of the layers is effected.

Transistor devices utilizing various different designs, diode resistors, and capacitors can be made in this fashion using properly designed layers of semiconductor, insulator, and conductor materials. A "junction field-effect transistor" (JFET) and a diffused-channel "insulated gate field-effect transistor" (IGFET) are shown in Figure 12.2.

The demand for pocket calculators and circuits for audio and video applications led to a widespread production of chips by a number of firms in the early 1960s and the technology advanced at a very rapid rate. Microchips of millimeter dimensions, having a circuit complexity ranging from a hundred to several thousand transistors and other devices were common by the late 1960s. Such chips were made on silicon wafers, 50–75 mm in diameter, using large-scale integration (LSI) manufacturing techniques.

A cross-sectional view of part of a microcircuit showing a resistor, two diodes, and a transistor is shown in Figure 12.3. The equivalent circuit is shown beneath the cross-sectional view (Ankrum, 1971).

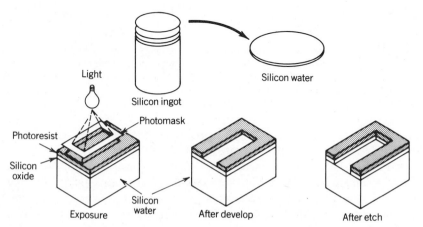

**FIGURE 12.1** Transfer of circuit pattern to a chip. Excerpted by special permission from *Chemical Engineering* (June 10, 1985) (c) (1985), by McGraw-Hill, Inc., New York, NY, 10020.

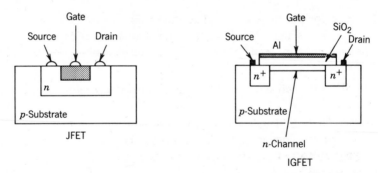

**FIGURE 12.2** A junction field-effect and an insulated gate field-effect transistor. Reproduced by permission from *Electronic Devices*, Allyn and Bacon, Inc., by F. R. Conner.

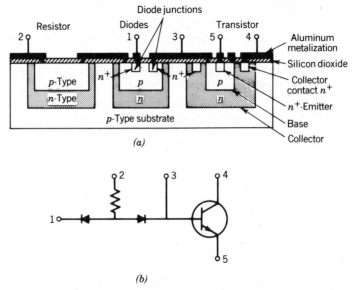

**FIGURE 12.3** A cross-sectional view of a microcircuit. (a) Single-crystal monolithic circuit containing a resistor, two diodes, and a transistor. (b) Schematic diagram of the monolithic circuit of (a). Paul D. Ankrum, *Semiconductor Electronics*, (c) (1971), p. 425. Reprinted by permission of Prentice-Hall, Inc., Englewood Cliffs, New Jersey.

    Microprocessors containing central processing units (CPU's) with one or more chips, made using LSI or VLSI (very large-scale integration) techniques were common by the mid-1970s. A chip made using VLSI manufacturing techniques contains tens of thousands to hundreds of thousands of transistors and/or other devices. Sixteen- and 32-bit microprocessors, the essential elements in computers, are all made using VLSI techniques; and active efforts aimed at putting ever more devices on the chip are a major part of many firms' research and development endeavors.

While various microcircuit products have been responsible for much of the explosive growth in the electronics industry, a number of other solid-state electronic products also have been developed as a result of the transistor and integrated-circuit inventions. Examples are a wide variety of sensors and detectors (infrared, ultraviolet, high-temperature, and biological), light-emitting diodes, solar cells, and lasers.

It is the purpose of this chapter to examine the role that chemical engineers can play as the electronics industry expands and creates new products using new materials. We do so by concentrating on an area of study which can best be termed semiconductor reaction and reactor analysis.

## 12.2. ELECTRONIC MATERIALS PROCESSING

### A. Economic Structure

The Department of Commerce provides us with a wealth of information on industrial activity in the United States. Their information helps to quantitatively define industrial groups such as those commonly referred to as "chemicals" and "electronics." The Department of Commerce does so by defining in considerable detail a Standard Industrial Classification (SIC) group.

There are 99 major groups defined for the United States. A SIC Classification Manual (1972) is available which contains the detailed definitions. What is commonly referred to as the "chemical process industries" consists of SIC groups 28 (Chemical and Allied Products) and 29 (Petroleum Refining and Related Industries). Although it could be argued that groups 20 (Food and Kindred Products), 26 (Paper and Allied Products), 30 (Rubber and Miscellaneous Plastic Products), and 32 (Stone, Clay, Glass, and Concrete Products) should be included. The major groups are broken down into three-digit subgroups (eight for SIC 28 and four for SIC 29). These subgroups are further subdivided into four-digit groups (e.g., 2834, Pharmaceutical Preparations). A detailed discussion for the chemical process industries is presented by Wei et al. (1979).

Like the chemical process industries, the "electronics" industry means different things to different people. In SIC terms, it is part of group 36 (Electrical and Electronic Machinery, Equipment, and Supplies). There are eight three-digit subgroups in SIC 36, but the most pertinent are SIC 367 (Electronic Components and Accessories) and SIC 366 (Communication Equipment).

Group SIC 367 (Electronic Components and Accessories) is the most valuable for "electronic" industry statistics. This group has the following four-digit subgroups:

> 3671 Radio and Television Receiving-Type Electronic Tubes,
> Except Cathode Ray
> 3672 Cathode Ray Television Picture Tubes

3673    Transmitting, Industrial, and Special-Purpose Electron
        Tubes
3675    Electronic Capacitors
3676    Resistors for Electronic Application
3677    Electronic Coils, Transformers, and Other Inductors
3678    Connectors for Electronic Application

If one wishes to see the products for which data are available on the number of manufacturing establishments, sales, people employed, capital investment, and so on, it is necessary to examine the four-digit group definitions. Table 12.1 shows the 39 products that make up SIC 3674 (Semiconductors and Related Devices). As can be seen from Table 12.1, SIC 3674 contains the very important category Microcircuits, integrated. SIC 3679 (Electronic Components Not Elsewhere Classified) has 50 product listings ranging from Antennas to Waveguides and Fittings.

A listing of total year-to-date shipments (sales) for all manufacturing, SIC 28, 29, and 36, is shown in Table 12.2 for 1984 and it gives us some perspective on the size of the electronics industry as compared with the chemical process industry.

The 1982 shipments for SIC 367 of some $34 million are broken down by percentages into the four-digit subgroups in Table 12.3. Groups 3679 and 3674 account for 78% of sales in SIC 367. The industry has been characterized by a phenomenal growth. This is illustrated in Table 12.4 for SIC 3674 (Semiconductors and Related Devices). Since 1967, when electronics was already a well-established industry, there has been a five-fold increase in the number of companies, a two-fold increase in the number of employees, and an eleven-fold increase in sales. Sales per employee, a measure of productivity, has increased by a factor of about 6. If we had gone back to the early 1950s to develop statistics, we would have found that many of the products listed in Table 12.1 did not exist. Some examples are light-emitting diodes, infrared sensors, "metal-oxide semiconductor" (MOS) devices, microcircuits, and solar cells. Part of the growth shown in Table 12.3 is due to the introduction of new products, and part is due to vastly improving such products as microcircuits. The SIC Manual was published in 1972. In an industry like electronics, which is constantly developing new products, a 1972 listing leaves out products of importance to us today. An important example is optical fibers, now approaching a billion-dollar-per-year business, but which only became commercially significant in the early 1980s.

A synopsis of international semiconductor production is given in Table 12.5 for 1985. The data were adapted from the March 1, 1985 issue of *Electronic Business*. North-American firms have about 60% of the business and Japanese firms have about 30%. The major firms involved are tabulated in Table 12.6. Of the 10 top firms (ranked by dollars of production), 5 are American and 4 are Japanese. Since there were 685 companies in 1982 (Table 12.4), it is obvious that there are many small firms in the U.S. electronics business whose sales are much less than those shown in Table 12.5.

Table 12.1 identifies the products known in 1972 which made up one part of the electronics industry. We examine some features of (1) the fabrication process

**TABLE 12.1   SIC Definition of Semiconductors and Related Devices**

*Semiconductors and Related Devices*

Establishments primarily engaged in manufacturing semiconductor and related solid-state devices, such as semi-conductor diodes and stacks, including rectifiers, integrated microcircuits (semiconductor networks), transistors, solar cells, and light-sensing and emitting semiconductor (solid-state) devices.

Computer logic modules
Controlled rectifiers, solid state
Diodes, solid state (germanium, silicon, etc.)
Electronic devices, solid state
Fuel cells, solid state
Gunn effect device
Hall effect devices
Hybrid integrated circuits
Infrared sensors, solid state
Light-emitting diodes
Light-sensitive devices, solid state
Magnetic-bubble memory device
Magnetohydrodynamic (MHD) devices
Memories, solid state
Metal-oxide silicon (MOS) devices
Microcircuits, integrated (semiconductor)
Modules, solid state
Molecular devices, solid state
Monolithic integrated circuits (solid state)
Nuclear detectors, solid state

Parametric diodes
Photoelectric cells, solid state (electronic eye)
Photovoltaic devices, solid state
Semiconductor circuit networks (solid state integrated circuits)
Semiconductors (transistors, diodes, etc.)
Solar cells
Solid-state electronic devices
Strain gages, solid state
Stud bases or mounts for semiconductor devices
Switches, silicon control
Thermionic devices, solid state
Thermoelectric devices, solid state
Thin-film circuits
Transistors
Tunnel diodes
Ultraviolet sensors, solid state
Variable-capacitance diodes
Zener diodes

**TABLE 12.2   Chemical and Electronic Industry Shipments**[a]

| | Value of Shipments (Millions of Dollars) | Percent of Manufacturing |
|---|---|---|
| All manufacturing | 2,273,301 | |
| Chemicals and allied products (SIC 28) | 211,306 | 9.3 |
| Petroleum and coal products (SIC 29) | 197,895 | 8.7 |
| Electronic machinery (SIC 36) | 181,630 | 8.0 |
| Communication (366) | 54,980 | |
| Electronic Components (367) | 44,756 | |
| 366 and 367 | 99,736 | 4.4 |

[a] Data from U.S. Department of Commerce, Bureau of the Census, Current Industrial Reports, December 1984.

**TABLE 12.3   Electronic Components and Accessories ( Standard Industrial Classification Group 367)**

| 1982 Shipments $34,500 Million | |
|---|---|
| Electronic Components, N.E.C. (3679) | 42% |
| Semiconductors and Related Devices (3674) | 36% |
| Tubes (3671) | 7% |
| Connectors (3678) | 7% |
| Capacitors, Resistors, Transformers (3675–3677) | 8% |

**TABLE 12.4   Semiconductors and Related Devices (Standard Industrial Classification Subgroup 3674)**

| | 1967 | 1982 |
|---|---|---|
| Number of companies | 141 | 685 |
| Number of employees | 85,400 | 166,500 |
| Shipments ($ million) | 1,141 | 12,429 |
| Sales/employee ($) | 13,400 | 74,700 |

for integrated circuits, a product with U.S. sales per year in billions of dollars, and (2) the manufacture of photovoltaic cells, a solid-state power source which presently has U.S. sales per year in the 100-million-dollar range, with the potential for very rapid growth. The fabrication processes for integrated circuit, photovoltaic or solar cells, and many other solid-state electronic devices have certain common features, shown in schematic form in Figure 12.4.

Raw materials with sufficiently low levels of impurities such that electronic properties are not adversely affected require extraordinary processing and care in handling. The chemical industry has played a major role in the multibillion-dollar-per-year business of supplying raw materials, and processing chemicals for both

**TABLE 12.5   Worldwide Semiconductor Production—1984[a]**

| $32,105 Million | |
|---|---|
| North America (61%) | $19,500 Million |
| Japan (29%) | $9,225 Million |
| Western Europe (8%) | $2,670 Million |
| Rest of world (2%) | $620 Million |

[a] Data from *Electronic Business*, March 1, 1985.

**TABLE 12.6   Worldwide Semiconductor Manufacturers—1984[a]**

| Firm | Estimated Production ($ Million) |
|---|---|
| Texas Instruments (USA) | 2350 |
| Motorola (USA) | 2255 |
| NEC (Japan) | 1985 |
| Hitachi (Japan) | 1690 |
| Toshiba (Japan) | 1460 |
| National (USA) | 1270 |
| Intel (USA) | 1170 |
| Philips (Europe) | 1150 |
| AMD (USA) | 935 |
| Fujitsu (Japan) | 815 |

[a] Data from *Electronic Business*, March 1, 1985.

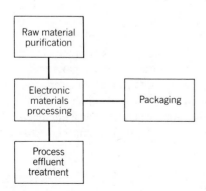

**FIGURE 12.4** Fabrication of microcircuits, photovoltaic panels, and other solid-state electronic devices.

TABLE 12.7　Electronic Materials for
Microcircuits

| Material | Approximate Percent of the Material Value in a Microcircuit |
|---|---|
| Semiconductors | 44 |
| Conductors | |
| Insulators | |
| Processing Chemicals | |
|    Photoresist | 6 |
|    Other Chemicals and Etchants | 9 |
| Packaging | 41 |
|    Ceramics | |
|    Pastes | |
|    Resins | |
|    Adhesives | |

electronic processing operations and packaging. Table 12.7 shows a material breakdown for microcircuit fabrication. A good deal of research, development, and engineering were carried out by the chemical process industries to develop electronic–chemical products such as photoresists for the patterning, etching, and masking operations used in the fabrication of both microcircuits and printed circuit boards. A good review of the electronics–chemicals business was recently published (Cox and Mills, 1985).

Annual growth rate of the materials designated in Table 12.7 is predicted to be around 20%, and 1987 world sales are predicted to be about 68 billion dollars. The United States' share of this market is expected to be about 40%; Japan's share 29%; and Western Europe's share 31%. Some of the larger U.S. companies involved are Eastman Kodak, Allied Chemical, E.I. DuPont, Union Carbide, Olin, and Rohm and Haas.

While the chemical process industry and the chemical professionals in it have played a major role as raw material suppliers for the electronics industry, they have not been very active in the electronic materials processing operation itself. In this chapter, the role that the chemical professionals, and chemical engineers in particular, can creatively and effectively play in electronic materials processing is examined. To do this in the space allowed, we concentrate on those parts of the fabrication process that require a chemical transformation to produce the semiconductor, insulator, or conductor materials that are the essential parts of an electronic device. There are many other areas in device fabrication where chemical engineers can contribute. Some of the most important, but which we are not able to discuss here, are:

- Bulk crystal growth
- Liquid-phase epitaxy

- Chemical and dry-plasma etching
- Doping by diffusion and ion implantation
- Resist processing
- Various control problems

Reactions are carried out many times in the fabrication of any microelectronic device and in the fabrication of larger electronic devices such as solar cells. For most electronic products, the process steps requiring reaction and reactor analysis are those that create the semiconductor insulator and conductor layers of a device (Figures 12.2 and 12.3). A reaction and reactor analysis is also needed if a quantitative understanding of the plasma-etching step in processing is to be achieved. Plasma etching is a reaction step in which material is selectively removed from the device.

## B. Semiconductor Layers

Table 12.8 lists some of the materials utilized in the production of semiconductor devices now in use or which may be in use over the next decade. Almost all microcircuits use single-crystal silicon, and about 60% of the solar cells are made of single-crystal silicon wafers. Many electronic devices and sensors use the other semiconductors shown.

Larrabee (1985) outlines how a silicon wafer is fabricated from single-crystal silicon and provides a good review of microcircuit fabrication procedures. A typical microcircuit-processing sequence starts with a wafer of single-crystal silicon and can have as many as 100–200 steps to define the devices and the interconnects.

**TABLE 12.8   Electronic Materials**

*Semiconductors*

Silicon
Gallium arsenide
Amorphous silicon
III–V compounds
II–VI compounds
Artificially structured materials (''superlattices'')

*Conductors*

Aluminum, copper, gold, molybdenum, nickel, tungsten
polysilicon, tin oxide, indium tin oxide

*Insulators or Dielectrics*

Silicon dioxide, $SiO_2$
Silicon nitride, $Si_3N_4$
Aluminum oxide, $Al_2O_3$
Polymers
$B_xN_y$

Those steps in which reactions to produce semiconductor layers are important include the following.

## 1.  Epitaxial Growth of Silicon

Two kinds of reactions are commonly carried out—pyrolysis of silane

$$SiH_4 \longrightarrow Si + 2H_2$$

and reduction of silicon tetrachloride

$$SiCl_4 + 2H_2 \longrightarrow Si + 4HCl$$

During the growth of the epitaxial film on a silicon wafer, dopant atoms are introduced with gases such as diborane and phosphine to give the desired electronic characteristics. These are the $p$ and $n$ regions shown in Figures 12.2 and 12.3. According to Hammond (1983), the number of "epi" reactors for epitaxial film growth needed per year will be around 150 by 1987. Current reactors are capable of processing between 0.3 and 0.5 $m^2/h$ of wafers in a batch operation.

Polysilicon (poly), which can have a range of electrical properties from semiconductor to conductor depending upon crystal structure and doping, is made by the following reaction:

$$SiH_4 + heat \longrightarrow Si\,(poly) + 2H_2$$

## 2.  Silicon Doping

Doping, usually with diborane or phosphine, to produce the $p$ and $n$ regions in the single-crystal silicon is also an important semiconductor reaction step in microcircuit fabrication. For tetravalent silicon, addition of a trivalent boron produces a $p$-type material (i.e., a "hole" or unfilled bond is created). The addition of a pentavalent phosphorous provides an extra electron ($n$-type material). In a doping furnace or reactor, the silicon wafers are exposed to a vapor phase containing the dopant atoms. The atoms diffuse into the exposed silicon surface, causing a change in electrical properties, while the masked areas remain unaffected (Figure 12.1). The dopant profile in the silicon is controlled by the temperature, the time of exposure to the dopant atom, and the temperature and time that the wafers stay in the annealing or so-called "drive-in" step. Ion implantation is a more recent technique used for doping and is becoming widely accepted, since very good control of the dopant-concentration profile can be achieved (Elliot, 1982). A quantitative analysis of doping and some of the history leading up to today's mathematical models is given by Dutton (1983).

## 3. Other Semiconductors

Creation of semiconductor layers with material other than silicon is beginning to become important in microcircuit fabrication with the use of compounds based on group III and group V elements, especially gallium arsenide. Gallium arsenide is used in discrete devices and integrated circuits for microwave, optoelectronic, and high-speed digital applications. There are some 1400 papers and patents published each year which deal with GaAs and the devices made with it (Hollan, 1980). Like silicon, bulk single crystals of GaAs are sliced into wafers that form a substrate from which a microelectronic circuit is built. A good review of the bulk-growth process is given by Ferry (1985). Epitaxial layers of GaAs are grown on the single-crystal wafer in molecular beam epitaxial (MBE) reactors, or in chemical vapor deposition (CVD) reactors using organometallic compounds (trimethyl or triethyl gallium and arsine). Some gallium arsenide is grown from molten gallium using liquid-phase reactors. There appears to be a market for some 10–20 CVD reactors per year for the epitaxial growth of GaAs (Little, 1985).

Nonsilicon semiconductors are becoming widely used in solar cells because it is felt that inexpensive large areas of semiconductor can only be made with thin films of amorphous silicon or polycrystalline compounds based on group II and group VI elements. Amorphous silicon solar cells with efficiencies in the 10% range are now being produced commercially (about 30% of the terrestrial market). Compounds based on II–VI elements are also used to make polycrystalline heterojunction solar cells of over 10% efficiency. The following types of cells have been made and extensively tested: $CdS/CuInSe_2$, $CdS/CdTe$, and $CdS/Cu_2S$. The GaAs-type solar cells with efficiencies in the 20% range are important for space and concentrator applications.

Amorphous silicon solar cells are produced in commercial quantities from monosilane in a plasma-enhanced CVD reactor. A typical amorphous silicon solar cell is shown in Figure 12.5 (DOE, 1983) and is made in layers using diborane or phosphine in the gas feed for the $p$ and $n$ layers of the cell. A number of cells connected in a series and parallel arrangement make up a module designed to produce power at some convenient voltage (Figure 12.6). The solar cell module is a large-scale version of a microcircuit. The individual devices in a solar cell module are several centimeters in dimension compared with the devices on a chip, which are a few microns in length and width. Masking and etching operations are much easier to carry out in making a solar module, but it is difficult to make cheaply large areas of semiconductor layers which will produce high-efficiency solar cells.

A typical polycrystalline cell using CdS as the wide-band-gap window material is shown in Figure 12.7 (Coutts and Meakin, 1985). These cells are made using sources containing CdS, CdTe, and Cu, In, and Se in physical vapor deposition (PVD) reactors. The individual layers are deposited on a substrate above the sources by using shutters on the source bottle or by moving the substrate to a different reactor.

Sensors and other solid-state electronic devices are made using compounds of

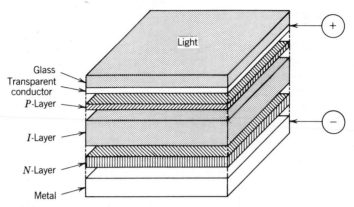

**FIGURE 12.5** Amorphous silicon solar cell. Reprinted from DOE, "Five Year Research Plan 1984–1988, Photovoltaics: Electricity from Sunlight," Solar Energy Research Institute, Department of Energy (May, 1983).

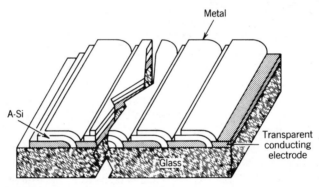

**FIGURE 12.6** Amorphous silicon solar cell module. Reprinted from DOE, "Five-Year Research Plan 1984–1988, Photovoltaics: Electricity from Sunlight," Solar Energy Research Institute, Department of Energy (May, 1983).

III–V and II–VI elements. Table 12.9 shows some of the compounds employed. Their fabrication does not require the submicron pattern definition and etching of today's microcircuit, but there are difficult and complex problems in consistently obtaining materials with the desired properties for effective sensing or imaging.

Research also is being done to develop semiconductor polymers and artificially structured materials. The latter are now a very active topic for research, since they offer the possibility of designing electronic materials to meet specific tasks. An artificially structured material or superlattice is composed of many thin layers (40–60 Å) of materials such as GaAs and AlGaAs. They are difficult to make and very good reactor design and control is required.

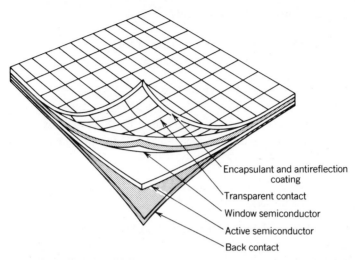

**FIGURE 12.7**   Polycrystalline solar cell. Reprinted from "Semiconductor Chemical Reactor Engineering and Photovoltaic Unit Operations," *Chem. Eng. Educ.*, 72–77 and 106–108 (Spring 1985).

**TABLE 12.9   Some III–V and II–VI Semiconductors**

| *III–V Semiconductors* | | |
|---|---|---|
| GaAs | InAs | $GaAs_{1-x}P_x$ |
| GaN | InP | $GaAs_{1-x}Sb_x$ |
| GaP | InSb | $Ga_{1-x}Al_xAs$ |
| GaSb | | |
| *II–VI Semiconductors* | | |
| ZnS | CdS | HgTe |
| ZnSe | CdSe | HgCdTe |
| ZnTe | CdTe | |

## C.   Insulating or Dielectric Materials

In microcircuits using silicon, the most common insulators are $SiO_2$ and $Si_3N_4$ (Elliot, 1982). The $SiO_2$ insulating layer is deposited in a reactor or furnace with the following chemistry using oxygen or steam over pure silicon:

$$Si + O_2 \longrightarrow SiO_2$$

$$Si + 2H_2O \longrightarrow SiO_2 + 2H_2$$

An $SiO_2$ layer can also be made using silane and carbon dioxide with a nitrogen carrier gas at 500–900°C.

$$SiH_4 + 4CO_2 \longrightarrow SiO_2 + 4CO + 2H_2O$$

A silicon nitride dielectric layer can be made in a CVD reactor using ammonia and monosilane in either a hydrogen or nitrogen carrier gas.

$$SiH_4 + 4NH_3 \longrightarrow Si_3N_4 + 12H_2$$

If nitrogen is used, the reaction takes place at 650°C; if hydrogen is used, 1000°C is required. Low-pressure and plasma-assisted CVD reactors are also used to deposit $SiO_2$ and $Si_3N_4$ layers.

## D.  Conducting Materials

Aluminum is widely used as the interconnection layer on integrated circuits, but many solid-state devices and solar cells utilize other conductors as shown in Table 12.8. Conductors are put onto a surface with various types of PVD reactors. Elliot (1982) lists the following techniques: flash evaporation, filament evaporation, electron-beam evaporation, sputtering, and induction evaporation. Chemical vapor deposition reactors are also used for some transparent conductive oxide layers such as tin oxide.

## 12.3.  CHEMICAL ENGINEERING ANALYSIS

## A.  Semiconductor Reaction and Reactor Analysis

The creation of a semiconductor, insulator, or conductor layer in a micro- or larger electronic device (such as a photovoltaic cell) requires careful design and operation of reactors to achieve certain optical and electronic properties. The general logic for doing so is simple (Figure 12.8), but the detailed implementation is complex. The widespread use of silicon is due, in part, to our ability to relate doping profiles

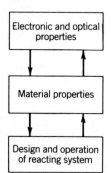

**FIGURE 12.8** Electron properties and reactor performance. Reprinted from "Semiconductor Chemical Reactor Engineering and Photovoltaic Unit Operations," *Chem. Eng. Educ.*, 72–77 and 106–108 (Spring 1985).

(a material property) to emitter-collector currents (an electronic property) on one hand and the performance of the doping operation (design and operation of the reactor) on the other (Dutton, 1983). The industrial and academic research effort to achieve the present state of model development described by Dutton took well over two decades and much remains to be done, particularly in relating material properties to reactor design and operation.

Chemical engineers have made a major contribution to process design and development in the chemical process industries by developing and effectively applying reaction and reactor analysis to processing steps in which some chemical transformation takes place. They can make a major contribution to electronic material processing by applying such skills to obtain a quantitative understanding of the deposition of the semiconductor, insulator, and conductor layers of electronic devices. The procedures that must be developed were termed ''semiconductor reaction and reactor engineering'' by the author in his 1984 3M Award Lecture of the Chemical Engineering Division of the American Society for Engineering Education (ASEE) (Russell, 1985).

Semiconductor reaction and reactor analysis needs to create procedures that improve the design, operation, and product quality of laboratory and commercial-scale reactors. Although a number of reactor types are designated by various names in the literature, they all have similar characteristics and a key part of the engineering analysis can be formulated which is not equipment-specific. Different reactor designations have developed based upon the different ways that molecules are delivered to the substrate on which a layer is to be grown. Molecules can reach the surface by direct line-of-sight impingement or by some form of diffusive, convective mass transfer. When they reach the surface directly in a molecular beam from some source in which raw material is vaporized, the term physical vapor deposition historically has been applied (Figure 12.9). At the pressures commonly used (less than $10^{-6}$ Pa), the material evaporated encounters few intermolecular collisions while traveling to the substrate. Physical vapor deposition reactors may use a solid, liquid, or vapor raw material in a source bottle with energy for evaporation supplied in a variety of ways. Resistive heating is common, and induction heaters or electron beams are also employed. Sputtering systems are PVD reactors in which a plasma supplies energy to the source which is a target somewhat bigger than the substrate. A very sophisticated PVD reactor known as a molecular beam epitaxy (MBE) reactor is commonly used for research and fabrication of some

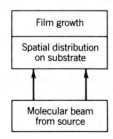

**FIGURE 12.9**  Physical vapor deposition.

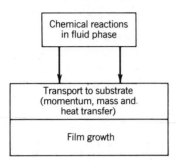

**FIGURE 12.10**   Chemical vapor deposition.

special devices. A "Citation Classic"* on MBE by Cho and Arthur of Bell Laboratories gives a very good review of the early development of this type of reactor (1975). The MBE systems operate at ultrahigh vacuum with considerable care devoted to in situ material characterization. Rates of film growth are very slow and highly controlled.

When the molecules that contain elements needed to form the film are supplied by a vapor phase flowing continuously over a substrate, the term chemical vapor deposition has historically been used (Figure 12.10). Chemical vapor deposition reactor analysis requires a quantitative understanding of transport phenomena, particularly fluid mechanics and mass transfer. The CVD reactors are widely used for semiconductors such as GaAs, other III–V compounds, some II–VI compounds, and amorphous silicon. In making amorphous silicon from silane or disilane, analysis of the gas-phase reactions is important.

Once the molecule has reached the surface, analysis of the molecular phenomena is the same for PVD- or CVD-type reactors. The phenomena are similar to that encountered for a typical surface catalytic reaction (Figure 12.11). After molecules reach the surface, they must adsorb, diffuse, and/or react to produce a layer that will possess the required electronic properties. In catalytic reaction analysis, we are concerned with the product that desorbs from the surface into the vapor phase. In semiconductor reaction analysis, we are concerned with the product that stays on the surface as a thin film.

## B. Analysis of a Growing Thin Film

An analysis for any growing film can be formulated using the control volume illustrated in Figure 12.12 (Jackson, 1984). Regardless of its source (e.g., line-of-sight molecular beam impingement or mass transfer from the gas phase), the rate at which a molecule impinges on the film is represented by $r(i)$ with units of mass per unit area per time, ($kg$-$moles/m^2$-$min$). The molecule may be reflected from the surface or be evaporated, with molar rates per unit of film surface area

*Citation Classics are published in *Current Contents*, Institute for Scientific Information, Philadelphia, PA.

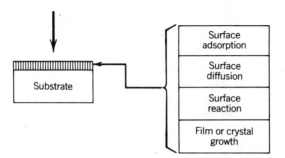

**FIGURE 12.11** Formation of a thin semiconductor, insulator, or conductor layer. Reprinted from "Semiconductor Chemical Reactor Engineering and Photovoltaic Unit Operations," *Chem. Eng. Educ.*, 72–77 and 106–108 (Spring 1985).

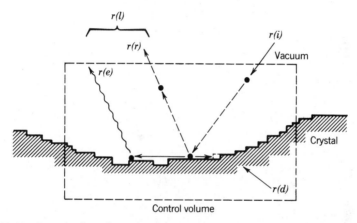

**FIGURE 12.12** Control volume for analysis of a growing thin film: $r(i)$, $r(e)$, and $r(r)$ are, respectively, molar fluxes at which species $j$ impinges on the substrate surface, reevaporates from the substrate surface, and reflects from the substrate surface; $r(i)$ is the sum of $r(e)$ and $r(r)$.

of reflection and evaporation denoted by $r(r)$ and $r(e)$, respectively. The total rate at which material leaves the control volume is the sum of these two terms, $r(1)$. The rate at which film is deposited is $r(d)$.

Using the law of conservation of mass, the following expression can be written for any species $j$ (Jackson, 1984):

$$\frac{1}{A_s M_W(j)} \frac{dM_j}{dt} = [r(i, j) - r(r, j) - r(e, j) - r(r \times t, j)] \quad (12.1)$$

The left-hand side is the molar rate of accumulation or deposition of adsorbed species $j$ per unit area:

$$r(d, j) = \frac{1}{A_s M_W(j)} \frac{dM_j}{dt} \quad (12.2)$$

This rate of film growth (Equation 12.2) can be obtained from measurements of film thickness versus time and knowledge of the film surface area ($A_s$) and is the quantity almost always measured in experiments.

Equation (12.1) is based on a control volume small enough that the incident flux from each source is uniform within it. The net rate of surface diffusion into the control volume is assumed to be negligible compared with this incident flux. The incident atoms of each element, $r(i, j)$, are either adsorbed or reflected from the surface where the rate of reflection is $r(r, j)$. The adsorbed species may react at a rate $r(rxt, j)$ to form a compound, reevaporate from the surface at a rate $r(e, j)$, or be codeposited with the compound in an elemental form at a rate $r(d, j)$. Solutions of Equation (12.1) with appropriate constitutive rate expressions for $r(i, j)$, $r(r, j)$, $r(e, j)$, and $r(rxt, j)$ for all elemental species and compounds within the control volume yield the deposition rate and composition of the film as a function of the component incident rates or fluxes, the temperature of the incident species, and the substrate temperature.

Using Equation (12.1) for the II–VI semiconductor, CdS, yields the following equations (Jackson, 1984):

Cadmium (Cd)

$$r(rxt, Cd) = r(i, Cd) - r(e, Cd) - r(r, Cd) \tag{12.3}$$

Sulfur (S)

$$r(rxt, S) = r(i, S) - r(e, S) - r(r, S) \tag{12.4}$$

Cadmium Sulfide (CdS)

$$\frac{1}{A_s M_W(CdS)} \frac{dM(CdS)}{dt} = -r(rxt, CdS) \tag{12.5}$$

The rate of reaction to form the CdS film is assumed to be:

$$r(rxt, Cd) = k(CdS) [Cd^s] [S^s] \tag{12.6}$$

Substitution of the rate expression (12.6) into Equation (12.5), rearrangement, and inclusion of molecular-mechanics constitutive relations for $r(e)$ allows Figure 12.13 to be prepared. The rate of deposition of CdS, $r(d, CdS)$, is plotted versus the rate at which cadmium is delivered to the surface, $r(i, Cd)$. Jackson (1984) measured both $r(i, Cd)$ and $r(d, CdS)$. His data are shown in Figure 12.13. The model predictions using Equations (12.3)–(12.6) and the constitutive relation for $r(e)$ are shown as solid lines. The impingement rate of sulfur, $r(i, S)$, (Equation 12.4), is varied as shown by the values given beside each curve. Device-quality

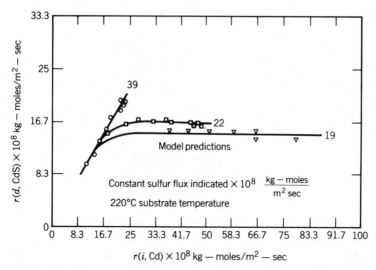

**FIGURE 12.13** Deposition versus incident rate of cadmium in cadmium sulfide. Reprinted from "Semiconductor Chemical Reactor Engineering and Photovoltaic Unit Operations," *Chem. Eng. Educ.*, 72–77 and 106–108 (Spring 1985).

CdS can be made over the range of values of $r(i, \text{Cd})$ and $r(d, \text{CdS})$, where the curves shown in Figure 12.13 begin to bend. This is in the region where the reaction represented by Equation (12.6) appears to control. Before the curves bend, the rate of deposition of CdS film is directly proportional to the rate at which cadmium can be delivered. At values of $r(i, \text{Cd})$ greater than $25 \times 10^{-8}$ kg-moles/m²-sec, the rate of deposition of CdS film is constant. The Cd atoms cannot find a surface site to react and they simply reflect from the surface. Graphs similar to Figure 12.13 are often seen in the literature describing thin-film growth.

This type of analysis can be carried out for any of the materials shown in Table 12.9 and Jackson et al. (1985) have done so for the general case. Their development is applicable for all the binary and ternary alloy semiconductors. The II–VI semiconductors, $(\text{CdHg})\text{Te}$, $(\text{CdZn})\text{S}$, and $\text{CuInSe}_2$, are used to demonstrate the analysis.

The epitaxial growth of gallium arsenide on single-crystal wafers of gallium arsenide can be modeled the same way as CdS with Ga substituted for Cd in Equation (12.3), as substituted for S in Equation (12.4), and with GaAs substituted for CdS in Equation (12.5).

The expressions for film growth rate show that the analysis of the growing film can be carried out without having to consider whether the reactor is a CVD or PVD reactor. The surface molecular phenomenon analysis (reaction engineering) is coupled to the transport phenomenon analysis (reactor engineering) through the rate of impingement $r(i)$. When gas-phase reactions are important, the coupling is more complex; this is illustrated by considering amorphous silicon film growth (Section 12.3D).

## C. Reactor Analysis—Physical Vapor Deposition Reactors

The CdS film growth represented by Equations (12.3)–(12.6) requires that $r(i, Cd)$ and $r(i, S)$ be known if the film thickness as a function of time is to be predicted. These terms are reactor-specific for they represent the rate at which a species is delivered to the substrate. For a PVD system, it is possible to obtain $r(i)$ by analysis of the source bottle evaporation process and the molecular beam distribution on the substrate. Such an analysis has been carried out by Rocheleau et al. (1982) and Jackson et al. (1985). A typical PVD reactor is shown in Figure 12.14. A summary of the analysis has been given by Russell (1985).

The rate of evaporation of any material is determined by the surface temperature of the source material. For thermal evaporation, this is a function of bottle geometry, the material surface area, and the design of the source heater. To make a semiconductor film, the material of interest is placed in the source bottle, heated to the point at which it evaporates or sublimes, flows out of the source bottle, and then condenses on the substrate, the temperature of which is carefully controlled.

The development and experimental verification of a model describing the rate of effusion for CdS that dissociates and sublimes has been thoroughly discussed by Rocheleau et al. (1982). The mass and energy balance equations are (see No-

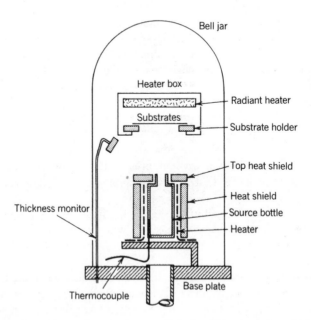

**FIGURE 12.14** Physical vapor deposition reactor. Reprinted from Rocheleau, R. E., B. N. Baron, and T. W. F. Russell, "Analysis of Evaporation of Cadmium Sulfide for Manufacture of Solar Cells," *AIChE J.*, **28** (1), 656 (1982). Reproduced by permission of the American Institute of Chemical Engineers.

menclature for explanations of notations):

$$\frac{\partial \rho V}{\partial t} = -\rho_g q \tag{12.7}$$

$$\rho V C_p \frac{\partial T_1}{\partial t} = -\rho_g q \Delta H_R + F_v F_\epsilon \sigma (T_2^4 - T_1^4) A_s \tag{12.8}$$

These equations can be solved numerically, given the initial dimensions of the material in the source bottle and the appropriate constitutive equations for the flow through the orifice in the source bottle. Table 12.10 gives the required equations in terms of the mass flux, $r$, related to $\rho_g q$ through the area available for flow. The solution method is somewhat complex and complete details are given by Rocheleau et al. (1982) and Rocheleau (1981). Solving the equations yields the rate of effusion versus charge temperature, $T_1$. A comparison of model prediction (solid lines) and experimental data (horizontal bars marked with the wall temperature ($T_2$) are shown for two different orifices in Figure 12.15. The heat transfer from the source bottle walls to the evaporating surface is the key issue in predicting rate of effusion from the source bottle.

This analysis yields rate of effusion from the bottle. To obtain $r(i)$, it is necessary to know where the species impinges on the substrate. This requires experiments in which the film thickness is measured as a function of spatial position on the substrate. Such experiments have been performed for Cd, Zn, and S by Jackson (1984). A typical set of such data is shown in Figure 12.16. Jackson et al. (1985) show that such data can be expressed in terms of a beam distribution function, $F(\phi)$.

$$F(\phi) = 8 \left[ \frac{\cos (\phi)}{\pi} \right]^{3/2}$$

$$\times \int_0^1 (1 - z^2)^{1/2} \int_0^{kz} \exp (-y^2) \, dy \, dz \tag{12.9}$$

$$k = \frac{A}{\tan (\phi) \left[ \sin^{1/2}(\phi) \cos^{1/2} (\phi) \right]}. \tag{12.10}$$

The parameter $A$ can be obtained from Figure 12.17. This figure shows a maximum or peak value of $A$ greater than 1.7 at a Knudsen number (the ratio of the mean free path to the nozzle diameter) of 0.10. The correlation falls off abruptly on both sides of this peak. The peak occurs in the transition regime for mass flow. The flow becomes increasingly laminar to the left and free molecular to the right of the peak.

Cadmium sulfide has been selected as an example because it illustrates the growth of a compound film without great algebraic complexity and because of its

**TABLE 12.10  Constitutive Flow Equations**[a]

| Flow Regime | Orifice | Pipe |
|---|---|---|
| Free molecular $(\lambda_m/R > 1)$ | $r = \left(\dfrac{\rho_1}{2\pi}\right)^{1/2}(p_1 - p_2)$ | $r = \dfrac{R^2\rho_1}{16\mu L}(p_1^2 - p_2^2)\left[4\left(\dfrac{2}{f} - 1\right)\lambda_m/R\right]$ |
| Viscous $(\lambda_m/R < 0.01)$ | $r = C_0 Y\left[2\rho(p_1 - p_2)\right]^{1/2}$ | $r = \dfrac{R^2\rho_1}{16\mu L}(p_1^2 - p_2^2)$ |

[a]Reprinted from Rocheleau, R. E., B. N. Baron, and T. W. F. Russell, "Analysis of Evaporation of Cadmium Sulfide for Manufacture of Solar Cells," *AIChE J.*, **28** (1), 656 (1982). Reproduced by permission of the American Institute of Chemical Engineers.

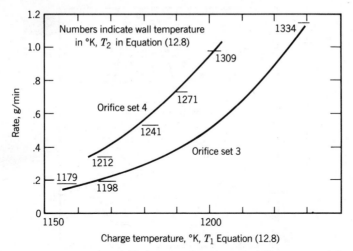

**FIGURE 12.15** Comparison of model behavior with experimental data. Reprinted from Rocheleau, R. E., B. N. Baron and T. W. F. Russell, "Analysis of Evaporation of Cadmium Sulfide for Manufacture of Solar Cells," *AIChE J.*, **28** (1), 656 (1982). Reproduced by permission of the American Insitute of Chemical Engineers.

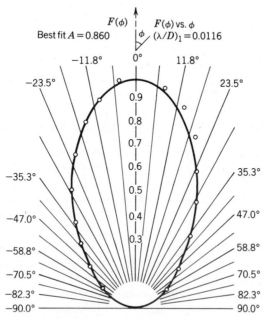

**FIGURE 12.16** Measured molecular beam distribution. Open circles represent the measured values. The curve represents predicted distribution. Reproduced by permission from S. C. Jackson, B. N. Baron, R. E. Rocheleau, and T. W. F. Russell, "Molecular Beam Distributions from High Rate Sources," *J. Vac. Sci. Technol.* **A3** (5), 1916–1920 (1985) by the American Institute of Physics.

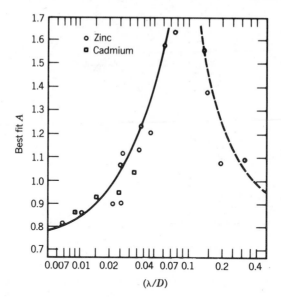

**FIGURE 12.17**   Parameter $A$ in Equation (12.10) vs. Knudsen number. Reproduced by permission from S. C. Jackson, B. N. Baron, R. E. Rocheleau and T. W. F. Russell, "Molecular Beam Distributions from High Rate Sources," *J. Vac. Sci. Technol.* **A3** (5), 1916–1920 (1985) by the American Institute of Physics.

importance as the wide-band-gap window material in polycrystalline solar cells. The CdS analysis has been used to design and operate unit-operations-scale equipment in which CdS was deposited continuously on a moving substrate (Russell, 1985; Rocheleau et al., 1982).

## D.   Reactor Analysis—Chemical Vapor Deposition Reactors

A comprehensive and valuable summary of the reactor analysis of various CVD reactors has been completed by Jensen (1986) and is reproduced with permission as Table 12.11. The table is self-explanatory and illustrates the importance of the gas-phase fluid mechanics and mass transfer in the reactor analysis.

An engineering analysis for amorphous silicon illustrates the procedure for CVD reactors and also shows how gas-phase reactions need to be considered. Bogaert et al. (1985) investigated thermal decomposition of both silane and disilane to produce amorphous silicon films. A sketch of the reacting system is shown in Figure 12.18. The $B_2H_6$ and $PH_3$ are used to dope the film. The film is deposited on substrates in the quartz tube. Film growth rates from 1–30 Å were achieved by varying the reactor temperature (360–485°C), reactor pressure (2–48 Torr), and gas holding time. The gas-phase reactions are dominated by the decomposition of saturated silanes and the insertion of silylene ($SiH_2$).

$$Si_nH_{2n+2} \longrightarrow Si_{n-1}H_{2n} + SiH_2$$

Bowery and Purnell (1971) and Ring (1977) showed that the silane-decomposition and silane-insertion reactions are first order. The gas-phase chemistry and polymer- and film-formation reactions are shown in Figure 12.19. The mathematical model equations for the reactor in which axial diffusion and radial dispersion are negligible are:

$$\frac{u dC_i}{dz} = -\Sigma\, r_i(rxt,\ i) - k_i C_i \qquad (12.11)$$

$$\frac{1}{MW_f} \frac{d\rho V_f}{dt} = \gamma_i k_i C_i \qquad (12.12)$$

where

$u$ = average linear velocity

$C_i$ = bulk gas concentration for species i;

$z$ = axial position

$r_i(rxt,\ i)$ = rate of gas-phase reaction for species $i$

$k_i$ = film-formation rate constant for species $i$

$\rho$ = film density, 2350 kg/m$^3$

$t$ = deposition time

$V_f$ = volume of the film

$\gamma_i$ = stoichiometric coefficient

Bogaert (1985) measured the gas-phase composition at the reactor exit and film growth rates. A comparison of the model predictions and experiments is shown in Figure 12.20, where $a$ shows the exit gas-phase composition versus gas-holding time, and $b$ shows the film growth rate at various holding times, versus axial position in the reactor. The solid lines are model predictions, and the points are experimental data. The study shows that silane pyrolysis and subsequent film growth can be represented by a series of gas reactions having a termination step at the substrate. The main film precursor appears to be a silicon hydride polymer. The by-products of the polymer reactions are hydrogen, monosilane, and solid amorphous silicon hydrogen film. The coupling between the gas-phase reactions and the film-formation reactions does not allow us to solve the problem in two uncoupled parts, as we did with the PVD reactor analysis.
PVD reactor analysis.

**TABLE 12.11  CVD Reactor Models**[a]

| Reference | Deposition System | Model Dimension and Solution | Flow and Transport | Chemical Kinetics |
|---|---|---|---|---|
| | | *Horizontal Reactors* | | |
| Eversteijn et al. (1970a, b) | $SiH_4/H_2$ | 2D Analytical | Stagnant layer adjacent to deposition surface / Linear temperature profile in layer / Plug flow with uniform temperature | Zero $SiH_4$ concentration at the surface / Mass-transfer-controlled deposition rate |
| van de Ven et al. (1987) | $Ga(CH_3)_3$ $AsH_3$ | 2D Analytical | | Deposition rate controlled by mass transfer of $Ga(CH_3)_3$ |
| Kuznetsov and Belyi (1970) | $GeI_4/H_2$ $SiCl_4/H_2$ | 2D Analytical | Boundary-layer treatment with $Pr = S_c = 1$ | Mass-transfer-controlled deposition rate |
| Rundle (1971) | $SiCl_4/H_2$ $SiH_4/H_2$ | 2D Analytical | Constant gas temperature / Plug flow / No natural convection | Mass-transfer-controlled film growth |
| Takahasi et al. (1972) | $SiCl_4/H_2$ $SiH_4/H_2$ | 3D Concentration / 2D Flow and temperature–numerical solution by finite difference | Constant physical properties / Well-developed laminar flow in forced-convection study / Natural convection included | Mass-transfer-controlled deposition rate |
| Berkman et al. (1977) Ban (1978) | $SiCl_4/H_2$ | 2D Analytical | Developing boundary layer / Turbulent bulk-flow region / linear temperature profile in layers / Thermodiffusion included | Mass-transfer-controlled deposition rate |

484

| Reference | System | Method | Flow assumptions | Kinetics / Comments |
|---|---|---|---|---|
| Westphal et al. (1982) | GaAs deposition by the AsCl$_3$/Ga/H$_2$ process | 2D Finite-difference solution | Boussinesq liquid Natural convection included | No deposition kinetics |
| Coltrin et al. (1984, 1986) | SiH$_4$/H$_2$/He | 2D Numerical solution boundary-layer equations Finite difference | No natural convection Varying physical properties Developing flow Thermodiffusion included (1986) | Detailed set of gas-phase reactions based on sensitivity analysis of detailed kinetic scheme |
| Moffat and Jensen (1986) | Ga(CH$_3$)$_3$, AsH$_3$ in H$_2$ SiH$_4$, H$_2$ | 3D Finite elements Parabolized fluid flow equations | Mixed convention Varying physical properties Developing flow | First-order surface reactions Ga incorporation mass transfer limited SiH$_2$, SiH$_4$, Si$_2$H$_6$ kinetics as from Coltrin and coworkers (1986) |

*Barrel Reactors*

| Reference | System | Method | Flow assumptions | Kinetics / Comments |
|---|---|---|---|---|
| Fujii et al. (1972) | SiCl$_4$/H$_2$ | 2D "Separation of variables" solution | Fully developed temperature and flow fields Constant physical properties No natural convection | mass-transfer-controlled deposition rate |
| Dittman (1974) | SiH$_4$/H$_2$/AsH$_3$ | 1D Analytical | Boundary-layer treatment | mass-transfer-controlled deposition rate |
| Manke and Donaghey (1977a, b) | SiCl$_4$/H$_2$ | 2D Separation of variables (a) Finite-difference solution (b) | Fully developed laminar flow No natural convection | 70% of SiCl$_4$ flux to surface is converted into Si(s) |
| Juza and Cermak (1980, 1982) | SiCl$_4$/H$_2$ | 2D Finite difference | Developing flow, temperature, and concentration | One reaction: $2H_2 + SiCl_4 \rightleftharpoons Si + 4HCl$ |

**TABLE 12.11** *(Continued)*

| Reference | Deposition System | Model Dimension and Solution | Flow and Transport | Chemical Kinetics |
|---|---|---|---|---|
| | | | profiles<br>Thermal diffusion included<br>No natural convection | Fitted-rate expression |
| *Vertical Reactor—Rotating Disk* | | | | |
| Sugawara (1972) | $SiCl_4/H_2$ | 2D<br>Finite difference<br>1D for infinite disk | No natural convection | Equilibrium of $SiCl_4$, HCl, and Si(s) at the deposition surface |
| Pollard and Newman (1980) | $SiCl_4/H_2$ | 1D<br>Finite-difference solution | Classical infinite rotating-disk analysis<br>Varying physical parameters<br>Multicomponent diffusion<br>No natural convection | Gas-phase and surface reactions<br>Mass-action kinetics |
| *Impinging-Jet Reactors* | | | | |
| Michaelidis and Pollard (1984) | $BCl_3/H_2$ | 1D<br>Finite difference | 1D impinging-jet analysis<br>Varying physical properties<br>Multicomponent diffusion<br>No natural convection | Gas-phase and surface reactions<br>Mass-action kinetics |
| Jenkinson and Pollard (1984) | $SiCl_4/H_2$<br>$BCl_3/H_2$ | As from Michaelidis and Pollard (1984)<br>Detailed investigation of thermodiffusion effects | | |
| *Stagnation Point-Flow Reactors* | | | | |
| Wahl (1984) | Arbitrary | 2D axisymmetric<br>Finite difference | Varying physical properties<br>Mixed convection included<br>Thermal diffusion | Several different mechanisms involving surface and gas-phase reactions, and mass-transfer-controlled deposition |

| | | | | |
|---|---|---|---|---|
| Kusumoto et al. (1985) | Ga(CH₃)₃/AsH₃/H₂ | 2D axisymmetric Finite difference | Varying physical properties Mixed convection Effect of walls | Mass transfer Limited growth |
| Houtman et al. (1986) | Arbitrary | 1D Finite elements 2D axisymmetric | Varying physical properties Mixed convection | Mass-transfer, first-order surface reaction |
| *LPCVD Reactors* | | | | |
| Ritchman et al. (1979) | SiH₄ pyrolysis | 1D Analytical | Plug flow No diffusion limitations Volume change included | Different models for surface decomposition of SiH₄ |
| Kuiper et al. (1982) | SiH₄ pyrolysis | 1D Analytical | Plug flow No diffusion limitation | SiH₄ decomposition mechanisms involving gas-phase and surface reactions |
| Jensen and Graves (1983) | SiH₄ pyrolysis | 2D Numerical solution by orthogonal collocation | 1D axial-convection diffusion model 1D reaction-diffusion model of wafer region No multicomponent diffusion | SiH₄ decomposition modeled by a Langmuir–Hinshelwood rate expression |
| Roenigk and Jensen (1985) | As above, but extended to include multicomponent-diffusion and multireaction schemes | | | |

"Reprinted with permission from "Micro-Reaction Engineering: Applications of Reaction Engineering to Processing of Electronic and Photonic Materials," Klavs F. Jensen (1987) by Pergamon Press, Inc.

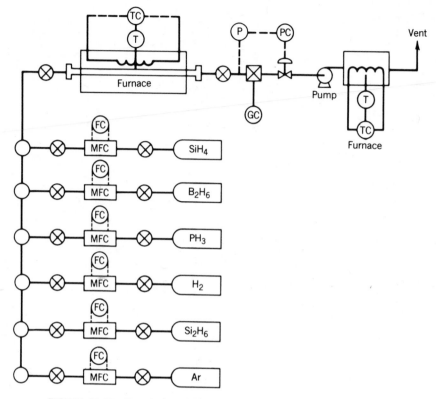

**FIGURE 12.18**   Chemical vapor deposition reactor (a-Silicon from disilane).

## 12.4   CONCLUDING REMARKS

An essential part of electronic materials processing is the creation of the semiconductor, insulating, and conducting layers of a device. These layers are made in
reactors which can be broadly classified into:

- Physical vapor deposition reactors. A PVD reactor is one in which there may
  be line-of-sight delivery of the species to the substrate.
- Chemical vapor deposition reactors. A CVD reactor is one in which the molecular species is delivered to the substrate by a convective–diffusive mass
  transfer through the vapor phase.

A reaction analysis of the film growth on the substrate surface is not equipment-
specific and has many features common to surface catalytic reaction analysis. With
some semiconductors, it may also be necessary to perform a reaction analysis on
the gas phase. An order-of-magnitude increase in our capability to measure species
present in the gas and on the surface must be achieved to see real progress in
reaction analysis.

## GAS PHASE REACTIONS

$$\text{SiH}_4 \underset{3}{\overset{2}{\rightleftharpoons}} \overset{\text{SiH}_2}{\text{Si}_2\text{H}_6} \underset{5}{\overset{4}{\rightleftharpoons}} \overset{\text{SiH}_2}{\text{Si}_3\text{H}_8} \underset{7}{\overset{6}{\rightleftharpoons}} \overset{\text{SiH}_2}{\text{Si}_4\text{H}_{10}} \underset{9}{\overset{8}{\rightleftharpoons}} \overset{\text{SiH}_2}{\text{Si}_5\text{H}_{12}}$$

$$\downarrow 1$$

$$\text{SiH}_2, \text{H}_2$$

## POLYMER FORMATION REACTIONS

$$\text{Si}_3\text{H}_8 \xrightarrow{k_{11}} \frac{2}{n} \text{Si}_n\text{H}_{2n} + \text{SiH}_4$$

$$\text{Si}_4\text{H}_{10} \xrightarrow{k_{11}} \frac{3}{n} \text{Si}_n\text{H}_{2n} + \text{SiH}_4$$

$$\text{Si}_5\text{H}_{12} \xrightarrow{k_{11}} \frac{4}{n} \text{Si}_n\text{H}_{2n} + \text{SiH}_4$$

## FILM GROWTH REACTIONS

$$\text{Si}_n\text{H}_{2n} \xrightarrow{k_{12}} n\text{SiH}_{0.08} + 0.96n\text{H}_2$$

$$\text{SiH}_2 \longrightarrow \text{SiH}_{0.08} + 0.96\text{H}_2$$

**FIGURE 12.19** Amorphous silicon reactions.

The reactor analysis is also badly hampered by a lack of information with which to test model predictions. Film growth rate data and electro-optical properties of the film are usually given with some information available on film uniformity as a function of spatial position. It is very rare to find an article that defines reactor geometry in the detail one needs and rarer still to find literature in which the key temperatures and/or gas composition have been measured.

It is easier to carry out the reactor analysis for PVD systems since more experimental information is available. It is usually possible to calculate the rate of impingement if reactor and source bottle geometry and temperature are given. Almost no information is available on gas-phase composition in CVD reactors, usually because very low conversions are attained (i.e., the reactor is operated as a differential reactor with no measurable concentration gradients).

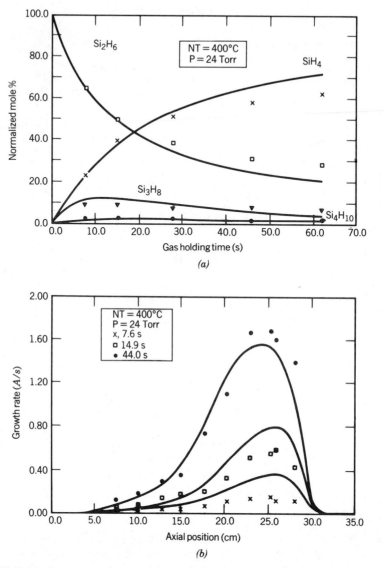

**FIGURE 12.20** Comparison of model prediction with data for amorphous silicon deposition. (a) Exit gas-phase composition vs. holding time. (b) Film growth rate vs. axial position in the reactor.

A major research effort to relate electronic and optical properties to the design and operations of the reactor will allow us to "tailor make" electronic materials for specific needs. This will require a multidisciplinary research team of material scientists, physicists, chemists, and electrical and chemical engineers. A reaction and a reactor analysis is a necessary part of such research but it is not, in itself, sufficient to achieve the creative breakthroughs. It will, however, provide a structure that allows a multidisciplinary team of researchers to more effectively communicate with each other, to plan experiments more creatively and more efficiently, and to design new innovative reactors.

## ACKNOWLEDGMENTS

The author would like to thank his colleagues, Bill N. Baron, R. E. Rocheleau, and S. C. Jackson for many valuable discussions and for their careful experimental and theoretical work. Thanks are also due to the University of Hawaii and the Hawaiian Natural Energy Institute for sabbatical support during manuscript preparation and for allowing the author to attend J. Holm-Kennedy's excellent microcircuit laboratory course.

A special debt is owed Paula Newton for her care in manuscript preparation and proofreading.

## NOMENCLATURE

| | |
|---|---|
| $A$ | Parameter in Equation (12.10), Figure 12.17 |
| $A_s$ | The surface area of substrate, Equation (12.1), $m^2$ |
| $C_i$ | The bulk gas-phase concentration, Equation (12.11), kg-mole/$m^3$ |
| $C_0$ | Orifice coefficient, Table 12.10 |
| $C_P$ | The heat capacity of CdS, Equation (12.8), J/kg-mole $\cdot$ K |
| $F_\epsilon$ | The effective emissivity, Equation (12.8), dimensionless |
| $F_\nu$ | The view factor, Equation (12.8), dimensionless |
| $L$ | The pipe or orifice length, Table 12.10, m |
| $MW(j)$ or $MW_j$ | The molecular weight, species $j$, dimensionless |
| $p_1$ | The pressure upstream of orifice or pipe, Table 12.10, Pa |
| $p_2$ | The pressure downstream of orifice or pipe, Table 12.10, Pa |
| $q$ | The volumetric flow rate of the exit stream from a PVD reactor of Figure 12.14, Equation (12.7), $m^3/s$ |
| $R$ | The ideal gas constant, Table 12.10, 847.8 kg$_f$-m/kg-mk $\cdot$ K |
| $r(d,j)$ | The molar flux (rate per unit area) at which species $j$ is deposited on the substrate to form a thin film, Equation (12.2), kg-mole/$m^2$-s |
| $r(e,j)$ | The molar flux at which species $j$ reevaporates from the substrate surface, Equation (12.1), kg-mole/$m^2$-s |
| $r(i,j)$ | The molar flux at which species $j$ impinges on the substrate surface, Equation (12.1), kg-mole/$m^2$-s |
| $r(r,j)$ | The molar flux of reflection from the substrate surface for species $j$, Equation (12.1), kg-mole/$m^2$-s |
| $r(rxt,j)$ | The net ratio of reaction of adsorbed species $j$ on the substrate surface, Equation (12.1), kg-mole/$m^2$-s |

| | |
|---|---|
| $T_1$ | The temperature of packed CdS in the PVD reactor of Figure 12.14, Equation (12.8), °K |
| $T_2$ | The temperature of bottle wall in the PVD reactor of Figure 12.14, Equation (12.8), °K |
| $t$ | Time, s |
| $u$ | The average linear velocity, Equation (12.11), m/s |
| $V$ | The volume of packed CdS powder in a PVD reactor of Figure 12.14, Equation (12.8), m³ |
| $V_f$ | The volume of thin film, Equation (12.12), m³ |
| $Y$ | Gas density expansion factor |
| $z$ | The axial coordinate |
| $\sigma$ | Stefan–Boltzmann constant |
| $\epsilon$ | The Boltzmann constant, Equation (12.8), $5.669 \times 10^{-8}$ w/m²-k⁴ or a $1714 \times 10^{-8}$ Btu/hr-ft²-°R⁴ |
| $\gamma_i$ | The stoichiometric coefficient, Equation (12.12), dimensionless |
| $\lambda_m$ | The mean free path, Table 12.10, m |
| $\rho$ | The density of thin film, Equation (12.12), kg/m³ |
| $\rho_g$ | The density of gas phase, Equation (12.7), kg/m³ |
| $\mu$ | The viscosity of gas at the average pressure, Table 12.10, kg/m-s |
| $\phi$ | Angle from the nozzle centerline |
| $\Delta H_R$ | The latent heat of vaporization, Equation (12.8), J/kg-mole |

## REFERENCES

Ankrum, P. D., *Semiconductor Electronics*, Prentice-Hall, Englewood Cliffs, NJ, 1971.

Ban, V. S., "Transport Phenomena Measurements in Epitaxial Reactors," *J. Electrochem. Soc.*, **125**, 317 (1978).

Berkman, S. V., V. S. Ban, and N. Goldsmith, "An Analysis of the Gas Flow Dynamics in a Horizontal CVD Reactor," in *Heteroepitaxial Semiconductors for Electronic Devices*, C. W. Cullen and C. C. Wang (eds.), Springer-Verlag, New York, 1977.

Bogaert, R. J., "Chemical Vapor Deposition of Amorphous Silicon Films," PhD Thesis, University of Delaware, Newark, DE, 1985.

Bogaert, R. J., and R. E. Rocheleau, B. N. Baron, M. T. Klein, S. C. Jackson, and T. W. F. Russell, "Chemical Reaction Analysis of Amorphous Silicon Deposition by CVD Using Disilane," AIChE Annual Meeting, Chicago, IL (November 1985).

Bowery, M. and J. H. Purnell, "The Pyrolsis of Disilane and Rate Constants of Silene Insertion Reactions," *Proc. Roy. Soc. Lond.*, **A321,** 341 (1971).

Braun, E. and S. MacDonald, in *Revolution in Miniature*, Cambridge University Press, Cambridge, England, 1982.

Coltrin, M. E., R. J. Kee, and J. A. Miller, "A Mathematical Model of the Coupled Fluid Mechanics and Chemical Kinetics in a Chemical Vapor Deposition Reactor," *J. Electrochem. Soc.*, **131**, 425 (1984).

Coltrin, M. E., R. J. Kee, and J. A. Miller, "A Mathematical Model of Silicon Chemical Vapor Deposition: Further Refinements and the Effects of Thermal Diffusion," *J. Electrochem. Soc.*, **133**, 1206 (1986).

Cho, A. Y. and J. R. Arthur, "Molecular Beam Epitaxy," *Prog. Solid State Chem.*, **10**, 157 (1975).

Connor, F. R., *Electronic Devices*, Allyn and Bacon, Boston, MA, 1980.

Coutts, T. J. and J. D. Meakin, *Current Topics in Photovoltaics*, Academic Press, New York, 1985.

Cox, D. S. and A. R. Mills, "Electronic Chemicals: A Growth Market for the 80's," *Chem. Eng. Prog.*, **81**, 11 (January 1985).

Dittman, F. W., "Epitaxial Deposition of Silicon in a Barrel Reactor," *Adv. Chem. Ser.*, **133**, 463 (1974).

DOE, "Five Year Research Plan 1984–1988, Photovoltaics: Electricity from Sunlight," Solar Energy Research Institute, U.S. Department of Energy, May 1983.

Dutton, R. W., "Modeling of the Silicon Integrated Circuit Design and Manufacturing Process," *IEEE Trans. Electron Dev.*, **30** (9), 968 (1983).

Elliott, D. J. *Integrated Circuit Fabrication Technology*, McGraw-Hill, New York, 1982.

Eversteijn, F. C. and H. L. Peek, "Design Considerations on the Epitaxial Growth of Silicon From Silane in a Horizontal Reactor," *Philips Res. Rep.*, **25**, 472 (1970a).

Eversteijn, F. C., P. J. W. Severin, C. J. H. van den Brekel, and H. L. Peek, "A Stagnant Layer Model for the Epitaxial Growth of Silicon From Silane in a Horizontal Reactor," *J. Electrochem. Soc.*, **117** (7), 925 (1970b).

Ferry, D. K. *Gallium Arsenide Technology*, Sams & Co., Inc., Indianapolis, IN, 1985.

Fujii, E., H. Nakamura, K. Haruna, and Y. Koga, "A Quantitative Calculation of the Growth Rate of Epitaxial Silicon From SiCl in a Barrel Reactor," *J. Electrochem. Soc.*, **119**, 1106 (1972).

Hammond, M. L., "Silicon Epitaxy: A 1983 Perspective," *Semicond. Int.*, **6**, 58 (October 1983).

Hegedus, S. S., R. E. Rocheleau, and B. N. Baron, "CVD Amorphous Silicon Solar Cells," Proceedings of the 17th IEEE Photovoltaic Specialists Conference, Orlando, FL, 1984, p. 239.

Hitchman, M. L., J. Kane, and A. E. Widmer, "*Thin Solid Films*," **59**, 231 (1979).

Hollan, L., J. P. Hollais, and J. C. Brice, in *Current Topics in Materials Science*, Vol. 5, E. Koldis (ed.), North Holland, New York, 1980, pp. 137–148.

Houtman, C., D. B. Graves, and K. F. Jensen, "CVD in Stagnation Point Flow," *J. Electrochem. Soc.*, **133** (5), 961 (1986).

Jackson, S. C., "Engineering Analysis of the Deposition of Cadmium–Zinc Sulfide Semiconductor Film," PhD Thesis, University of Delaware, Newark, DE, 1984.

Jackson, S. C., B. N. Baron, R. E. Rocheleau, and T. W. F. Russell, "Molecular Beam Distributions from High Rate Sources," *J. Vac. Sci. Technol.*, **A3** (5), 1916 (1985).

Jenkinson, J. P. and R. Pollard, "Thermal Diffusion Effects in Chemical Vapor Deposition Reactors," *J. Electrochem. Soc.*, **131** (12), 2911 (1984).

Jensen, K. F., "Micro Reaction Engineering—Applications of Reaction Engineering to the Processing of Electronic and Photonic Materials," *Chem. Eng. Sci.*, **42**, 923 (1987).

Jensen, K. F., and D. B. Graves, "Modelling and Analysis of Low Pressure CVD Reactors," *J. Electrochem. Soc.*, **130**, 1950 (1983).

Juza, J. and J. Cermak, "Model of the Epitaxial Chemical Vapor Deposition Reactor for Design and Performance Optimization," *Chem. Eng. Sci.*, **35**, 429 (1980).

Juza, J. and J. Cermak, "Phenomenological Model of the CVD Epitaxial Reactor," *J. Electrochem. Soc.*, **129** (7), 1627 (1982).

Kuiper, A. E. T., C. J. H. van den Brekel, J. de Groot, and F. W. Veltkamp, "Modelling of Low Pressure CVD Processes," *J. Electrochem. Soc.*, **129** (10), 2288 (1982).

Kusumoto, Y., T. Hayashi, and S. Komiya, "Numerical Analysis of the Transport Phenomena in MOCVD Process," *Jap. J. Appl. Phys.*, **24** (5), 620 (1985).

Kuznetsov, F. A. and V. I. Belyi, "Vapor Deposition and Etching Open Tube Kinetics under Diffusion Controlled Conditions," *J. Electrochem. Soc.*, **117**, 785 (1970).

Larrabee, G. B., "Microelectronics: A Challenge to Chemical Engineering," *Chem. Eng.*, **92**, 51 (June 10, 1985).

Little, R. C., personal communication, Spire Corporation, Bedford, MA, 1985.

Manke, C. W. and L. F. Donaghey, "Numerical Simulation of Transport Processes in Vertical Cylinder Epitaxy Reactors," Proceedings 6th International Conference on Chemical Vapor Deposition, Electrochemical Society, Pennington, New Jersey, 1977a, pp. 151–165.

Manke, C. W. and L. F. Donaghey, "Analysis of Transport Processes in Vertical Cylinder Epitaxy Reactors," *J. Electrochem. Soc.*, **124** (4), 561 (1977b).

Michaelidis, M. and R. Pollard, "Analysis of Chemical Vapor Deposition of Boron," *J. Electrochem. Soc.*, **131** (4), 861 (1984).

Moffat, H. and K. F. Jensen, "Complex Flow Phenomena in MOCVD Reactors. I. Horizontal Reactors," *J. Cryst. Growth*, **77**, 108 (1986).

Pollard, R. and J. Newman, "Silicon Deposition on a Rotating Disk," *J. Electrochem. Soc.*, **127** (3), 744 (1980).

Ring, M. A. in *Homoatomic Rings, Chains and Macromolecules of Main Group Elements*, Elsevier, New York, 1977, Ch. 10.

Rocheleau, R. E., "Design Procedures for a Commercial Scale Thermal Evaporation System for Depositing CdS for Solar Cell Manufacture," PhD Thesis, University of Delaware, Newark, DE, 1981.

Rocheleau, R. E., B. N. Baron, and T. W. F. Russell, "Analysis of Evaporation of Cadmium Sulfide for the Manufacture of Solar Cells," *AIChE J.*, **28**, 656 (1982).

Roenigk, K. F. and K. F. Jensen, "Analysis of Multicomponent LPCVD Processes. Deposition of Pure and *in situ* Doped Poly-Si," *J. Electrochem. Soc.*, **132**, 448 (1985).

Rundle, P. C., "The Growth of Silicon in Horizontal Reactors," *J. Cryst. Growth*, **11**, 6 (1971).

Russell, T. W. F., "Semiconductor Chemical Reactor Engineering and Photovoltaic Unit Operations," *Chem. Eng. Educ.*, 72-77 and 106-108 (Spring, 1985).

Sugawara, K., "Silicon Epitaxial Growth by Rotating Disk Method," *J. Electrochem. Soc.*, **119**, 1749 (1972).

Takahasi, R., Y. Koza, and K. Sugawara, "Gas Flow Pattern and Mass Transfer Analysis in a Horizontal Flow Reactor for Chemical Vapor Deposition," *J. Electrochem. Soc.*, **119** (10), 1406 (1972).

van de Ven, J., G. M. J. Rutten, M. J. Raaijmakes, and L. J. Giling, "Epitaxial Growth of GaAs by MOCVD," *J. Cryst. Growth*, in press (1987).

Wahl, G., "Theoretical Description of CVD Processes," Proceedings of the Ninth International Conference on Chemical Vapor Deposition, Electrochemical Society, Pennington, New Jersey, 1984, pp. 60–77.

Wei, J., T. W. F. Russell, and M. W. Swartzlander, in *Structure of the Chemical Processing Industries: Function and Economics*, McGraw-Hill, New York, 1979.

Westphal, G. H., D. W. Shaw, and R. A. Hartzell, "A Flow Channel Reactor for GaAs Vapor Phase Epitaxy," *J. Cryst. Growth*, **56,** 324 (1982).

## FURTHER READING

Braun, E. and S. MacDonald, *Revolution in Miniature: The History and Impact of Semiconductor Electronics*, 2nd ed., Cambridge University Press, Cambridge, England, 1982.

Conner, F. R., *Electronic Devices*, Edward Arnold Publishers Ltd., London, England, 1980.

Ferry, D. K. (ed.), *Gallium Arsenide Technology*, Howard W. Sams & Co., Inc., Indianapolis, IN, 1985.

Jackson, S. C. and T. J. Anderson (eds.), "Engineering Research Needs for Electronic Materials Processing," Proceedings of a National Science Foundation Workshop, University of Delaware, Newark, DE, 1985.

Jenson, K. F., "Micro-Reaction Engineering," *Chem. Eng. Sci.*, **42,** 923 (1987).

Kerridge, C., *Microchip Technology*, Wiley, New York, 1983.

*National Materials Advisory Board*, "Advanced Processing of Electronic Materials in the United States and Japan," State-of-the-Art Reviews, Panel on Materials Science, National Materials Advisory Board, Commission on Engineering and Technical Systems and the National Research Council, National Academy Press, Washington, D.C., 1986.

*The Federation of Materials Societies*, "Electronic Materials: A Key to U.S. Competitiveness?," Workshop Summary, The Federation of Materials Societies, Washington, D.C., February 26–27, 1986.

# INDEX